HOTEL
Engineering

SUJIT GHOSAL

Professor
Mechanical Engineering Department
Jadavpur University, Kolkata

OXFORD
UNIVERSITY PRESS

OXFORD
UNIVERSITY PRESS

Oxford University Press is a department of the University of Oxford.
It furthers the University's objective of excellence in research, scholarship,
and education by publishing worldwide. Oxford is a registered trademark of
Oxford University Press in the UK and in certain other countries

Published in India by
Oxford University Press
22 Workspace, 2nd Floor, 1/22 Asaf Ali Road, New Delhi 110 002

First published 2011
10th impression 2023

ISBN-13: 978-0-19-806291-2
ISBN-10: 0-19-806291-5

Typeset in Times New Roman
by Shubham Composer
Printed in India by Manipal Technologies Limited, Manipal

For product information and current price, please visit www.india.oup.com

Dedicated
to
my parents

Preface

The hospitality industry is second only to the global oil industry in terms of turnover, and is, by far, the largest employer around the world. Ten per cent of the world's workforce is associated with this industry, and ten per cent of the world's total GNP comes from tourism. As a result of its global boom, the hospitality industry, along with the tourism industry, has emerged as a key contributor to the overall economy of the world. This surge has been the outcome of burgeoning business and leisure travel, both inland and overseas. Rising income levels and spending power coupled with the governmental open sky policy have provided a major boost in this sector of business.

Engineering systems are vital for performance of hotels. For example, a failure of the AC machine or fire in the hotel can be devastating for the business. High energy loss and improper selection and operation of equipment can lead to excessive revenue drain. Hence, students and professionals in hotel management should have a fair knowledge of engineering systems employed in hotels. As a matter of fact, in big hotels, engineering systems are very elaborate, sophisticated, and complex.

Hotel engineering has been included as a compulsory subject in the course curriculum of the Institute of Hotel Managements (IHMs) under the Council of Hotel Management at the undergraduate level. Some of the topics of hotel engineering are also taught in facility planning courses in many institutes. This subject is also included in the certificate courses in craftsmanship in food production. Topics of hotel engineering are also included in many postgraduate diploma courses related to hotel management and catering technology.

About the Book

Hotel Engineering has been designed with the aim of providing a proper mix of the fundamentals required for the subject vis-à-vis the description of the working of engineering systems.

The book discusses the basics of all engineering systems used in the hotel industry, including science of illumination, motion, electricity, refrigeration systems, utility systems, and so on. It covers the basics of heat and the different types of fuels and heat appliances used in hotels. Topics related to water hardness and its treatment, water distribution systems, refrigeration, ventilation, fire safety norms, and air conditioning are elucidated. The maintenance requirements of buildings, utility systems, and the security of a hotel have also been dealt with extensively.

A few topics which are not explicitly mentioned in the syllabus or usual engineering practices, such as the use of solar cooker or rainwater harvesting, have also been included in the chapters making the coverage very comprehensive. Relevant references have been made to industry standards and norms, as well as statutory requirements. The maintenance and troubleshooting guidelines of various engineering systems in the hotel industry is an added feature of this book.

Apart from the students of hotel management, practising engineers in the hotel industry will also find the book a handy reference.

Pedagogical Features

The book has been developed with the following structural and pedagogical features in each chapter:

- The chapters start with learning objectives and end with key terms, review questions, and review problems, wherever relevant.
- Each chapter explains the concepts along with relevant industrial applications to help students understand them better.
- The basic principles of science behind various engineering concepts have been covered in each chapter. While the emphasis is on explaining the engineering systems, the underlying physical principles have been explained so that students of all backgrounds are able to understand them.
- Review questions and numerical problems provided in the chapters will test the students' understanding of the topics as well as prepare them to answer related questions and problems in the examinations.

Other key features of the book are listed as follows:

- Provides numerous photographs, tables, and figures to explain the engineering systems employed in the hotel industry
- Contains many industrial examples along with simple and solved problems to explain the concepts
- Includes maintenance and troubleshooting guides of engineering systems in the hotel industry

Coverage and Structure

The book is divided into sixteen chapters devoted to the various areas of the subject. Efforts are made to see that the reader gets a fairly comprehensive idea about the fundamental principles of science as applicable for understanding the engineering systems.

Chapter 1 deals with the importance and description of various engineering systems used in a hotel, in general.

Chapter 2 introduces the basics of motion, electricity, and electrical machines needed for the overall understanding of electrical systems in a hotel.

Chapter 3 provides coverage on elementary electrical wiring and safety precautions in dealing with electrical appliances with special reference to the hotel industry. The chapter also gives information on how to read an electric metre and do some elementary calculations on energy bill and electric tariff.

Chapter 4 introduces the science of illumination, a few common lighting devices and lighting requirements for different functional areas in a hotel.

Chapter 5 discusses the fundamentals of heat, fuel, and combustion. It also deals with the different types of fuels and heat appliances used in the hotel industry.

Chapter 6 covers water hardness and the methods for its treatment.

Chapter 7 provides a description of cold- and hot-water distribution systems in the hotel industry and a few components used therein.

Chapter 8 addresses the issue of waste disposal and pollution, including the legal requirements.

Chapter 9 is about fire prevention and control as applicable to the hotel industry. It also discusses care and maintenance of fire control equipment.

Chapter 10 focuses on the various refrigeration systems employed in a hotel.

Chapter 11 is about ventilation and air-conditioning systems.

Chapter 12 deals with the maintenance procedures and contract maintenance in the hotel industry.

Chapter 13 discusses equipment replacement policy.

Chapter 14 is about treatment of miscellaneous utility systems, including audio-visual equipment, computers, and sensors and detectors.

Chapter 15 deals with the safety and security aspects of a hotel.

Chapter 16 contains useful information on the maintenance and troubleshooting of some engineering systems in the hotel industry.

Acknowledgements

Goethe thought prefaces are useless, but an ordinary person may feel inclined to state somewhere in between the covers of a book, his motivation and thinking behind writing what could be just another book trying desperately to justify that it is not so. As such I do not claim myself to be an exception.

It is a pleasure to remember with gratitude all the people who have helped me in the preparation of this book. I should particularly thank Mr Ranjit Chowdhury, Principal, Institute of Hotel Management, Catering Technology and Applied Nutrition, Kolkata, who has always shown great interest in the book. I am indebted to Sri Sujit Nath, Research Scholar, Mechanical Engineering Department, Jadavpur University, Kolkata, for preparing many figures carving out time from his busy schedule. I would also like to thank Mr Tarun Chatterjee, Chief Engineer, ITC, Sonar Bangla Hotel, Kolkata, for his valuable suggestions and information. I also remember many useful suggestions made by Prof. Dipten Misra of Mechanical Engineering Department, Jadavpur University, Kolkata.

I run short of words to express the contribution of my wife, Archana Ghosal, and son, Joyjit Ghosal, who helped me in writing the book and at times suffered silently due to my long working hours.

Last but not the least, I express my sincere appreciation towards the editorial team of Oxford University Press for its support, fruitful suggestions, and encouragement.

Suggestions for further improvement of the book are welcome.

Sujit Ghosal

Contents

1 Chapter

Hospitality Industry and Engineering Systems

Learning Objectives

Modern hospitality industry can now be called an engineering industry considering the multitude of engineering systems employed therein for running the business.

The principle objectives of this chapter are to:

- emphasize the complexity and importance of engineering systems and equipment in the hospitality industry
- identify different kinds of engineering systems and equipment used in the hospitality industry
- help the readers to appreciate the role of engineers in helping managers run the business in a competitive environment

1.1 | INTRODUCTION

For a long time, hospitality industry had not given much thought to operation and economics of engineering services and utilities employed by it. Hospitality managers had very little knowledge about engineering systems. Engineering utilities and systems were considered to be complex. But, however complex and mysterious engineering systems might seem to the hospitality managers, they have, of late, come to realize that the efficiency of hotel operation critically depends upon smooth and economic functioning of engineering utility services. Earlier, the building manager undertook the engineering and maintenance of hospitality property. But in most of the cases, the building managers did not possess special skills to keep the utilities in good condition. Later, the job was transferred to the chief engineer of the hotel. Until recently, not much attention was given to the department of engineering and maintenance. The recent sky-rocketing price of fuel and ever-increasing maintenance costs of sophisticated engineering equipment have changed this situation altogether. Earlier, many general managers did not have the opportunity to discuss operational matters with the chief engineer; some even did not like to visit the maintenance department, as they had very little knowledge about the engineering services. The chief engineer, on the other hand, also knew very little about management and cost control. However, this scenario is undergoing a drastic change and managers who do not have working knowledge and appreciation of engineering services will soon become obsolete. Managers who have some skill and knowledge of engineering services will definitely have an edge over others in the field. Managers of today should be able to discuss the problems of operation vis-à-vis engineering requirements in technical language. The status of chief engineer has also elevated to that of a departmental head, such as director of sales or food and beverages manager, because of complexity and critical importance of the engineering department in the operation of business.

1.2 | ENGINEERING SYSTEMS IN HOSPITALITY INDUSTRY

Modern hospitality industry facilities, including those in the hotel industry, have a complex array of very sophisticated engineering systems ranging from complex property building, sophisticated water treatment plant, sewage treatment plant to the state-of-the-art sensors and detectors. Any hospitality facility also has boilers, diesel-generating units, transformers, air-conditioning and refrigeration units, sophisticated kitchen appliances, laundry equipment, fire safety equipment, to name a few. Right from the time a guest enters a hotel, goes to the reception, walks through the lobby, and retires in his/her room, he/she encounters various engineering systems. The building, the front glass door in the lobby which opens automatically when it senses the appearance of the customer, the sophisticated internal communication system the receptionist uses for arranging stay of the customer, the computer the receptionist uses to scan through the room availability and updating status, the fire sprinkler system in the corridor the guest moves through, the hot water system, cold conditioned air, the channel music, the television (TV), the smoke detector inside the lodging room, all are the products of engineering systems installed in the hotel. The customer wants his/her dirty but favourite apparel washed, dried, and ironed within hour, so that he/she can attend a conference in his/her favourite dress. The laundry operator gets the dress ready with the state-of-the-art washing machine and dryer. When the electricity supply goes off, the in-house generator takes over. Dishwashers do wonders in the kitchen. Hospitality industry's facilities of today can be really called engineering units doing hospitality business. As such, the managers of a hotel or a hospital should have a basic understanding of the science of engineering and working knowledge of the systems for efficient operations of the business.

All the engineering systems and equipment in a hotel or similar industry can be categorized into various broad engineering divisions such as civil engineering, mechanical engineering, and electrical engineering. This division arises out of technical requirement because specialist engineers and technicians must deal with the specific systems and equipment. Here, we bring in to the fore various engineering systems and facilities commonly used in the modern hospitality industry vis-à-vis their commonly assigned engineering categories.

1.2.1   Facilities under the Scope of Civil Engineering

There are many facilities in a hotel that can be categorized under civil engineering. A few of them are listed below.

1. Buildings
2. Road
3. Parking
4. Lawn and landscape
5. Drinking and sanitation water supply lines
6. Sanitation facilities
7. Drainage
8. Sewage-treatment plant
9. Swimming pool

While some of these facilities are discussed in the later chapters, buildings, road, parking, lawn, and landscape, etc. are kept aside the scope of this book as they may come under the specialized topic of facility planning.

1.2.2   Facilities under the Scope of Mechanical Engineering

Modern hotel industry uses a lot of equipment essentially involving mechanical engineering. Some of them are as follows:

1. Hot water calorifier
2. Boiler and accessories
3. Diesel engines for power generation
4. Gas bank
5. Ventilation and air conditioning
6. Refrigeration
7. Cooking equipment such as mixers, grinders, ovens, ranges, meat saw, etc.
8. Dishwashers

9. Dryers
10. Washing machine and dryer
11. Water-pumping system
12. Water-treatment plant
13. Firefighting system

1.2.3 Facilities under the Scope of Electrical and Electronics Engineering

As hospitality industry is adopting advanced technology, a number of electrical and electronic equipment and their complexities are increasingly growing. A few such equipment installed in modern hotels are as follows:

1. Transformer
2. Electric motor and starter
3. Alternator/generator for backup power supply
4. Lifts
5. Hotel illumination system
6. Electrical wiring
7. Electrical controls and switch gears
8. Inverter
9. Uninterruptible power supply (UPS)
10. Electric ranges in the kitchen
11. All the controls in kitchen and laundry equipment
12. Audio-visual equipment
13. Public address and internal communication systems
14. Data communication and computer systems
15. Sensors and detectors

It is evident that many of the equipment shown under one category have components to be serviced by people specializing in other categories. For example, almost all the rotating mechanical equipment will have motors and electronic control systems. Similarly, a water-treatment plant and swimming pool have chemical engineering and chemistry associated with them. However, the enormity and complexity of the engineering systems and equipment used in the hotel industry is clear from the above-mentioned lists. As a matter of fact, the business of the property is critically dependent on the efficient functioning of the engineering systems. Proper understanding of all these sets of equipment is vital for the operational success of the business. As such, a full-fledged department of engineering service and maintenance having qualified and experienced engineers, supervisors, and technicians has been set up in all medium- and large-sized hotels and most of the small hotels. The chief engineer is the departmental head and enjoys a very high position in the management structure. The engineering department has the responsibility of maintaining the facilities to the top level of functional fitness. This department has to look into measures for energy conservation and pollution control. It has to also select and purchase/install new equipment in consultation with various managers who are involved in operations such as food and beverages manager, chef, banquet hall manager, etc. As already indicated, the operation managers of today must have some working knowledge of the equipment and engineering systems, their operations, functions, and importance so that they can interact with the engineering department as well as top management. In subsequent chapters, the basic principles of science and engineering for understanding the working of engineering systems as needed by students of hotel management and managers of hotel industry have been introduced. A few important engineering systems are also discussed in some detail so that students of hotel engineering who will be future managers, get some idea about engineering systems and their importance in hotel business.

It may be said that hospitality industry is, to a large extent, an integrated form of many engineering systems employed for different kinds of essential and support services. The success of even those departments seemingly remotely connected with engineering facilities, such as food and beverages, front office, housekeeping, etc. also depends upon the performance of many different engineering systems, which they require to operate and which may not be always very apparent. Thus, engineering system is inseparable from any activity in a hotel and is all-pervading in nature. Proper selection/design, operation, and maintenance of engineering systems are, thus, very important and all managers in a hospitality industry must appreciate the critical role of engineering systems and equipment in the success of the business. The managers should have a good knowledge and information about the engineering systems in order to be able to interact with personnel working in different departments and discharge their duties as managers more effectively.

1.3 | CONCLUSION

A guest staying or visiting a hotel comes across a host of engineering systems. These systems play a vital role in proper functioning of the various departments of a hotel. From the kitchen to the reception, every minor function is carried out with the support of these engineering systems. Therefore, managers of today should be able to discuss the problems of operation vis-à-vis engineering requirements in technical language. Keeping this in view, this chapter listed the various engineering systems and facilities commonly used in modern hospitality industry. All these and many more aspects of hotel engineering have been covered in the subsequent chapters.

KEY TERMS

Civil engineering	It is a branch of engineering dealing primarily with analysis, design, construction and maintenance of systems that include foundations for building and other structures, bridges, dams, buildings, steel structures, river hydraulics, roads, highway, public health engineering such as sewerage, drinking-water supply, irrigation, etc.
Electrical engineering	It is a branch of engineering which deals with generation, transmission, distribution of electricity; analysis, design, operation and maintenance of electrical machines such as generator, electric motor, transformer, electrical wiring, illumination systems, etc.
Electronics engineering	It is a branch of engineering dealing with analysis, design, operation, and maintenance of various electronic equipment and systems such as constant voltage supply system, sensors and detectors, audio-visual equipment, etc.
Engineering system	It is a functional system comprising various engineering components to achieve the satisfaction of some specific need.
Hospitality industry	It consists of hotels, motels, inns, or such businesses that provide transitional or short-term lodging, with or without food services, as also recreation, and entertainment sectors.
Mechanical engineering	Mechanical engineering involves the analysis, design, manufacturing, installation, operation and maintenance of various systems such as engines, pumps, power plants (involving boiler, turbine, etc.), heating equipment, refrigeration, air conditioning, automobiles. It is one of the oldest and broadest engineering disciplines.

REVIEW QUESTIONS

1.1 Write about the relationship between hotel managers and engineering systems and equipment.

1.2 Categorize different engineering systems and equipment, as used in the hotel industry, citing examples for each of them.

1.3 List a few systems and equipment that are categorized under mechanical and civil engineering, as generally used in the hotel industry.

1.4 List a few engineering systems and equipment that are categorized under electrical and electronics engineering, as generally used in the hotel industry.

2
Chapter

Basics of Motion, Electricity, and Electrical Machines

Learning Objectives

A multitude of equipment and utility systems used in households and the hospitality industry, among others, are essentially engineering systems and involve some kind of motion, thermal effect, and the use of electricity for their operation. Working with them efficiently and safely, thus, requires a basic understanding of their principles of working.

The objectives of the present chapter are as follows:

- recapitulating some of the very basic principles and terms associated with general mechanics and electricity
- discussing different units of measurement for different relevant quantities such as force, work, energy, power, electric current, etc.
- presenting the fundamentals of a few important and relevant physical quantities in an elaborate manner
- helping to understand the working of engineering systems that operate on electricity employed in houses and the hospitality industry
- supplementing the understanding of the principles by solving numerical problems on force, work, energy, electric-current flow, and power in simple electrical circuits

2.1 | INTRODUCTION

Modern hotels have a host of modern and classical engineering equipment and can be veritably called an engineering service industry. These include boilers, engines, air-conditioning and refrigeration units, sensors and detectors, and kitchen equipment such as electric ranges, mixers, choppers, etc. The description of quite a few of them will follow in subsequent chapters. Here, we introduce and explain a few scientific and engineering terms that are needed for understanding the basic engineering systems commonly encountered in the hospitality industry. These are related to general mechanics and electricity.

2.2 | MOTION—SPEED, VELOCITY, ACCELERATION

2.2.1 Speed (s)

It is the rate at which a body travels from one point to another. The speed of a moving object is the distance travelled divided by the time it takes to travel that distance. Its unit is, therefore, m/s, km/h, etc. If an object takes time, t, to move from A to B and the distance between A and B is x (Fig. 2.1), the speed, s, is given by $s = x/t$.

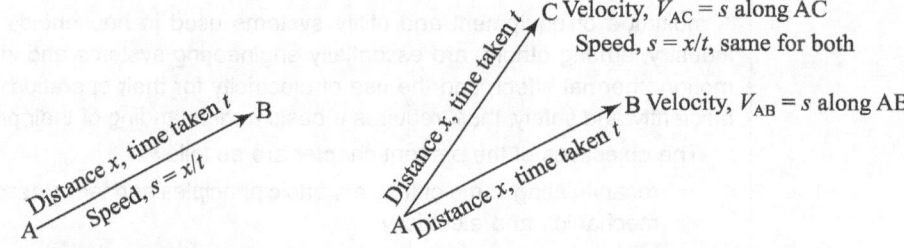

Fig. 2.1 Speed is distance covered divided by time taken

Fig. 2.2 Velocity is speed along with direction

If an object moves in one direction from A to B (Fig. 2.2) by some distance, x, taking time, t, while another object moves the same distance taking the same amount of time but moving in some other direction (A to C); both objects are said to have the same speed, *i.e.*, $s = x/t$. We do not attach any property of direction to it. Therefore, speed has only magnitude and does not indicate direction. Such a quantity is called a scalar quantity. Temperature and mass are two other examples of scalar quantities.

2.2.2 Velocity (V)

It is the rate at which a body changes its position. It has both magnitude and direction. Its unit is also m/s, km/h, etc.

If two objects are moving from A and take the same time and same path to reach B, we say that their speeds are same, as well as their velocities are same. But if an object is moving from A to B at speed of s and another from A to C at the same speed (Fig. 2.2), we say that the velocities are different in the two situations, as the directions of motion are different for them, although they have the same speed. In the first case, we say that the velocity is s along AB and in the second case, s along AC. Hence we note that velocity has two quantities. One is the speed, s, which is the magnitude of the velocity and the other is the direction. Such quantities which that require both magnitude and direction for their complete specification or identity are called vector quantities. Force and acceleration are two other examples of vector quantities.

So, the velocity of an object can be said to change if it

(a) Changes only its speed, without changing its direction, or

(b) Changes only its direction, without changing its speed, or

(c) Changes both speed and direction during its journey

The three different situations are shown in Fig. 2.3.

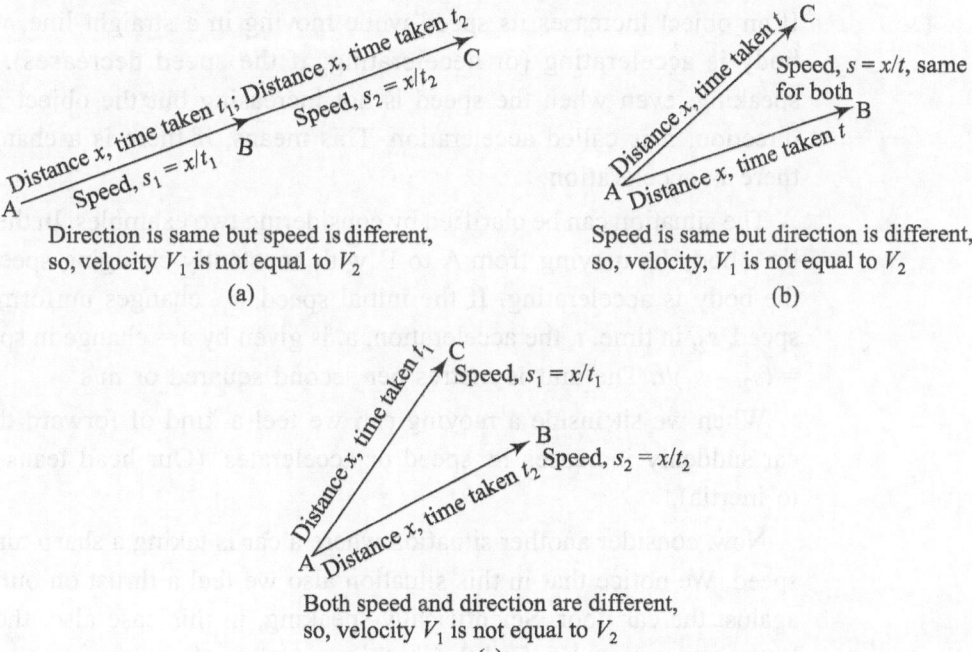

Direction is same but speed is different, so, velocity V_1 is not equal to V_2

(a)

Speed is same but direction is different, so, velocity, V_1 is not equal to V_2

(b)

Both speed and direction are different, so, velocity V_1 is not equal to V_2

(c)

Fig. 2.3 Velocity changes due to (a) change of speed, (b) change of direction, and (c) change in both speed and direction

Figure 2.4 shows how resultant velocity is computed when an object changes its velocity during its journey.

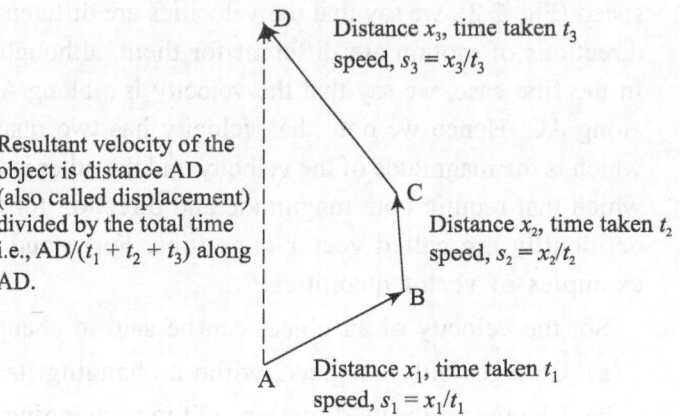

Fig. 2.4   Resultant velocity

2.2.3   Acceleration

If an object increases its speed while moving in a straight line, we say that the body is accelerating (or decelerating, if the speed decreases). Scientifically speaking, even when the speed is not increasing but the object is changing its direction, it is called acceleration. This means, if there is a change of velocity, there is acceleration.

The situation can be clarified by considering two examples. In the first example, let a body be moving from A to B with constantly changing speed. We say that the body is accelerating. If the initial speed, s_1, changes uniformly to the final speed, s_2, in time, t, the acceleration, a, is given by a = change in speed/time taken = $(s_2 - s_1)/t$. The unit is metres per second squared or m/s^2.

When we sit inside a moving car, we feel a kind of forward thrust when the car suddenly increases its speed or accelerates. (Our head leans backward due to inertia).

Now, consider another situation where a car is taking a sharp turn at a constant speed. We notice that in this situation also we feel a thrust on our side, if sitting against the car door. Scientifically speaking, in this case also, the car is said to have acceleration. We find that in this case, the velocity is changing (by changing direction), although the change in velocity here is to be found out in a little more involved manner.

Acceleration is, therefore, formally defined as the change in velocity per unit time. Since velocity is a vector quantity, acceleration is also a vector quantity.

2.3 | FORCE, WORK, AND POWER

2.3.1 Force

Force, when exerted on a body at rest, either changes or tends to change the position of a body at rest and when applied on a moving body, changes the velocity of a moving body. It has both magnitude and direction, and hence can be called a vector quantity. Examples of force are magnetic force, gravitational force, frictional force, human body force tending to push a pillar, force due to water on the sidewall of a dam, etc.

If a body is under the influence of a number of forces acting simultaneously on it, the balance of the forces (also called resultant force or net force) will determine the dynamic state of the body. If there is no net force on the body, the body will either be at rest or moving at a uniform velocity. So, we can say that if a body is at rest or moving at a uniform velocity, there is no net force on the body. If there is a net force on the body in a particular direction, the body will move with acceleration in that direction. The relationship between the net force and acceleration is written as (Newton's second law of motion):

$$F = ma \tag{2.1}$$

where F is the net force, m the mass of the object, and a the acceleration in the direction of F.

When a man jumps from a height, the only force acting on him is the gravitational pull and hence, the force system acting on him is not balanced. So the man will have acceleration, i.e., his speed will keep increasing till he hits the ground, which, by its reactions, balances the gravitational pull and reduces the acceleration of the man to zero and, additionally, brings the man to a zero speed. When a book is kept on the table, the gravitational pull is balanced by the reaction offered by the table that acts on the book in a direction opposite to the gravitational pull. Hence, the book remains at rest. When a car moves with uniform velocity, there is no acceleration and Newton's law states that there cannot be any net force on the car. But as we know, whenever something moves on a surface and/or through air, resistive frictional and pressure forces oppose the motion. When the car is moving with no acceleration, the engine force balances this resistive force of road and air friction. If we wish to accelerate the car, the engine force acting on the car has to be more than the opposing force of resistance so that there could be a net force acting on the car in the forward direction and we do this by increasing the fuel input to the engine.

Units of force

Absolute unit of force   The unit of force is defined in respect of Eqn 2.1. If a net force acting on a mass of 1 kg produces an acceleration of 1 m/s^2 (i.e., say a change of velocity from 6 m/s to 9 m/s in 3 s), we say that the force is having a magnitude of 1 Newton or 1 N.

Newton is the absolute unit of force in SI unit.

In FPS (foot-pound-second) system, the absolute unit of force is poundal. 1 poundal is the force, which, when acting on a body of mass 1 pound (1 lb or 453.6 g), produces an acceleration of 1 foot/s^2.

Let us solve a numerical problem.

A single force acts upon an object of mass 80 kg moving in a straight line. Initially, the object was moving with a velocity of 5 m/s before the application of the force. After the application of force, the velocity changes uniformly to 7.5 m/s in 2 s while moving in the same direction. Find the magnitude of the force acting on it.

We have

$$F = ma$$

$$m = 5 \text{ kg}$$

$$a = \text{change of velocity per unit time} = (7.5 - 5) \text{ m/s} / 2\text{s}$$

$$= 1.25 \text{ m/s}^2$$

Therefore, force, $F = 80 \times 1.25 \text{ N} = 100 \text{ N}$

Gravitational unit of force   However, when we say that a body is experiencing, suppose, a force of 100 N, we do not have a physical feeling of the magnitude of this force. But, if we say that there is a force that is equivalent to a feeling that we experience when we hold, suppose, a 10 kg mass kept on our palm, we have some idea of the magnitude of the force.

When a mass is falling freely under the action of gravitational pull, the acceleration is constant and equal to 9.81 m/s^2. Now if 1 kg mass is falling freely, the corresponding gravitational force on it is given by

$$F = ma$$

$a = 9.81$ m/s^2 (the acceleration due to gravity, which is the acceleration with which a body moves when falling freely under gravity)

Therefore, $F = 1 \times 9.81 \text{ N} = 9.81 \text{ N}$

Therefore, if we have a force of 9.81 N, we can say that it is equivalent to a feeling of force of 1 kg mass on the palm. Thus, if we have a force of 19.62 N, we say that it is equivalent

(Contd)

(*Contd*)

to a 2 kg mass on the palm. So, we call 19.62 N as 2 kilogram-force or 2 kgf. This latter unit of force is called the gravitational unit of force. So any force expressed in Newton can be converted to gravitational unit by dividing it by 9.81 and, conversely, any force expressed in kgf can be converted to Newton by multiplying it by 9.81. So, 100 N will be equal to 100/9.81 kgf or 10.2 kgf.

Following the above discussion, the gravitational unit of force in FPS system is pound-weight (lb-wt) or pound-force (lbf) and is obtained by dividing the quantity of force in poundal by 32.2 (acceleration due to gravity is 32.2 ft/s^2 in FPS system) and conversely.

Pressure

The term pressure is related to fluids (liquid or gas). Pressure is an inherent property of a fluid by virtue of the random movement of their rather loosely bonded molecules. Any solid surface in contact with a fluid will experience a force due to this pressure, which is calculated as

$$\text{Force} = \text{pressure} \times \text{area}$$

The direction of force is perpendicular to the area.

From the above relation, we find that the unit of pressure is Newton per square metre, known as pascal, abbreviated as Pa. Atmosphere exerts on us a pressure of about 101 kPa. Pressure is also expressed as bar and 1 bar is equal to 10^5 Pa. In FPS system of units, the unit of pressure is pound force per unit square inch or psi. Atmosphere has a pressure of about 14.7 psi. So, 101 kPa is equal to 14.7 psi.

Gas and liquid pressures are usually quoted as the values over and above atmospheric pressure and such values of pressure are called *gauge pressure* values. The pressure regulator of a gas cylinder has a dial, indicating pressure, called pressure gauge. This pressure gauge indicates gauge pressure. For example, if the dial indicates 25 psi, actual pressure (or, absolute pressure) of the gas inside is (25 + 14.7) psi or 42.7 psi. In common language, gas/liquid pressure is often quoted with a unit of Pounds. This is actually pounds per square inch or psi.

2.3.3 Work

Work is said to be done when a force is exerted on a body and the body changes its position. Following this definition, whatever amount of force we may apply and whatever amount of sweat we release in the process, if we cannot move a pillar or a big tree, scientifically speaking, we do not do any work. Work done by a force is given by,

$$W = F \times d \qquad (2.2)$$

where W is the work done, F is the force applied, and d is the displacement of the body in the direction of the force, F.

Unit of work

When F is in newton and d is in metres, the unit of work, W, becomes newton-metre (N–m) or joule (J), i.e., the SI unit of work is joule (named after English physicist James Prescott Joule).

Suppose, a force of 50 N is applied on a body, as a result the body moves by 10 m in the direction of the force. The work done on the body is, therefore, 50×10 N-m = 500 J.

If a car is moving with a speed of 36 km/h uniformly and the engine is producing a driving force (or motive force) of 400 N, the work done by the engine on the car during a run of 5 minutes will be as follows:

The distance moved by the car in 5 minutes is

$$d = \text{uniform speed} \times \text{time} = 36 \times 1000/3600 \text{ m/s} \times (5 \times 60)s$$

$$= 3000 \text{ m}$$

Therefore, the work done by the engine during this period is

$$F \times d = 200 \text{ N} \times 3000 \text{ m} = 60{,}000 \text{ N-m} = 60{,}000 \text{ Joule}$$

In FPS system, the unit of work is ft-poundal when force is in absolute unit (poundal) and ft-lb when force is expressed in gravitational unit (lbf). So, we say that work of 1ft-lb will be spent if a force of 1slbf (32.2 poundal) acting on a body moves it by 1 foot.

2.3.4 Power

Suppose a source of energy exerts a force 10 N on a body and moves it uniformly by 2 m in 5 s. Another source of energy exerts the same 10 N on a body and moves it by the same distance, i.e., 2 m, in 2 s. In both the cases, the work done is 10×2 J or 20 J. But in the latter case, this work is done more quickly than in the first. This permits us to say that the latter source is more powerful than the former.

Power is, thus, formally defined as the work done or energy expended or consumed in unit time. So the power, P is defined, symbolically, as

$$P = W/t \qquad (2.3)$$

SI unit of power is J/s or watt.

In the first case of the above-mentioned example, $W = F \times d = 10 \times 2$ J = 20 J and $t = 5$ s. So, power, $P = 20$ J/5 s = 4 J/s = 4 W = 4 W [1 J/second = 1 W].

In the second case, $W = F \times d = 10 \times 2$ J = 20 J and $t = 2$ s

So, power, $P = 20$ J/2 s $= 10$ J/s $= 10$ W $= 10$ W

A larger unit of power used in SI unit is kilowatt or kW. 1 kW is equal to 1000 W.

In FPS system, the unit of work, in gravitational units, is usually expressed as ft-lb.

So, the unit of power becomes ft-lb/s. If the power taken from a source or consumed by some object is 550 ft-lb/s, we say that the power is 1 horsepower or, in short, 1 hp. Although, internationally the units of physical quantities should be in SI, there is a parallel practice of using hp as the unit of power, particularly in mechanical systems such as pump power, engine power, etc.

1 hp = 550 ft-lbf = 550 ft $\times$ 1 lbf

$\qquad$ = 550 $\times$ 0.3048 m $\times$ 1 $\times$ 32.2 poundal = 550 $\times$ 0.3048 m $\times$ 1 lb $\times$ 32.2 ft/s^2

$\qquad$ = 550 $\times$ 0.3048 m $\times$ 1 $\times$ 0.4536 kg $\times$ 32.2 $\times$ 0.3048 m/s^2 = 746.3 J/s

$\qquad$ = 746.3 W

So, 1 hp is approximately equal to 746 W or 0.746 kW.

If power, P, delivered by an energy source (or power consumed by some object) and the period of time, t, for which the power is supplied (or consumed) are known, the energy, E, expended or consumed is found out by rewriting Eqn 2.3 as

$$E = P \times t$$

Let us now solve a composite problem.

A car is moving at a speed of 40 km/h. The driver wants to increase the speed to 60 km/h in 5 s. What will be the acceleration needed to achieve this condition? If the car has a mass of 800 kg, what will be the force needed to do this, considering no additional road/air resistance due to the increased speed?

We know that acceleration is rate of change of velocity, i.e., change in velocity divided by the time during which the change occurs.

Now speed in m/s = 5/18 $\times$ speed in km/hr

So, initial velocity, $V_1 = 5/18 \times 40 = 11.11$ m/s

Final velocity, $V_2 = 5/18 \times 60 = 16.67$ m/s

Therefore, by definition,

Acceleration, $a = (V_2 - V_1)/\text{time} = 5.56/5 = 1.11$ m/s^2

The corresponding force, which is needed to cause this acceleration, is given by the product of the mass on which this acceleration is to be produced and the acceleration, i.e.,

Force, $F = m \times a = 800 \times 1.11 = 888$ Newton

(Contd)

(*Contd*)

(In actual practice, the extra force will be more than this value because as the speed increases, the road friction and air resistance also increase, and so the force to be applied by the engine piston will be 888 N plus the additional forces of resistances to motion due to increased speed.)

If the distance covered by the car during the time is 69.4 m, we calculate the additional power requirement for covering the distance, which has to be supplied by the engine over and above what it was supplying earlier.

Whenever a force is applied on some body and the body moves some distance, some work is done. This work is given by the product of force and the distance moved. Hence, we first calculate the additional work as

$$W = F \times d = 888 \times 69.4 = 61627.2 \text{ J}$$

This work is done in 5 seconds. Therefore, the additional power the engine has to provide is 61627.2/5 W, or 12323.4 W, or 12.32 kW (approx.).

2.4 | ENERGY

Energy is the inherent ability of an object to do work. There are different forms of this energy inherent in a body. For example, mechanical energy (kinetic, potential), electrical energy, chemical energy (energy released by ordinary chemical reaction), nuclear energy (energy released by nuclear chemical reaction), etc. One form of energy can be converted to another form employing suitable machines/mechanism. Work is said to be done when energy is converted to kinetic energy, either as the end effect or intermediate effect. As a result, the unit of energy is the same as that of work, i.e., joule in SI unit.

Energy exists in many different forms. Examples of these are light energy, heat energy, mechanical energy, gravitational energy, electrical energy, sound energy, chemical energy, nuclear or atomic energy, and so on. These forms of energy can be transferred and transformed between one another.

In the car speeding up problem above, 61627.2 J mechanical work was done. The car engine did this work by using the heat it received from petrol when it burned. If 1 litre (l) petrol can produce 2,40,000 J upon burning, considering 100 per cent combustion efficiency and 30 per cent engine efficiency, only 72,000 J of energy will be converted to useful work by the engine. So, to achieve the speeding up, an additional amount of 0.86 l (61627.2/72,000) of petrol will be consumed by the car for the said journey (in actual practice, there will be transmission loss in the mechanical components that lie between engine and the wheels in a car as also increased road friction and air resistance will be present, and hence more than 0.86 litre of petrol would be necessary).

2.4.1 Kinetic Energy and Potential Energy

Out of the many forms of energy, we discuss here two important forms of energy, namely kinetic energy and potential energy.

Kinetic energy (KE)

The kinetic energy of an object is the extra energy that it possesses by virtue of its motion. It is defined as the work needed to accelerate a body of a specified mass from rest to its current state of velocity. Having gained this energy during its acceleration, the body maintains this kinetic energy unless its speed changes.

Potential energy (PE)

Potential energy can be thought of as energy stored within a physical system. This energy can be released or converted into other forms of energy, including kinetic energy. It is called potential energy because it has the potential to change the states of objects in the system, when the energy is released.

There are different types of potential energy. They are as follows:

Gravitational potential energy   Gravitational energy is the potential energy associated with gravitational force. For example, consider a book placed on top of a table. It will have more potential energy than a book on the floor. This fact forms the basis of work we get in a water turbine when water stored at a height is brought down to a lower level to drive the turbine.

Chemical potential energy   Chemical potential energy is a form of potential energy related to the structural arrangement of atoms or molecules that make up the matter. This structural arrangement (arising out of particular nature of chemical bonding) can undergo change through what is known as chemical reaction to release other forms of energy such as light, heat, sound, etc. For example, when fuel is burnt, the chemical energy is converted to heat.

Electrical potential energy   An object can also have potential energy by virtue of its electric charge. There are three main kinds of this type of potential energy, namely electrostatic potential energy, electrodynamic potential energy (sometimes called magnetic potential energy), and nuclear potential energy.

If an electric charge is placed in a space, the space is said to have a property called electric field. This electric field exerts a force on other electrically charged objects placed in its domain of influence. So, if another charge is brought from outside the domain of influence of the field (say, from infinite distance away) into the field, the charge will experience a force acting on it and either some work will have to be done (if the force is repulsive in nature) on the charge or the charge will do some work (if the force is attractive in nature).

Electrostatic potential energy Electrostatic potential energy is the energy associated with an electrically charged particle (at rest) kept in an electric field. It is defined as the work that must be done to move it from an infinite distance away to its present location, in the absence of any nonelectrical forces on the object.

Electrodynamic potential energy (or magnetic potential energy) In case a charged object or its constituent charged particles are not at rest but in motion, it gives rise to a magnetic field giving rise to yet another form of potential energy, often termed as magnetic potential energy. This kind of potential energy is a result of the phenomenon of magnetism.

Nuclear potential energy It is the potential energy of the particles inside an atomic nucleus, some of which are indeed electrically charged. This kind of potential energy is different from the previous two kinds of electrical potential energies because in this case the charged particles are extremely close to each other.

Elastic potential energy Elastic potential energy is the potential energy of an elastic object (for example, a rubber band or a catapult) that is stored inside when deformed by applying some kind of force.

Thermal potential energy Thermal energy of an object is the sum of average kinetic energy of particles constituting the object plus average potential energy of their displacement from their equilibrium positions.

We can say that potential energy is any form of energy, which has stored potential that can be put to future use as source of energy. For example, water stored in a dam for hydroelectricity generation is a form of potential energy and is in a state of relatively higher potential energy compared to a position at the base of the dam or further down. The force of gravity causes the water to begin to flow when the valves are opened. The gravitational potential energy of the water is converted to kinetic energy. The flowing water can turn a turbine (a rotor with blades having flow passages between them), which, in turn, converts the kinetic energy of the water into usable mechanical energy. An alternator or generator, mechanically connected (coupled) to the turbine, then rotates and converts the mechanical energy from the turbine into electrical energy. This electricity is then sent, via suitable transformers, to the electricity grid and to people's houses where it is converted into light energy (lights and television), sound energy (television, stereo), heat energy (hot water, toasters, ovens), mechanical energy (fans, vacuum cleaners, refrigerator compressors, air conditioners, water pumps), and so on.

2.5 | POTENTIAL

In continuation of the concept of potential energy, as outlined earlier, let us now discuss 'potential'.

Potential may be defined as the status of some entity (e.g., liquid in a tank, a metal plate at a given temperature or a terminal of an electric cell, etc.) in respect of its ability to do some work.

A body at a higher potential is said to be at a higher energy level than a body at a lower potential if there is no other form of energy except potential energy in it. So, when a body at a higher potential comes, by some mechanism, to a state of lower potential, there is flow of energy out of the body equal to the difference in the energy level between the two states of potential. This flow of energy can be gainfully utilized for the use of humankind through machines. A body is in a state of potential due to gravitational height, temperature, concentration of species, presence of positive electrical charges (positive potential), or presence of negative charges (negative potential), etc. Whenever a system at a state of higher potential is connected to another at a state of lower potential, energy will be exchanged and, eventually, potentials will become equal causing no further energy transfer, unless energy is added to maintain the potential difference.

Here we discuss the following potentials (Fig. 2.5):

1. Gravitational potential
2. Electrical potential
3. Temperature potential
4. Concentration potential

Gravitational potential It is the potential acquired by a body by virtue of its position with respect to the centre of the earth. A body at a higher height from the surface of the earth is at a higher potential than a body at a lower height.

Electric potential It is the electrical potential energy per unit of charge that is associated with a static (not changing with time) electric field. It is typically measured in volts. The difference in electrical potential between two points is known as voltage. Electric potential may be conceived of as 'electric pressure'. Where this 'pressure' is uniform (i.e., electrical potential is same everywhere), no current flows and nothing happens.

Temperature potential A body at higher temperature is at a higher thermal potential than a body at a lower temperature.

Concentration potential A space having a higher concentration of some species (say, carbon dioxide) is at a higher carbon dioxide potential than a space having lower concentration.

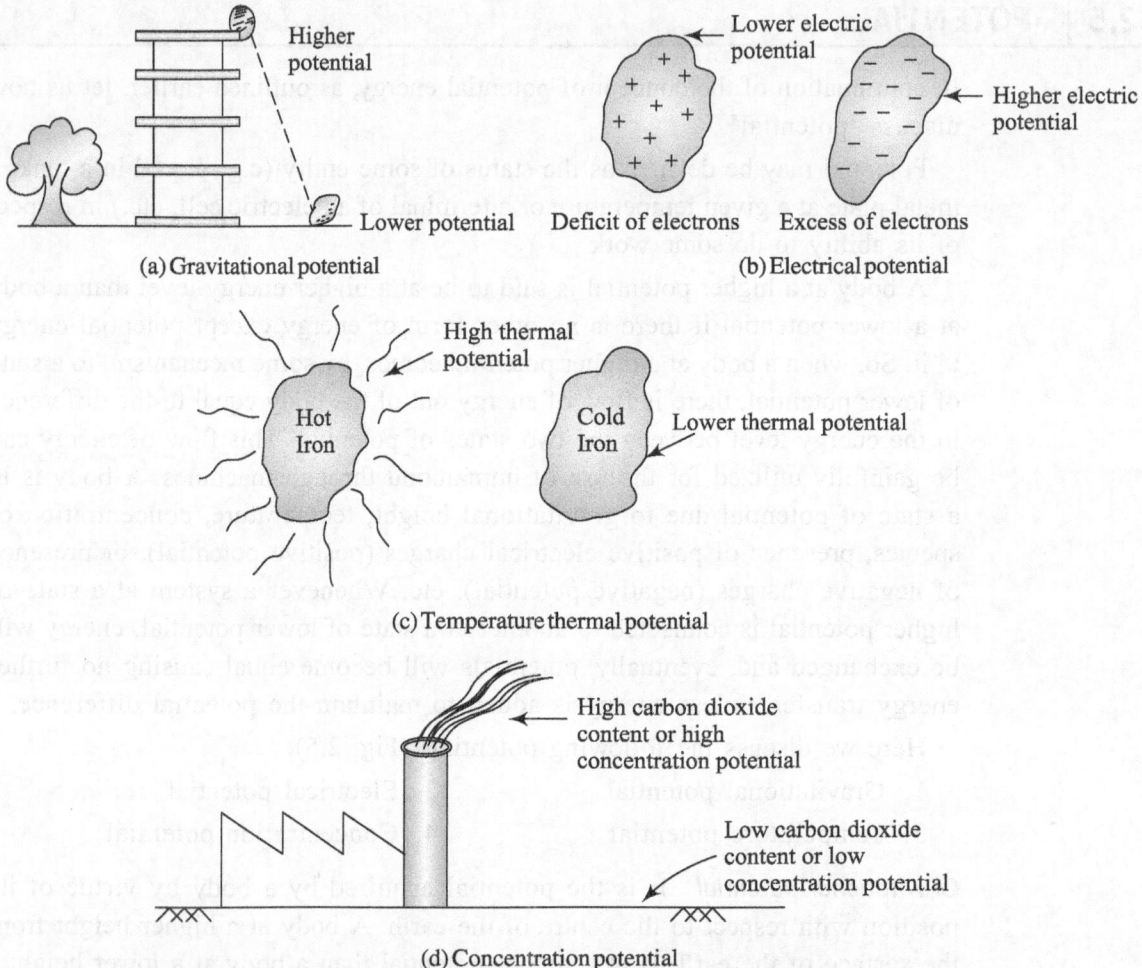

Fig. 2.5 Different types of potentials showing their high and low states

Figure 2.6 shows that water stored at a height can be connected to a reservoir at a lower level via a turbine (a rotating machine with blades inside through which water passes). When the water reaches the entry to the turbine, it is at a much lower gravitational potential and this loss of potential energy corresponds to a gain in kinetic energy of the water, i.e., the velocity of water increases at the cost of potential energy. This increase in kinetic energy of water is then transferred to the turbine shaft when the water passes through the turbine blades. Water eventually comes out of the turbine and discharges into the lower-level reservoir. So, two water reservoirs at different potentials, if connected by a conduit (conductor), will cause a flow of water.

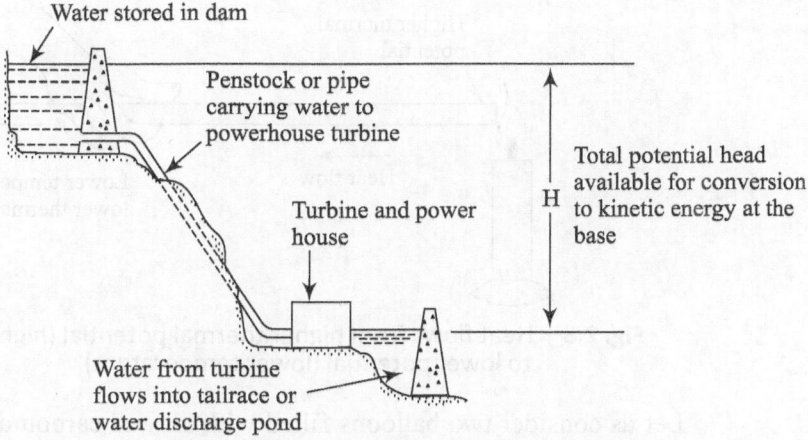

Fig. 2.6 Water turbine utilizes the potential energy of water in a dam to produce power in a generator

Similarly, we find that the two terminals of a battery which are at different electric potentials, if connected by copper wire (conductor), will cause flow of charges (or electrons or current) and transfer energy to the electric lamp in the circuit (Fig. 2.7).

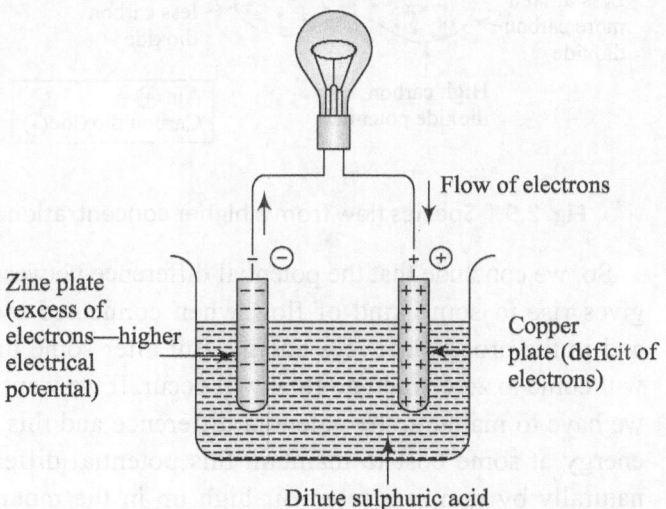

Fig. 2.7 Difference of electrical potential causes flow of electrons in a circuit

We also observe that if we connect one body at a higher temperature to a body at a lower temperature by means of a copper rod (conductor), heat flows from the body at a higher temperature to the body at a lower temperature (Fig. 2.8).

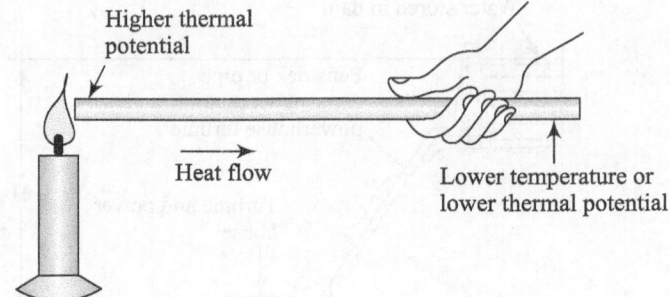

Fig. 2.8 Heat flows from higher thermal potential (higher temperature) to lower potential (lower temperature)

Let us consider two balloons filled with air and carbon dioxide, called species (Fig. 2.9). In exactly similar fashion, if one balloon has a higher concentration of carbon dioxide inside than another balloon and if both of them are connected by a tube, carbon dioxide will diffuse from the balloon with a higher concentration of carbon dioxide to the one with the lower concentration (Fig. 2.9).

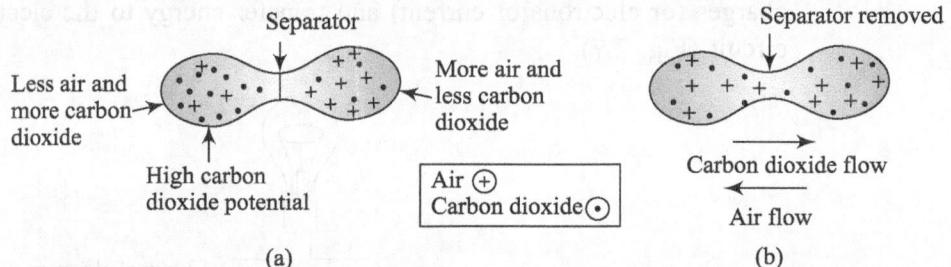

Fig. 2.9 Species flow from a higher concentration potential to a lower one

So, we conclude that the potential difference between two points in space always gives rise to some kind of flow when connected by an appropriate conductor, and in the process conveys energy. But after some time, the potential difference will come to zero when no flow will occur. If we want to maintain a constant flow, we have to maintain the potential difference and this is where we have to supply energy at some cost to maintain this potential difference (unless it is supplied naturally by rain at a reservoir high up in the mountain).

2.6 | ELECTRICITY AND FUNDAMENTALS OF ELECTRIC CIRCUIT

Electricity is the flow of electrical power or charge. It is a secondary energy source which means that we get it from the conversion of other sources of energy, such as coal, natural gas, oil, nuclear power, and other natural sources, which are called primary sources. The energy sources we use to make electricity can be renewable or non-renewable, but electricity itself is neither renewable nor non-renewable.

Whenever electric charge moves or flows, that is an electric current. The words 'electric current' are the same as the words 'charge flow'.

We can formally define current as the free flow of electrons (negative charge) in any conductor.

Whenever two points having electrical potential difference are connected by a conductor (a continuous material which can carry electrons for its flow), electricity flows. The potential difference may be likened to 'excess electrons to offer' at one point and 'an appetite for electrons' at the other; so when these two points are connected, there will be flow of electrons from one to the other till the number of free electrons at both the ends would equalize (same potential at the two ends or zero potential difference between them). So, if we want to maintain the flow, we have to create, by some mechanism expending energy (hence money), a permanent condition of electron excess at one and electron deficiency at the other end, i.e., maintain a potential difference between them. In a cell, this is achieved through chemical reactions.

Current flows from the point with excess electrons (points are given − sign) to one with deficiency of electrons (points are given + sign). But conventionally, it is shown to flow in the reverse direction (Fig. 2.7), i.e., from a positive terminal to a negative terminal.

2.6.1 Batteries Produce Electricity

Battery produces electricity using two different metals (called electrodes or more commonly terminals when they come out of the battery casing where we connect wires) immersed in a chemical solution. Chemical reactions between each of the electrodes and the chemicals cause excess of electrons to accumulate in one electrode than in the other. The end that has more electrons develops a negative charge since electrons are negatively charged (conventionally this end is called the positive terminal), while the other end develops a positive charge due to deficit of electrons (conventionally called the negative terminal). Thus, a potential difference is created between the two terminals. If a wire is attached from one end of the battery to the other, electrons flow through the wire to make the quantum of electrical charge same in both the terminals. In a Zn-Cu electric cell, one electrode is made of zinc (Zn) and the other is made of copper (Cu) and kept in a dilute solution of sulphuric acid (H_2SO_4), as shown in Fig. 2.7. Due to chemical reactions, positively charged hydrogen ions (H^+) come out of the solution and accumulate on the copper-plate electrode, making it positive (deficit of electrons). On the other side, positively charged zinc (Zn^+) ions come out of the zinc plate electrode and react with sulphate (SO_4^-) ions in the solution. As a result, the zinc plate becomes negatively charged (excess electrons). Thus, the two plates create a condition of what is called 'electric potential difference' between them. We note that if these two terminals are connected externally through a tiny bulb, the circuit is complete, current flows, and the bulb glows (Fig. 2.7 and Fig. 2.10). Dry-cell battery that we commonly use for various appliances work essentially on the same principle but the electrolytes in them are not in liquid state but are contained in the form of a low-moisture paste and hence the name 'dry cell'. The electrodes for a common dry-cell battery are zinc and carbon. Continuous chemical reaction has to take place to maintain the potential difference, and hence, current, through the circuit. Electrons flow from the negative end of the battery through the wire to the light bulb and back to the battery, thus completing the circuit. The chemical reaction inside the battery maintains the potential difference between the terminals and we note that during this time, the zinc plate gets constantly thinned because of departure of zinc ions from it. So, eventually, the plate has to be replaced, or more conveniently, the entire cell should be replaced. This means that to keep the flow of current continuing we have to expend energy (by replacing plate or cell) to keep the bulb glowing.

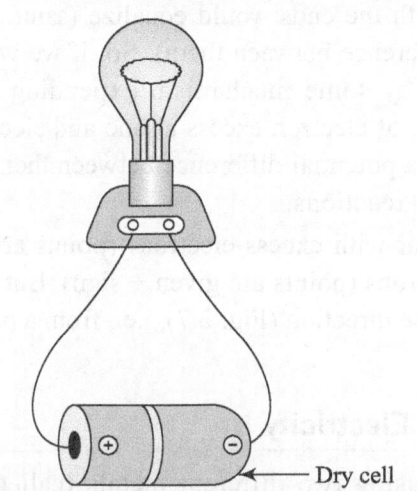

— Dry cell

Fig. 2.10 Battery produces electricity and lights a lamp

Electricity travels in closed loops

Electricity travels in closed loops or circuits (from the word 'circle'). If a circuit is open, the electrons cannot flow. When we flip on a light switch, we close a circuit and electricity flows from the electric wire through the bulb and back. When we flip the switch off, we open the circuit. Electricity ceases to flow through the bulb. When we turn a light switch on, electricity flows through a tiny wire inside the bulb and completes a path for continuous electric circuit. The wire gets very hot due to the resistance it offers to the path of electrons. The hot wire emits visible light. When the tiny wire has broken (Fig. 2.11), the path through the bulb becomes discontinuous. As a result, the whole circuit is broken and electrons can no longer reach from one terminal to the other. The bulb stops glowing and we need to replace the defective bulb with a new one. When we turn on the television, electricity flows through wires inside the set, producing pictures and sound using the components and circuitry inside. Electricity also runs motors, e.g., in ceiling fans, domestic pumps, washing machines, or mixers. Each of these systems (electric bulbs, fans, motors, etc.) is called load or resistance in a circuit. Electricity does a lot of work for us. So, it is quite legitimate to say that we live in a world of electricity.

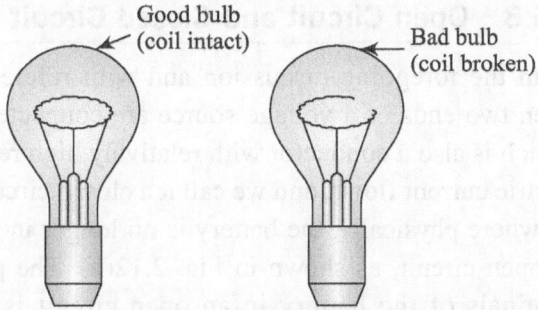

Fig. 2.11 Flow of electrons and hence electric current needs a continuous path to travel through

2.6.2 Symbolic Representation of the Electric Circuit

Let us now discuss a few electrical quantities.

Current It is the free flow of electrons in any conductor. It is represented by the symbol I. Ampere is the SI unit of current and it is represented by Amps and symbolically by α, e.g., a current of 5 Amps, or 5 A, or 5^{α}.

Volts It is the force or pressure (potential difference) that causes the flow of electrons in any closed circuit. The unit of this force is volt and is represented by V.

Resistance Any electrical appliance connected in an electric circuit is a load or resistance for the circuit against which the current has to flow, thereby consuming electric power. Resistance may be defined as that property of a substance which opposes the flow of electricity through it. It is represented by R. The unit of resistance is ohm (or Ω).

So the basic requirement in an electric circuit is a voltage source (a battery, etc.) and a load or resistance connected across the terminals (+ and −) of the voltage source by means of a conductor that carries the moving charge (or electrons). The battery-bulb circuit as shown in Fig. 2.10 can now be represented as an equivalent electrical circuit as follows (Fig. 2.12(a)).

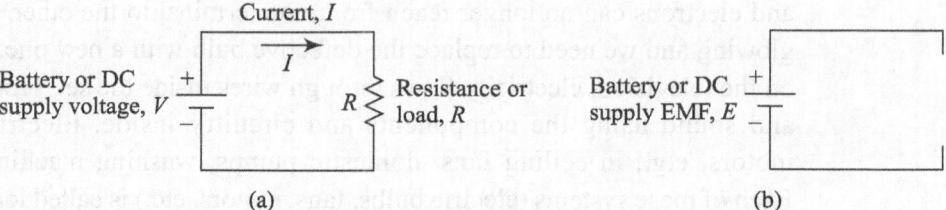

Current, I

Battery or DC supply voltage, V ⊣⊢

Resistance or load, R

Battery or DC supply EMF, E

(a)   (b)

Fig. 2.12 (a) Basic closed electrical circuit and symbols
(b) Open circuit configuration

2.6.3 Open Circuit and Closed Circuit

From the foregoing discussion and with reference to Fig. 2.12(a), we see that when two ends of a voltage source are connected via a conductor wire and load (which is also a conductor with relatively high resistance), the circuit is complete, electric current flows, and we call it a closed circuit. But, if we just snap the circuit anywhere physically, the battery is no longer in the closed circuit and it is called an open circuit, as shown in Fig. 2.12(b). The potential difference between two terminals of the battery in an open circuit is called the electromotive force (EMF). EMF is slightly higher than the potential difference, V, available for driving current through the external load because of internal losses of the battery as current flows through the battery. So, we formally define closed and open circuits as follows:

Closed circuit It is an electric circuit that provides an endless continuous path for uninterrupted supply of electric current.

Open circuit It is an incomplete electric circuit in which the normal path of current has been interrupted.

Distribution of voltage among loads

From a study of the circuit, we are permitted to say that the voltage available in the battery is necessary to overcome the resistance (the bulb in this case) in the circuit. Now we connect another bulb in series (to be discussed in some more details later) with the former bulb. We may redraw the circuit, as shown in Fig. 2.13.

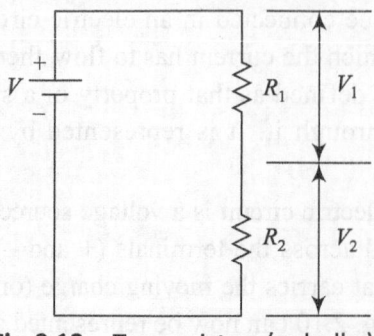

V ⊣⊢

R_1 V_1

R_2 V_2

Fig. 2.13 Two resistors (or bulbs) in series

In this case, we can say that the total voltage (V) available with the battery is shared by bulb 1 (resistance 1 or R_1) and bulb 2 (resistance 2 or R_2).

Or, $V = V_1 + V_2$

where V_1 and V_2 are the voltages needed for flow of current, I, through R_1 and R_2, respectively. They are also termed as voltage drop due to R_1 and R_2, respectively, to maintain a current, I, through them. So we find that in a circuit, voltage drop is equal to the voltage available. We shall discuss this in more detail later in this chapter.

2.7 | OHM'S LAW OF ELECTRICITY AND SIMPLE CIRCUITS

Ohm's law is the most fundamental law involving current and potential drop in a circuit. Two fundamental types of circuits, namely series and parallel circuits are introduced here. Simple circuit calculations will be shown with the help of Ohm's law to find out the current flowing through different electrical loads in a circuit and also the power consumed by them.

2.7.1 Ohm's Law

Ohm's law connects the voltage drop (voltage required to be supplied) across a resistance, the current flowing through it, and the resistance between points 1 and 2 in a circuit (Fig. 2.14). It is given by

$$V_{12} = R_{12}\, I_{12}$$

So, we can say that the voltage drop across a resistance is proportional to the current as well as resistance of the load.

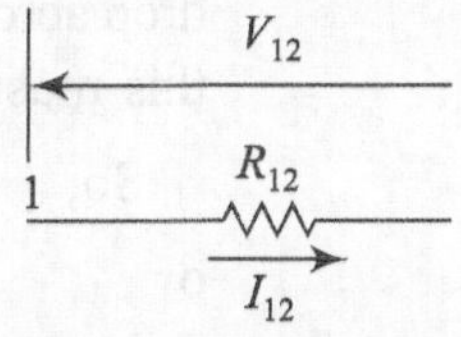

Formally, Ohm's law states the following:

Fig. 2.14 Definition sketch for Ohm's law

In an electrical circuit, the current passing through a conductor between two points is directly proportional to the potential difference (or voltage) across the two points, and inversely proportional to the resistance between them.

To put it mathematically,

$$V = RI \tag{2.4}$$

where potential difference, V, is in volts and current, I, is in ampere, and the resistance, R is in Ohms.

A closer look at Ohm's law shows that if the resistance is very small in a circuit, for a given voltage, the current will be very high, which may lead to a high temperature in the circuit, causing the melting of the wire and the associated fire hazard.

Power consumed by the lamp (or any resistor or load in the electric circuit)

Power is needed to be supplied (from the battery or domestic power supply points) constantly to maintain the voltage supply in order to maintain the flow of current. The power consumed by a load (or resistance) served by a battery or steady DC supply voltage source is given by

Power, $\qquad P = VI = I^2R = V^2/R$ (2.5)

where V is the voltage used by the load, I is the current through the load, and R is the resistance offered by the load to flow of current.

Any of the three forms may be used, as convenient. If voltage is in volts, current is in ampere, and resistance is in ohms, then power will be in watt.

Higher supply voltage means higher current for the appliance, and hence higher power output as also a risk of appliance burn out.

We refer to Fig. 2.12(a). Let R, which may be an electric bulb, be connected to a voltage source (battery or the domestic power supply, in which case of course the supply voltage will be much higher). Let the constant voltage supplied be 10 V. Let the resistance of the bulb be 2 ohms (2 Ω). We calculate the current flowing through the bulb.

If I be the current flowing through the bulb, according to Ohm's law, voltage drop across the bulb or the voltage required by the bulb will be $R \times I$ or $2 \times I$. Now, this must be supplied by the voltage source, which in the present case, is 10 V.

So, $\qquad\qquad 10 = 2 \times I$

or $\qquad\qquad\qquad I = 5$ A

So, for the present case, power consumed will be $V \times I = 10 \times 5 = 50$ W. We call the lamp a 50 W lamp under a supply voltage of 10 V. We get the same value if we calculate P as equal to V^2/R.

Now, suppose the lamp is connected to a 20 V supply. Will the power remain the same?

We must be clear about one thing—it is the resistance of the bulb (due to the long and fine wire inside the bulb, called the filament), which remains constant under different situations of supply voltage and is the characteristic property of the bulb. In the new situation, power will be $V^2/R = 200$ W. The current will be $V/R = 10$ A. So we find that both power consumed by the bulb and the current flow through the bulb have increased. The increased current may cause the bulb to burn out (a broken coil) due to the heat generated by the increased current.

Normally, therefore, we have a combination of wattage and the corresponding voltage supplied printed on the body of the bulb, the combination of which should be adhered to as far as practicable to get the safest and best performance.

Suppose we have a 100 W bulb at 230 V. We are to find out what will be the power consumed by the bulb and the current drawn by it if we connect it to a 110 V supply. Will the bulb glow brighter or dimmer?

From the given data, we can find out the resistance of the bulb, which is the only thing constant for the bulb and changes neither with the supply voltage nor with the way it is connected to a circuit. Power, P, resistance, R, and supply voltage (V) are connected by

$$P = V^2/R,$$

or $\qquad R = V^2/P = 230 \times 230/100 = 529\ \Omega$

Also, the current through the bulb is given by

$$V = RI$$

or $\qquad 230 = 529 \times I,$ or $I = 0.435\ \text{A}$

Now, let the bulb be connected to a 110 V source. So, the new quantity of power consumed by the bulb is given by

$$P = V^2/R = 110 \times 110/529 = 22.87\ \text{W}$$

[Resistance, R, of the bulb remaining the same, i.e., 529 Ω]

The corresponding current through the bulb is

$$V = RI$$

or $\qquad I = V/R = 110/529 = 0.208\ \text{A}$

We find that the power and current in the second case are less, and hence the bulb will be less bright in the second case. Conversely, if we connect it to a higher voltage line, the bulb will definitely glow much brighter but the current may be so high that it might heat up the filament to the point of melting.

2.7.2 Simple Circuits

Series Circuits

In a series circuit, resistances or loads are put one after another and current remains the same through all the resistances and equal to the main current. Voltage drops according to the particular resistance and total voltage drop will be equal to the applied voltage.

The circuit in Fig. 2.15 shows three bulbs in series and Fig. 2.16 shows three resistors connected in series, and the direction of current is indicated by the arrow. Figure 2.16 is the equivalent electric circuit.

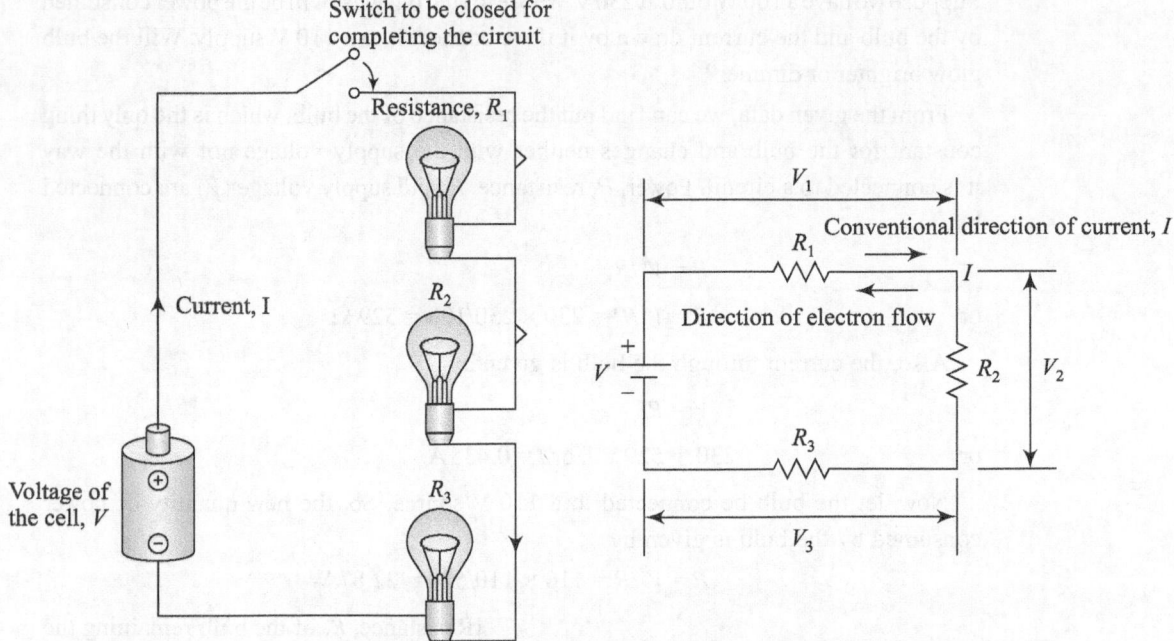

Fig. 2.15   Three bulbs in series   **Fig. 2.16**   Resistors in series (Equivalent circuit for Fig. 2.15)

Note that since there is only one path for the current to travel, the current through each of the resistors is the same which is equal to I, as shown in Fig. 2.16.

Total voltage drop (*i.e.*, voltage required by the resistors which is also the voltage to be supplied by the battery) is

$$V_1 + V_2 + V_3 = V_{req.} = V_{total}$$

Or $V = I R$ (Ohm's law),

$$IR_1 + IR_2 + IR_3 = V_{total} \qquad (2.6)$$

Here and in subsequent discussion, we neglect the voltage drop inside the battery.

We understand that since in the series circuit one resistor is put after another, the same current has to flow through them which is the circuit current.

$$V_{total} = I (R_1 + R_2 + R_3) \qquad (2.7)$$

Also, the voltage drops across the resistors must add up to the total voltage supplied by the battery, i.e.,

$$V_{total} = V_{supply} = I (R_1 + R_2 + R_3)$$

or current flowing through each resistor, I is given by

$$I = V_{supply}/(R_1 + R_2 + R_3) \qquad (2.8)$$

So we find that in the series circuit, as we add more devices, the denominator in Eqn 2.8 increases and the magnitude of the current through the circuit and hence the loads decrease, which is an obvious disadvantage because power that can be used by the devices goes on decreasing (power $= I^2/R$) with each additional load. So, if we use several electric bulbs, for example, in series, none will get full voltage and will glow less and each addition will further dim the light.

Parallel circuits

If several resistances R_1, R_2, R_3, ..., are joined in such a way that the positive (+) sides all resistances are connected at one point and the negative (–) sides at another point, the circuit is called a parallel circuit. Figure 2.17 shows three bulbs connected in parallel, and Fig. 2.18 and Fig. 2.19 show the equivalent electrical circuit for it.

In a parallel circuit: (a) voltage drop (voltage available) remains the same in each branch and (b) total current divides in separate branches.

Each of the three bulbs (or resistors) in Fig. 2.17 is one path for current to travel between points A and B and the current, in fact, flows through all the paths so that all the three lamps glow simultaneously.

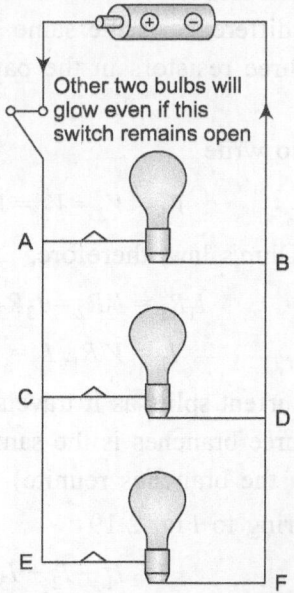

Other two bulbs will glow even if this switch remains open

Fig. 2.17   Three lamps in parallel

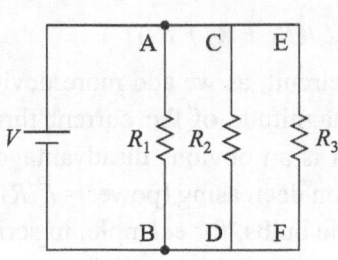 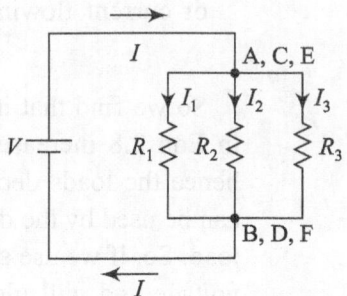

Fig. 2.18 Circuit containing resistors in parallel

Fig. 2.19 Example of a circuit containing three resistors connected in parallel, equivalent to Fig. 2.18

In Fig. 2.17, the junction points A, B, C, D, E, and F are called nodes. Figures 2.18 and 2.19 show the equivalent circuit for Fig. 2.17. Although Fig. 2.17 resembles the actual system closer, both will yield same analysis and results.

In a parallel circuit we observe that the magnitude of the current through each of them does not change, even if the order of connection (e.g., it could be in the order of R_3, R_2, and R_1 from left side of the figure) of the resistors are changed.

At A, the potential must be the same for each resistor. Similarly, at B, the potential must also be the same for each resistor. So, between points A, and B, the potential difference is the same whichever path we choose to follow, i.e., each of the three resistors in the parallel circuit must have the same voltage across them.

We can also write

$$V_1 = V_2 = V_3 = V_{AB} = V$$

Applying Ohm's law, therefore,

$$I_1 R_1 = I_2 R_2 = I_3 R_3 = V_{AB} = V$$

or $\qquad I_1 = V/R_1, I_2 = V/R_2,$ and $I_3 = V/R_3$

Also, the current splits as it travels from A to B. So, the sum of the currents through the three branches is the same as the current at A and at B (where the currents from the branches reunite).

Now, referring to Fig. 2.19

$$I = I_1 + I_2 + I_3 \qquad (2.10)$$

or $\qquad I = V/R_1 + V/R_2 + V/R_3 = V(R_1 + R_2 + R_3) \qquad (2.11)$

2.7.3 Appliances in Series Circuit

Three electrical appliances (one electric lamp, one electric range, and one room heater) are connected to a battery/DC supply voltage of 230 V, as shown in Fig. 2.20. We are to find the current flowing through each resistor and the power consumed by them. The power-voltage ratings of each appliance are given as follows:

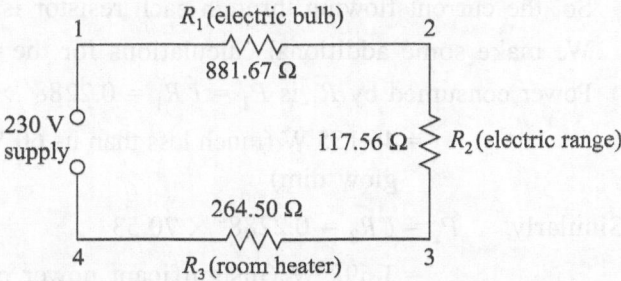

Fig. 2.20 Appliances (or loads) connected in series

Appliance 1 (electric lamp)   230 V/60 W

Appliance 2 (electric range)   230 V/750 W

Appliance 3 (room heater)   230 V/1000 W

We know that power = V^2/R, so the corresponding resistances (property of the appliance coils) are given by

$$R_1 = 230 \times 230/60 = 881.67 \ \Omega$$
$$R_2 = 230 \times 230/750 = 70.53 \ \Omega$$
$$R_3 = 230 \times 230/1000 = 52.9 \ \Omega$$

The first principle to understand about series circuits is that the amount of current is the same through all the resistors in the circuit. This is because there is only one path for electrons to flow in a series circuit, and because free electrons flow through conductors like water flowing in a tube. Just as the rate of flow (litre/minute, etc.) of water in the tube must be same everywhere along the tube, current should be also same in the series circuit everywhere along the length of the circuit. Secondly, the voltage available is distributed among the resistors as per requirement.

If the current be I, voltage required by each of the resistors will be, according to Ohm's law,

$$V_1 = IR_1 = I \times 881.67, \ V_2 = IR_2 = I \times 70.53, \ V_3 = IR_3 = I \times 52.9$$

Therefore, total voltage required,

$$V_{\text{total}} = V_1 + V_2 + V_3 = I \times (881.67 + 70.53 + 52.9)$$
$$= I \times 1005.1$$

This total voltage required (or voltage drop) must be equal to voltage supplied

or $\quad V_{\text{total}} = 230$ V,

or $\quad 1005.1 \times I = 230$ or $I = 230/1005.1$ Ampere $= 0.2288$ A

So, the current flowing through each resistor is 0. 2288 A

We make some additional calculations for the power consumed.

Power consumed by R_1 is $P_1 = I^2 R_1 = 0.2288^2 \times 881.67$

$\qquad\qquad = 46.155$ W (much less than its 60 W output and so, the bulb will glow dim)

Similarly, $\quad P_2 = I^2 R_2 = 0.2288^2 \times 70.53$

$\qquad\qquad = 3.692$ W (insignificant power output compared to its rated output of 750 W; the range will hardly heat the materials for cooking with this low wattage.

and $\quad P_3 = I^2 R_3 = 0.2288^2 \times 52.9$

$\qquad\qquad = 2.769$ W (rated output is 1000 W and the toaster will take possibly 360 times more time to toast)

So, we observe that all the appliances are operating at power level much below their individual rated output power level when connected in series.

Total power consumed by the loads is 51.616 W.

Total power could also be calculated as $P = V_{\text{supply}} \times$ Circuit current $= 230 \times 0.2288 = 52.624$ W (the small difference is due to round-off errors).

2.7.4 Parallel Circuit

We now place the same resistors in parallel across the same uniform voltage source of 230 V, as shown in Fig. 2.21. We have one source of voltage and three resistances in parallel. How do we use Ohm's law here?

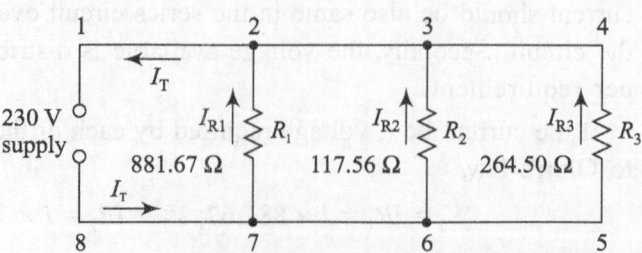

Fig. 2.21 Appliances (or loads) connected in parallel

Calculation of current and power

We find that points 1, 2, 3, and 4 are the same points electrically, and hence have the same electrical potential. Similarly, points 5, 6, 7, and 8 are also points having the same electrical potential. We are to find the electric currents, I_{R1}, I_{R2}, and I_{R3} and total current, I_T. From foregoing argument we can, therefore, write

$$V_{18} = 230 = V_{27} = V_{36} = V_{45}$$

Again from Ohm's law, $V_{27} = I_{R1}R_1 = 230$ or $I_{R1} = 230/881.67 = 0.261$ A

Similarly, $\qquad\qquad V_{36} = I_{R2} \times R_2$ or $I_{R2} = 230/70.53 = 3.261$ A

and $\qquad\qquad\qquad V_{45} = I_{R3} \times R_3$ or $I_{R3} = 230/52.9 = 4.348$ A

We see that maximum current flows through the smallest resistance, as expected (in the series circuit, the current through all the resistors were the same).

Total current drawn by the circuit from the battery is given by

$$I_T = I_{R1} + I_{R2} + I_{R3} = 7.870 \text{ A}$$

Note that in the series configuration, the total current drawn by the appliances were much less (only 0.2288 A).

Power drawn by each resistance is given by, $P = V^2/R$ (here, it is convenient to use this form)

$$P_1 = (V_{27})^2/R_1 = 230 \times 230/881.67 = 60 \text{ W}$$
$$P_2 = (V_{36})^2/R_2 = 230 \times 230/70.53 = 750 \text{ W}$$
$$P_3 = (V_{48})^2/R_3 = 230 \times 230/52.9 = 1000 \text{ W}$$

So, we observe that all the appliances operate at their desired rated output power level when connected in parallel.

Total power consumed is, therefore, $(60 + 750 + 1000)$ W = 1810 W.

Total power consumption can also be calculated as voltage supplied $\times$ total current drawn (using the form, $P = V_{\text{supply}} \times I_T$) = $230 \times 7.870 = 1810.1$.

We also observe that loads connected in parallel draw more current and power from the same voltage supply in comparison to when they are connected in series.

Now, we study the situation when the additional load of resistance 300 Ω is connected in parallel.

We observe that the full supply voltage of 230 V is available for each of the four loads, as before (in actual practice, it will fall by a very small value, as line loss will increase a bit due to increase in total current and hence, full 230 V will not be available but a little less, say 228 V or 225 V, depending on additional line loss). So, current through the earlier three appliances will remain as before (with no change in the power consumed) and the power consumed by the new appliance will be as per its rated power if the rated supply voltage is also 230 V.

From this example, we arrive at one very important conclusion that addition of new loads in a parallel circuit does not materially affect the power drawn by an appliance already in the circuit, whereas it is adversely and significantly affected in a series circuit, lowering the power output of each appliance in the circuit with each additional load being added.

Further, if we put a switch in a series circuit, putting this switch off means current flow through each of them will stop simultaneously and, thus, we cannot switch off an appliance of our choice. This defeats the very purpose of a switch, which should selectively put appliances off according to our choice. But in a parallel circuit, we can put an individual switch in each parallel path (e.g., in paths 2-7, 3-6, and 4-5) so that switching off a particular path will stop flow of current in that path, leaving the other part of the circuit as it is. So we can switch off either the bulb or toaster or range in any combination.

We can list the reasons for putting electrical appliances in a parallel circuit in an electrical installation system as follows:

- Loads connected in parallel draw more current and power from the same voltage supply in comparison to when they are connected in series.
- Loads connected in parallel work to their specified power output, while in series, they work much below their capacity, hence desired output effect will not be achieved.
- Addition of new loads in parallel circuits does not materially affect the power drawn by an appliance already in the circuit, whereas it is adversely and significantly affected in series circuit.
- Loads can be selectively switched off in parallel circuits without disturbing other loads, but, in series circuits, switching off one switch in the circuit stops current through the circuit and all the loads are, thereby, put off simultaneously.

We put domestic electrical loads in parallel not in series because in the former case, full supply voltage is available to each load, and hence they operate very near to their rated power (or wattage). Further, switching off one load will put off all the other loads in the house.

2.8 | PRINCIPLES OF GENERATION OF ELECTRICITY

2.8.1 How Electricity is Commercially Produced/Generated?

A generator is a device that converts mechanical energy into electrical energy. The process is based on the relationship between magnetism and electricity. In 1831, Michael Faraday discovered that when a magnet is moved inside a coil of wire, electrical voltage is induced in the wire (Fig. 2.22a) and if a circuit is made, current flows in the wire. In other words, when a wire made of conducting material cuts through a magnetic field, an electrical voltage is created across the wire ends.

Voltage can also be generated (or induced) in the coil if the magnet is kept fixed and the coil is moved (Fig. 2.22b). Moving the wire through the magnetic field creates an electric potential difference between the ends of the wire.

Note that the wire must be a part of an electrical circuit to produce current, and hence energy and power. If the ends are attached to a light bulb or to an electrical motor, the circuit is complete and electrical current flows and we get power. However, in practice, complex mechanical systems are needed to get stationary terminals from these rotating system of wire (also called armature) and then further elaborate and complex electrical arrangements before we get the supply lines at our houses.

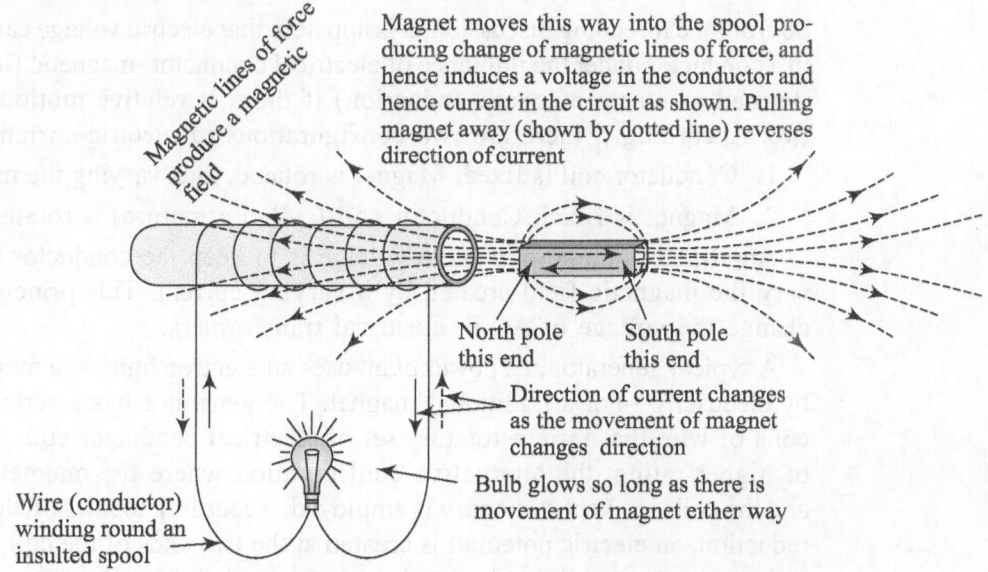

Magnetic lines of force produce a magnetic field

Magnet moves this way into the spool producing change of magnetic lines of force, and hence induces a voltage in the conductor and hence current in the circuit as shown. Pulling magnet away (shown by dotted line) reverses direction of current

North pole this end   South pole this end

Direction of current changes as the movement of magnet changes direction

Bulb glows so long as there is movement of magnet either way

Wire (conductor) winding round an insulted spool

(a) Principle of voltage generation by electromagnetic induction (when the conductor is fixed and the magnet moves)

The conductor moves towards the magnet, producing change in magnetic lines of force cut by it, and hence is induced by voltage being generated in the conductor and hence current in the circuit as shown. Moving the conductor away from the magnet (shown by dotted line) reverses direction of current

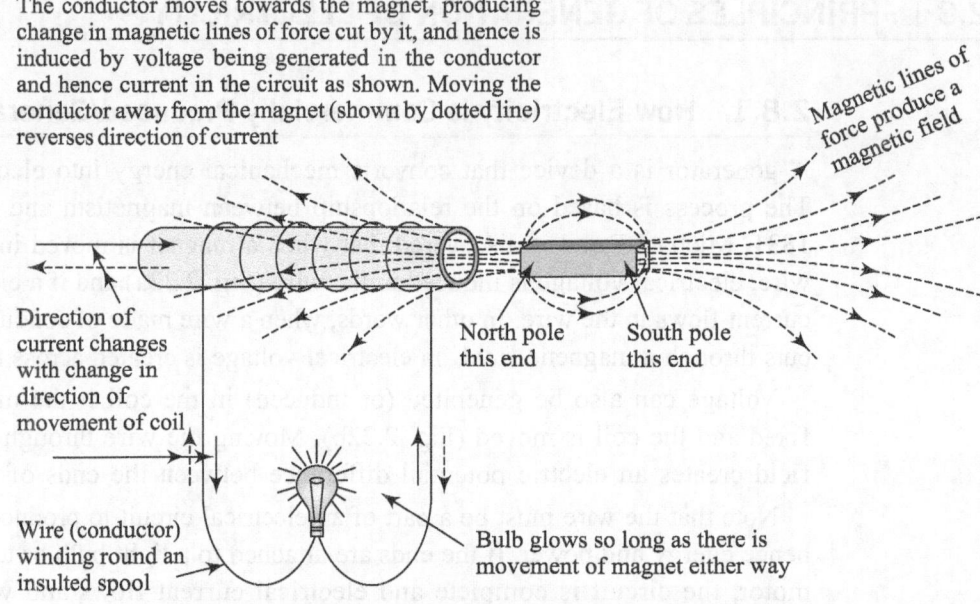

Magnetic lines of force produce a magnetic field

Direction of current changes with change in direction of movement of coil

North pole this end

South pole this end

Wire (conductor) winding round an insulted spool

Bulb glows so long as there is movement of magnet either way

(b) Principle of voltage generation by electromagnetic induction
(when the magnet is fixed and the conductor moves)

Fig. 2.22 Principles of generation of electricity

2.8.2 Basic Configurations of Electric Current Generators

So from the foregoing discussion, it is apparent that electric voltage can be generated in a conductor under the influence of electrical conductor–magnetic field interaction (termed as electromagnetic induction) if there is relative motion between the two. Accordingly, there are two configurations of electric-current generator:

1. Conductor coil is fixed: Magnet is rotated, thus varying the magnetic field.
2. Magnet is fixed: Conductor coil (called armature) is rotated.

Another technique to create a voltage is to keep the conductor stationary but vary the magnetic field created by a varying current. This principle is used to change the voltage of AC in electrical transformers.

A typical generator at a power plant uses an electromagnet—a magnet produced by electricity—not a traditional magnet. The generator has a series of insulated coils of wire that form a rotating set of electrical conductor coil. In a generator of higher rating, the alternative configuration where the magnets are moving and the coils are kept stationary is employed. According to laws of electromagnetic induction, an electric potential is created at the two ends of the coil and if the two ends are connected through an external resistance (or load), just as in case of two terminals of a battery are connected to an electric bulb (a load), electric current flows through it. This voltage is the source of electric power that is transmitted from the power house to the consumer's premises. We shall discuss further about generators in subsequent chapters.

2.9 | FUNDAMENTALS OF AC AND DC SYSTEM OF ELECTRICITY

There are two systems of electric generation and supply, namely,

1. Direct current or DC
2. Alternating current or AC

2.9.1 Fundamentals of DC

Direct current does not change its direction, i.e., electrons flow through the circuit in one direction only. This is because the polarities of terminals of the voltage source do not change. The positive polarity of the voltage source remains positive and negative polarity remains negative always, so that the current at all times flows in the same direction.

Direct current is produced by sources such as batteries, solar cells, and commutator-type electric generator of the dynamo type (also called DC generator), the working principle of which is described later. But the types of voltage (hence current) generated in case of a battery and in case of a DC generator are a little different. In case of a battery, the voltage generated is direct and steady, and hence the current in the circuit connected to the battery is almost uniform, but in case of a DC generator, the voltage produced is direct (+ve) but not steady (Fig. 2.23) and it rises and falls as the rotating armature inside the generator (discussed later) moves through its cycle, creating a pulsating direct current. However, in both the cases, current comes out of the generator/battery through the +ve terminal, flows through the +ve wire and returns through the (–ve) wire and goes back to the generator through the (–ve) terminal.

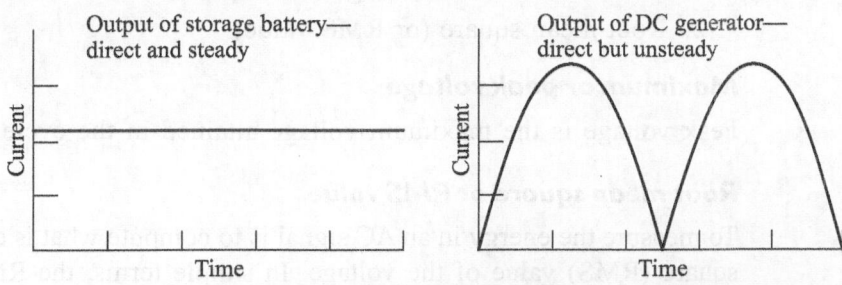

Fig. 2.23 Nature of direct current

2.9.2 Fundamentals of AC

The other type of electricity is called alternating current or AC. This is the electricity that we get from the wall plug base in our houses nowadays. We use this type of electric current to power most of our electrical appliances. The electricity is not provided as a single, constant voltage, but rather as a sinusoidal (sine) wave that varies in magnitude as well as polarity (+ve to –ve and so on) thus changing direction over time. It starts at zero, increases to a maximum value and then decreases to a minimum negative value; the cycle then repeats itself (Fig. 2.24).

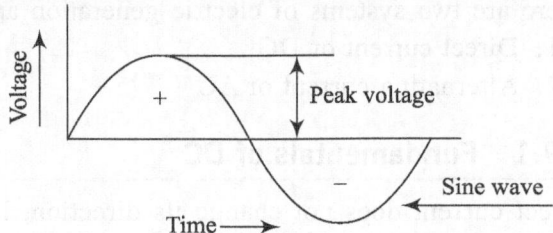

Fig. 2.24   Variation of AC voltage over time
(the sine wave)

We have noted that AC means alternating current or voltage, which alternates its direction and magnitude with the passage of time. It passes through a definite cycle consisting of two half-cycles. During one of these half-cycles, the relevant electrical quantity (voltage and current) acts in one direction in the circuit and during the other, in the opposite direction. Thus, current first flows in one direction (called +ve direction), then in the reverse direction (called -ve) direction, and this phenomenon repeats in all subsequent cycles in passage of time.

While simple direct current sources are generally described only by their voltage, alternating current circuits require more detail. First of all, if the voltage goes from a positive value to a negative value and back again, what do we say is the voltage? Is it zero, because it averages out to zero or is it the peak or maximum voltage?

Normally, the voltage of an AC is defined through the following specifications:
1. Maximum or peak voltage (or amplitude)
2. Root mean square (or RMS value)

Maximum or peak voltage

Peak voltage is the maximum voltage attained in the cycle (Fig. 2.24).

Root mean square or RMS value

To measure the energy in an AC signal is to compute what is called the root mean square (RMS) value of the voltage. In simple terms, the RMS value of an AC electrical voltage represents an equivalent DC voltage supply of that value, which will produce the same energy that the AC voltage would produce when connected across the same resistive load. It indicates, in essence, an average of the alternating current waveform. Whenever we see an AC voltage specification, it gives us the RMS number unless they say otherwise specifically. For example, in India, most homes have 230 V AC electricity. This means that if a heater is connected to this supply, it will have the same heating effect as if it is connected to a 230 V DC supply. Similarly, we have the RMS value for the AC current also. So, if we say that a load is connected to a 230 V AC and the current is 5 A, both the figures represents equivalent values in DC system, and hence power consumed can easily be calculated as $VI = 230 \times 5 = 1150$ W. Again, supposing we have a 230 V AC supply to bulb of resistance 880 Ω, we consider a 230 V DC supply connected to the bulb and calculate the power as V^2/R or about 60 W.

RMS value is 70.7 per cent of the peak value. So, 230 V AC has a peak voltage of 230/0.707 or 325 V, i.e., the AC will alternate between +326V and −325 V. Hence, we note that 230 V AC is more dangerous than 230 V DC to work with.

Different parts of the world use different voltages ranging from 100 V AC to 240 V AC for domestic uses, and of course, heavy equipment, such as heavy-duty electric motors, etc., can use much higher voltages for efficient running.

Frequency

The other key characteristic of AC is its frequency. It describes how many times in a second the voltage alternates from positive to negative and back again, completing one cycle. Consequently, it is measured in cycles per second (cps) or, more commonly, hertz (Hz). In India and the rest of world (except the USA), the standard is 50 Hz, meaning the voltage changes sign 50 times in a second from positive to negative. In the USA, the standard is 60 Hz.

It is quite evident that the AC voltage passes through zero conditions twice every cycle. So if a bulb is connected to it, it should become off when the voltage becomes zero. But we cannot perceive this, as this automatic switching off–on occurs so fast (AC has 50 cycles per second and hence passes through zero voltage 100 times or the circuit switches 100 times off in one second) that our eyes cannot distinguish it (our eyes can distinguish between two events only if one occurs at an interval of more than 1/16 seconds after the other).

2.9.3 Principles of AC Generator (Alternator)

Unlike the constant polarity of the two terminals of a battery, and hence a constant direction of current through the circuit connected to it, an AC generator creates a voltage and hence current through the circuit connected to it, which changes its magnitude as well direction. Figure 2.25 shows the basic arrangement of an AC generator. It consists of a coil of wire called armature, the magnets called poles, the two rotating rings to which the ends of the coil are rigidly fixed, called slip rings, S_1 and S_2, and two carbon brushes, which are fixed and do not rotate with the rings. The rings have always positive contact with the brushes. The brushes, therefore, form the two ends of the generator like the terminals of a battery to which the external circuit is connected, as shown in Fig. 2.25.

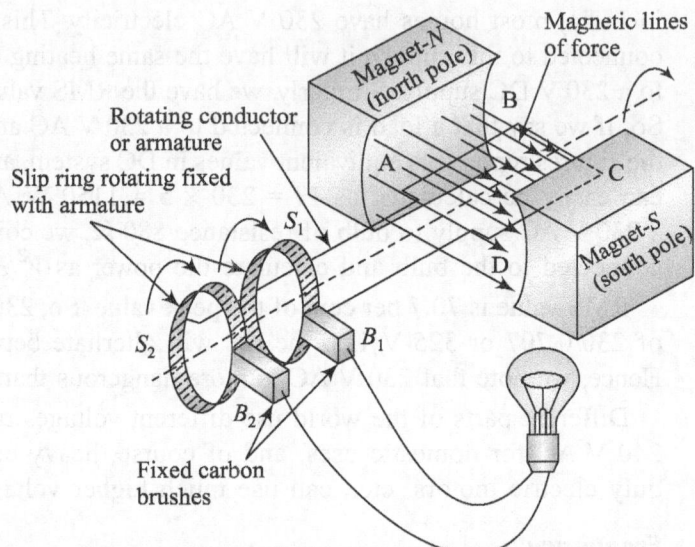

Fig. 2.25 Basic working of an alternator (or AC generator)

For each half rotation of the armature (Fig. 2.26), the current starts at zero with the wire moving parallel to the invisible magnetic lines of force (pointing from north pole to south pole) as shown. The current increases to its maximum level at the point where the wire cuts through the lines of flux at a 90° angle. The current then returns to zero a quarter of a rotation later when the wire again moves to become parallel to the lines of flux.

At this point, the armature has completed one-half rotation through the magnetic field, starting at zero current, moving to its maximum current and then returning to zero. Now, however, things get a little interesting as the rotor begins its second half of the rotation.

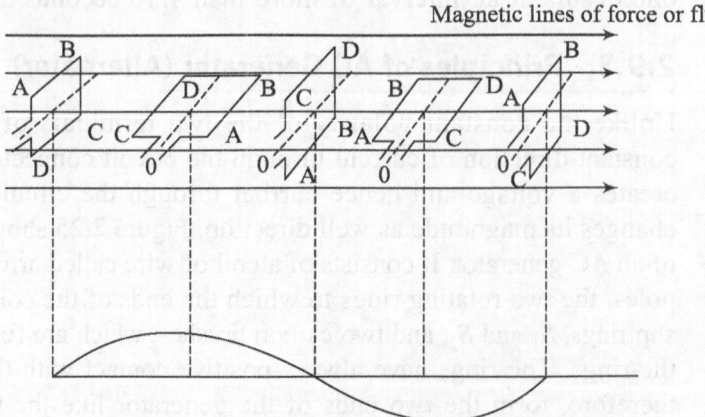

Fig. 2.26 Development of alternating voltage

The wire loop, through which current passed in one direction during the first half rotation, will now be 'upside down' relative to the magnetic field for the next half of rotation. So the electromagnetic forces responsible for inducing the voltage, which, in turn, caused current to flow through it in one direction during the first half rotation will now be causing current to flow in the opposite direction for the other half rotation.

In case of an AC generator, the ends of the armature coils stick to the same slip ring while rotating (Fig. 2.25). Now one end of the coil will have positive polarity for 180° rotation (half of the rotation) and will have negative polarity for the other half. So the corresponding slip ring and the terminal carbon brush attached to it will have +ve and –ve polarity, alternately producing alternating current in the circuit connected to the generator. Thus, when an alternator produces AC voltage, the voltage switches polarity over time, but does so in a very particular manner. When graphed over time, the pattern or 'wave' traced by this voltage of alternating polarity takes on a distinct shape, known as a sinusoidal or sine wave. Figure 2.27 shows a typical variation of voltage available at the slip-ring brush of the AC generator with angular position of the rotating coil of the alternator (the other name of AC generator).

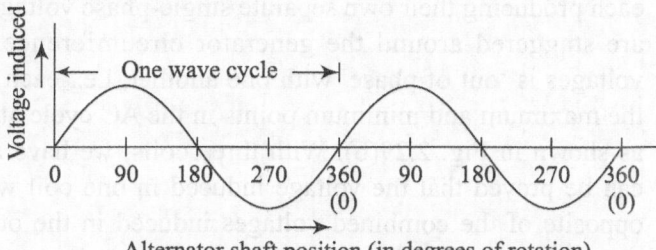

Fig. 2.27 Alternator voltage as function of shaft position

Since the horizontal axis of this graph can be thought as the passage of time as well as shaft position in degrees, the dimension marked for one cycle is often measured in a unit of time, most often seconds or milliseconds. The pattern would be exactly similar, and hence we can think of the plot in Fig. 2.27 as a variation of AC voltage with the passage of time.

2.9.4 Phase

We consider two alternating voltages, having waveforms A and B, produced by some means. They have the same peak and frequency but still they may differ in one very important aspect. We take the help of Fig. 2.28 to examine that aspect. Upon close observation of the two AC waveforms, we find that they do not reach their peaks or zero simultaneously, i.e., they are not synchronized, or to put it otherwise, they are out of phase with each other.

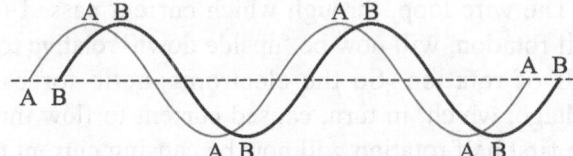

Fig. 2.28 Out of phase waveforms

In technical terms, this is called a phase shift. If, in an alternator/generator, instead of only one coil, we put more coils placed at some angles with one another, we get AC voltages from each of them but with mutual phase shifts. Accordingly, AC system can be single-phase, two-phase, or three-phase configuration.

However, because of technical and other compatibility reasons, only single-phase and three-phase electricity is supplied to consumers and the electrical appliances and machines are also designed on the basis of either single-phase or three-phase supply. Normally, electrical motors are three-phase machines while lighting and heating/cooking appliances use a single-phase supply.

We will discuss in brief only about three-phase electricity generation principles.

Three-phase generators usually have three separate armature coils (Fig. 2.29), each producing their own separate single-phase voltage. Since these coil windings are staggered around the generator circumference, each of the single-phase voltages is 'out of phase' with one another, i.e., each of the three-phases reaches the maximum and minimum points in the AC cycle at different times (or phases), as shown in Fig. 2.29(b). With three coils, we have a three-phase system and it can be proved that the voltage induced in one coil will always be equal and the opposite of the combined voltages induced in the other two coils. This permits us to join one end of each coil together to produce a zero voltage, which is taken out of the generator as neutral line (Fig. 2.29(a) and Fig. 2.30). The other three ends of the three coils form the three live lines in the distribution system. L_0 represent neutral line while L_1, L_2, and L_3 denote the three live lines or three-phases. The voltage between L_0 and L_1 is called phase voltage, while that between L_1 and L_2, is called line voltage. Line voltage is 1.73 times the phase voltage for sine wave. In domestic supply, we get L_0 and any one of L_1, L_2, and L_3. In India, voltage between one-phase and neutral (or phase voltage) is 230 V and between any two-phases or line-to-line or, in short, line voltage is 1.73×230 or 400 V. So we get 230 V supply in our domestic connection, while industrial establishments, where three-phase supply is provided, get 400 V as well as 230 V supply. Of course, this supply voltage combination system is not obtained right at the terminals of the generator but at the local street transformer which is finally distributed to the local street poles for commercial use.

We observe that single-phase and two-phase systems can be obtained only from a three-phase system, and hence AC generator produces only a three-phase AC.

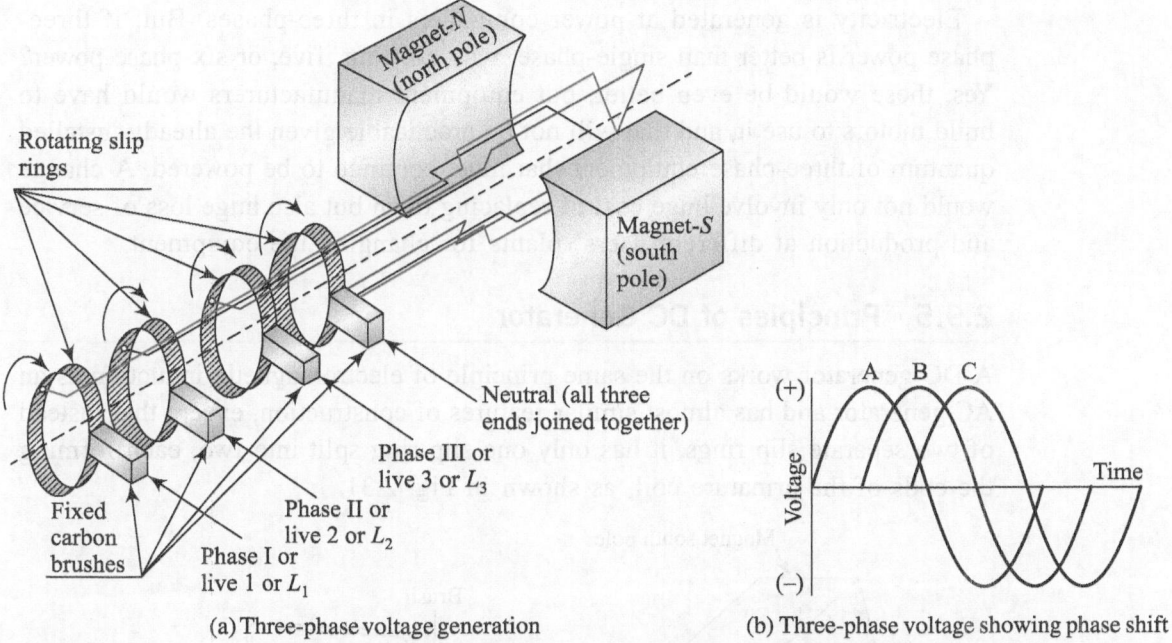

(a) Three-phase voltage generation

(b) Three-phase voltage showing phase shift

Fig. 2.29   (a) Three-phase voltage generation and (b) phase shift

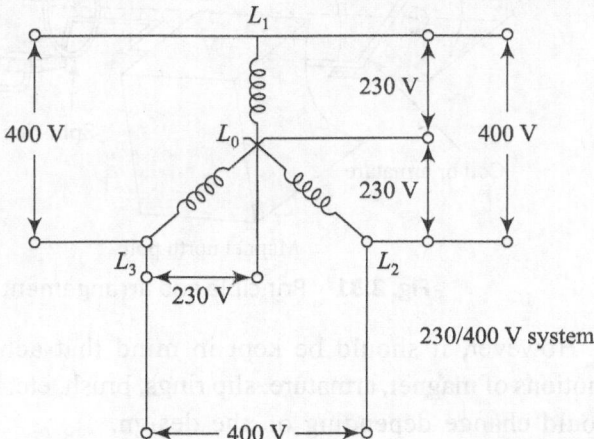

230/400 V system

Fig. 2.30   Three ends of the three coils meet at a point to make a point of zero potential and neutral line is taken out of it (refer to Fig. 2.29)

Three-phase power is designed especially for large electrical loads where the total electrical load is divided among the three separate phases. As a result, the wires and transformers will be less expensive than if these large loads were carried on a single-phase system.

Electricity is generated at power companies in three-phases. But, if three-phase power is better than single-phase, why not four, five, or six-phase power? Yes, these would be even better, but equipment manufacturers would have to build motors to use it, and that will not be practicable given the already installed quantum of three-phase equipment that must continue to be powered. A change would not only involve huge cost of replacing them but also huge loss of service and production at different users' plants for changing the equipment.

2.9.5 Principles of DC Generator

A DC generator works on the same principle of electromagnetic induction as an AC generator and has almost similar features of construction, except that instead of two separate slip rings, it has only one slip ring split into two, each forming the ends of the armature coil, as shown in Fig. 2.31.

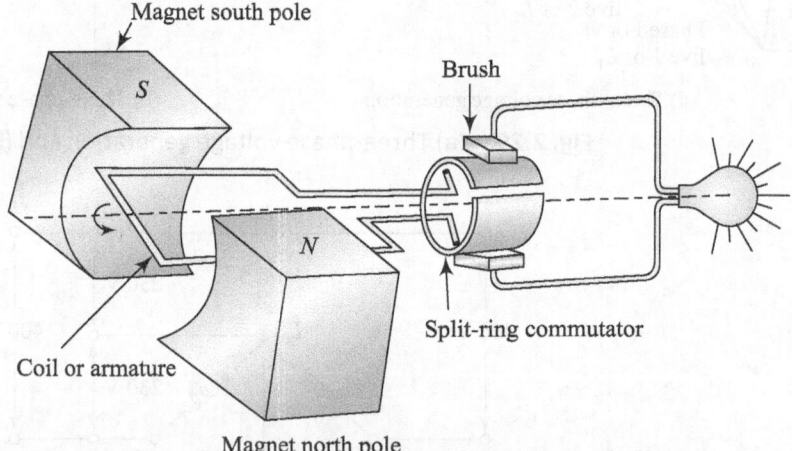

Fig. 2.31   Principle and arrangement of DC generator

However, it should be kept in mind that actual arrangement and relative motions of magnet, armature, slip rings, brush, etc. for both AC and DC generators could change depending on the design.

These connections pass current to a different contact (or brush) for each half of the armature cycle, i.e., each brush makes contact with each end of the wire for only half a rotation. So, current reaching each brush always flows in the same direction, regardless of the position of the armature (Fig. 2.32).

Thus, a DC generator produces direct current but unlike the constant potential of a battery, a DC generator creates a series of current fluctuations. This is known as pulsating direct current.

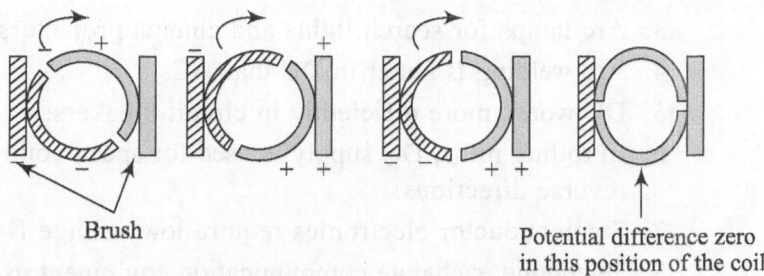

Brush

Potential difference zero
in this position of the coil

Fig. 2.32 Polarity at the carbon brushes always remain the same
due to split-ring design

2.9.6 Advantages of AC

Why does standard electricity come only in the form of alternating current?

There are a number of reasons, but one of the most important characteristics of AC is that it is relatively easy to change voltages from one level to another using a transformer. Transformers work with voltages that change with time and this is the reason why it works for AC but not for DC. This capability allows the companies that generate and distribute electricity to do it in a more efficient manner, by transmitting it at high voltage for long lengths, which reduces energy loss due to the resistance in the transmission wires. Another reason is that it could be easier to mechanically generate alternating current electricity than direct current. We summarize briefly the advantages of AC as follows:

1. Higher voltages can be generated (up to 1,32,000 V or even more), when DC can go up to 650 V only.
2. AC voltage can be increased or decreased with the help of transformers.
3. AC transmission and distribution is more economical, as the material (say, copper) can be saved by transmitting power at higher voltage.
4. For the same power, AC motors are cheaper, light in weight, require lesser space, and require less attention in operation and maintenance as compared to DC motors.
5. AC can be converted to DC easily, when and where required, but the reverse cannot be done.

2.9.7 Advantages of DC

DC supply system has its own fields of application with very distinct advantages as listed below:

1. DC series motors are most suitable for traction purposes in tram cars, railways carriages, and lifts.
2. DC is required for electroplating, electrolytic, and electrochemical processes.

3. Arc lamps for search lights and cinema projectors work in DC.

4. Arc welding is better in DC than AC.

5. DC works more efficiently in circuit breakers.

6. In rolling mills, DC supply is used for speed control in both forward and reverse directions.

7. Semiconductor electronics require low voltage DC supply.

8. Telephone exchange communication equipment use standard low voltage DC power supply.

9. Solar cells and hydrogen fuel cells all produce DC.

2.10 | DISTRIBUTION OF ELECTRICAL ENERGY AND THE TRANSFORMER

Electricity is produced in power plants and it has to be carried to the consumers' premises (industrial and domestic), long distances away. Transmission lines are used to carry electricity (Fig. 2.33). It can be shown that if electricity is carried at high voltage, the loss of energy in transit is much less compared to when it is transported at low voltage via transmission lines. The voltage of the electricity produced in a generating station is normally 13 kV in India.

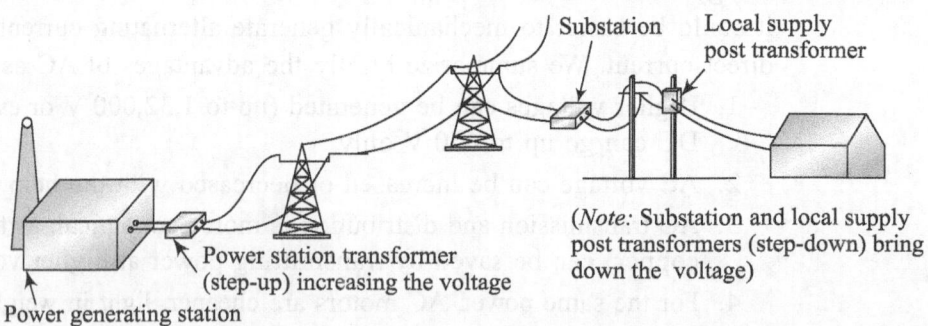

Substation   Local supply post transformer

(*Note:* Substation and local supply post transformers (step-down) bring down the voltage)

Power station transformer (step-up) increasing the voltage

Power generating station

Fig. 2.33   Electricity is carried at high voltage to locations away from the generation station

To solve the problem of sending electricity over long distances, William Stanley developed a device called a transformer, which increases (or steps up) the voltage to any desired level. The transformer allows electricity to be efficiently transmitted over long distances, making it possible to supply electricity to homes and businesses located far from the electricity-generating plant. Recently, it has been possible to send electricity at a voltage of 750 kV although 400 kV, 132 kV, 66 kV, and 33 kV transmission lines are more common. Transformers can also decrease the voltage as required and these transformers are called step-down transformers.

Upon reaching the consumers' locality, the voltage again needs to be decreased (or stepped down) because industrial motors and domestic electrical appliances work with 230V/400V lines. Usually, this is done in the substations at two stages. First, the high-voltage input is stepped down to 6 kV (or 3 kV) and then to 230/400 V. So, at these localities, step-down transformers (situated in what is called a substation) are again used. From the substation, distribution lines carry the electricity to industrial establishments and houses (Fig. 2.33)

2.11 | CONCLUSION

From the discussion in this chapter, it is quite clear that elementary knowledge of basic mechanics and electricity is vital for common people's appreciation of the working marvels of modern engineering equipment in everyday use. It is apparent from what has been discussed that understanding the basics of mechanics and electricity is helpful for working with the equipment, as used in the hotel industry, in a safe and efficient manner. A discussion of electrical machines was also included here which gave some idea about electrical generators and motors generously used in rotary machines everywhere. The essential concepts of electrical power transmission and distribution were also touched upon for the sake of completeness in the basic understanding.

KEY TERMS

AC	AC is defined as the kind of electrical voltage/current, which not only changes its magnitude but also its sign as time passes, in a definite manner.
Acceleration	It is the rate of change of velocity.
Amplitude	It is the maximum value reached in a waveform.
Current	It is the continuous flow of electrons between two points in an electric circuit.
DC	DC is defined as the kind of electrical voltage/current that may or may not change the magnitude but the direction of current (i.e., the sign of polarity of the voltage source terminals) will never change.
Electric potential	It is the electrical potential energy per unit of charge that is associated with a static (not changing with time) electric field. It is typically measured in volts.
Energy	It is the ability of an object to do work by virtue of its physical state of rest, or of motion, or both.
Force	It is a physical quantity that either tends to displace an object or actually changes its position from its present state.

	Examples are gravitational attraction force, magnetic force, muscular force of human being, etc.
Frequency	It is the number of times a cycle is repeated in one second.
Generator	It is a device that generates electricity on a large-scale basis, commercially.
Neutral	This is a line in electrical distribution system that has zero potential.
Peak voltage/current	This is the maximum voltage/current in an AC waveform.
Phase	This is the shift in the time of occurrence of similar conditions in two or more waveforms of same frequency and amplitude.
Potential	Potential may be defined as the status of some entity (e.g., liquid in a tank, a metal plate at a given temperature, or a terminal of an electric cell, etc.) in respect of its ability to do some work.
Potential difference	The difference in electrical potential between two points is known as potential difference or popularly as 'voltage'.
Potential energy	Potential energy can be thought of as energy stored within a physical system.
Power	It is the rate at which some work is done.
Resistance	It is the opposition offered by any electrical appliance or motor to the flow of current in a circuit.
RMS current	It is the DC equivalent current for an AC voltage system.
RMS voltage	It is the DC equivalent voltage for an AC voltage system that produces the same heating effect to a load as the AC system.
Scalar	A quantity is said to be scalar if it has only magnitude such as mass, temperature, etc.
Speed	It is the rate at which a body traverses actual distance.
Transformer	It is an electrical device which increases or decreases electrical voltage.
Turbine	Turbine is a mechanical rotary device which moves the armature coil of the generator by getting energy from high-pressure and high-temperature steam (steam turbine) or high-velocity water (water turbine).
Vector	A quantity is said to be vector if, in addition to magnitude, direction is also needed to fully define it such as velocity, acceleration, force, etc.
Velocity	It is the rate of change of position of a body.
Work	It is the product of force and displacement in the direction of force.

REVIEW QUESTIONS

2.1 Distinguish between speed and velocity.

2.2 What are the two different systems of units for force? How is kgf related to Newton as a unit of force?

2.3 Define work and power.

2.4 State the units of force, work, and power.

2.5 Explain what do you mean by electricity.

2.6 Define potential and potential energy.

2.7 What is electrical potential and what is its unit?

2.8 What are the common commercial sources of potential difference for continuous supply of electric current?

2.9 State Ohm's law.

2.10 Explain, with sketches, the terms AC and DC.

2.11 Define the terms peak voltage, RMS voltage, frequency, and phases in relation to AC system.

2.12 Explain why 230 V AC supply is more dangerous to work with than 230 V DC supply.

2.13 State the advantages and disadvantages of AC and DC system of electric voltage supply.

2.14 Sketch a simplified generator system to show its working.

2.15 What is the basic principle on which generators work?

2.16 Explain single-phase and three-phase AC systems.

2. 17 Name a few sources of electrical power other than conventional fossil fuel thermal power.

2. 18 What is the function of a transformer in an electric transmission and distribution system?

2.19 Explain why electricity is transmitted over long distance at a very high voltage.

2.20 What is a step-up transformer? Why is it used in electrical transmission system?

2.21 What type of transformer would you expect to be present in a locality substation?

REVIEW PROBLEMS

2.1 A car is moving at a speed of 30 km/h. The driver wants to increase the speed to 60 km/h in 10 s. What will be the acceleration needed to do this? If the car has a mass of 600 kg what will be the force needed to achieve this speed? Neglect any change in resistance to motion one to this acceleration. $[0.833 \text{ m/s}^2, 500 \text{ N}]$

2.2 A force of 100 N is applied to a body of mass 20 kg moving at a uniform speed of 2 m/s. What will be the acceleration developed on the body? What will be its speed after 3 s, if the force continues to act? $[5 \text{ m/s}^2, 17 \text{ m/s}]$

2.3 What is the kW equivalent of a power of 2 hp. $[1.492 \text{ kW}]$

2.4 You buy a 1500 W, 230 V electric kettle from the market. What is the coil resistance for the kettle? The kettle is connected to a supply of 230 V but on a particular time, the voltage drops to 200 V. What will be the current passing through the kettle ? What power will it consume? Will the kettle boil water faster or slower?

$[35.267 \ \Omega, 5.671 \text{ A}, 1134.19 \text{ W}, \text{It will boil slower.}]$

3

Chapter

Basic Electrical Wiring, Safety, and Electrical Tariff

Learning Objectives

This chapter deals with the fundamentals of electrical wiring and related elements found in domestic buildings as well as commercial establishments. The objectives of the chapter are to:

- make the readers familiar with the methods by which a connection is obtained from the electric supply post in the street
- help understand the basic philosophies and circuit arrangement of wiring and safety precautions to be taken while handling electricity
- make the readers understand the basic components of an electrical wiring system such as fuse, earthing system, switches, etc. as installed in a hotel
- introduce the principles of electric energy metre and electrical tariff system and help make simple calculations for electricity bill

---

3.1 | INTRODUCTION

Electrical wiring in a domestic set-up is a familiar experience. It is used to carry electric supply to the premises and distribute it at various locations where many different electrical equipment and appliances (or electrical load) get connected via plugs and sockets and receive the necessary power to run. After going through the chapter, the reader should be able to understand the basics of electrical wiring and other features in the distribution system and to make electrical tariff calculation for simple combinations of domestic loads. The reader should also be able to handle the appliances safely keeping in mind the parameters influencing both personal safety and condition of the appliance.

3.2 | CONDUCTORS AND INSULATORS—PROPERTIES AND APPLICATIONS

A good knowledge of the properties and applications of electricity conductors and insulators is important for the basic understanding of electrical wiring and safety. The following gives a brief introduction to these materials.

3.2.1 Electrical Conducting Property of Materials

The current flowing through a substance placed in a circuit for a given potential difference is determined by the number of free electrons in the substance. Substances like metals, which have a large number of free electrons, help in moving charge from one end of the substance in the circuit to the other end and are said to offer low resistance to the flow of current. They are, therefore, called conductors. Some materials such as wood have very few free electrons in them, so they offer high resistance to the flow of current. This material property of matter offering resistance to flow of current is called resistivity or specific resistance. If we have a length of wire, its resistance depends upon the length (higher the length, higher the resistance), cross-sectional area (higher the area, lower the resistance), and material property (a copper wire of the same length and cross-sectional area as an aluminium wire will have lower resistance than the aluminium wire) called resistivity. The dependence of resistance on different factors is shown in Fig. 3.1. This relationship may be expressed mathematically as

$$R = r\, l/A \tag{3.1}$$

where R is the resistance, r is the resistivity or specific resistance of the material, l is the length of the wire, and A is the cross-sectional area of the wire.

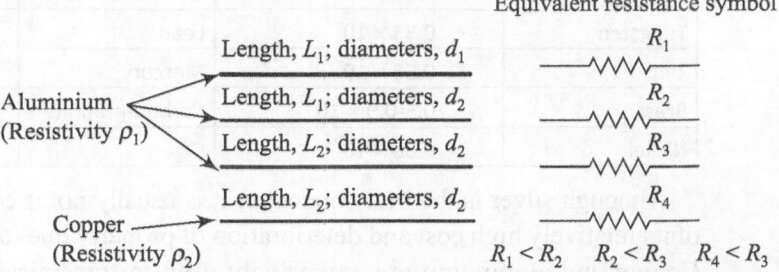

Fig. 3.1 Wire resistance changes with length, diameter, and nature of material

So, two pieces of electrical wires having the same diameter (or cross-sectional area of the wire) and length will offer different resistance to electrical current flow if the resistivity (i.e., material of the wire) of the two pieces of wires are different. So, a material having low resistivity for carrying electrical current must be chosen.

The unit of resistivity is clearly ohm-m (Ω-m).

Resistivity of a material is, thus, defined as the electrical resistance offered by a piece of wire of the material of unit length and unit cross-sectional area.

The reciprocal of resistivity is conductivity. Materials generally vary widely in their electrical conducting properties, i.e., in their magnitudes of resistivity. Accordingly, they are classified as

- Conductors (very low resistivity or high conductivity)
- Insulators (very high resistivity or poor conductivity)

Conductors

Conductors are substances through which electricity can pass easily without much resistance. Copper, aluminum, iron, etc. are examples of conductors. Silver is the best conductor of all metals.

We can further classify conductors as

- Good conductors
- Moderately good conductors

Tables 3.1 and 3.2 provide a list of these substances in descending order of conductivity (reciprocal of resistivity). The properties are approximate and may vary a little in different sources.

Table 3.1 List of good and moderately good conductors (in descending order of conductivity)

Material	Resistivity (Ω-m) at 20°C	Material	Resistivity (Ω-m) at 20°C
Silver	0.165×10^{-7}	Iron	0.91×10^{-7}
Copper	0.17×10^{-7}	Platinum	1.05×10^{-7}
Gold	0.2×10^{-7}	Tin	1.15×10^{-7}
Aluminium	0.266×10^{-7}	German silver	$1.6\text{-}4.0 \times 10^{-7}$
Tungsten	0.53×10^{-7}	Lead	1.9×10^{-7}
Zinc	0.58×10^{-7}	Mercury	9.56×10^{-7}
Brass	$0.5\text{-}0.9 \times 10^{-7}$	Carbon-graphite	7.837×10^{-6}
Nickel	0.68×10^{-7}		

Although silver is the best conductor, it is usually not used commercially because of its relatively high cost and deterioration of property due to atmospheric oxidation. Copper and aluminium are extensively used in transmission lines, while copper is used for windings in electrical machines due to its better current-carrying capacity leading to less weight and cooler operation. Domestic wiring also uses copper because of its better current-carrying capacity (and smaller diameter for a given electrical load) and good mechanical flexibility, although it is costlier.

Brass is used in electrical contact elements in various electrical fittings and appliances.

Let us solve a problem to calculate the resistance of the coil in a bulb.

Let the straightened length of the coiled coil in the bulb be 1.3 m. The material of the coil is tungsten whose specific resistance is 0.53×10^{-7} ohm-m (Table 3.1). The diameter of the coil wire is 0.01 mm. We have to calculate the resistance of the coil. If we use this coil in the bulb and fit it to a 230 V supply what will be power output (wattage) of the bulb?

We have from Eqn 3.1, resistance, $R = r\,l/A$

Given $\quad\quad\quad r = 0.53 \times 10^{-7}$ ohm-m, $l = 1.3$ m,

A (area of the circular cross-section of the wire)

$\quad\quad\quad\quad = (\pi/4)\, d^2 = (3.14/4) \times (0.01 \times 10^{-3})^2$ square metre

$\quad\quad\quad\quad = 0.785 \times 10^{-10}$ square m

Therefore, $\quad R = 0.53 \times 10^{-7} \times 1.4/(0.785 \times 10^{-10})$ ohm = 877.7 ohm

$\quad\quad\quad\quad R = 878$ ohm (rounded off)

Power $= V^2/R = 230 \times 230/878 = 60.25$ watt or it makes a 60 W bulb

Insulators

Insulators are substances which offer tremendous resistance so that they allow practically no electricity to flow through them. Materials such as rubber, asbestos, bakelite, mica, ebonite, etc. fall in this category. Plastic and rubber are often used to insulate household wiring. Ceramics are used as insulators on utility poles. Table 3.2 gives a list of materials used as insulators.

Table 3.2 List of materials used as insulators

Insulator	Resistivity at 0°C (ohm-m)	Insulator	Resistivity at 0°C (ohm-m)
Bakelite	1	Perspex	10^{13}
Distilled water	10^5	Polystyrene	10^{14} to 10^{16}
Glass	5×10^9 to 10^{13}	Porcelain	10^{12} to 10^{13}
Marble	10^8	Pressed amber	10^{16}
Mica	10^{11} to 10^{15}	Vulcanite	10^{14}
Paraffin oil	10^{16}	Dry air	3×10^{17}
Paraffin wax	10^{16}		

3.3 | BASIC COMPONENTS OF ELECTRICAL WIRING—WIRES, CABLES, SWITCHES, AND FUSES

The basic components of electrical wiring are wires and cables, switches, and fuses, among others. The following sections discuss various such components.

3.3.1 Wires

Wires are used for carrying current from one point to another. Before final entry and after exit out of appliances the wiring has to travel distance from the supply point and return to supply point. During this travel, the wires can not be laid bare. To prevent fatal accidents and fire hazards, wires must be coated with insulating materials.

Sometimes the term 'cable' is also used to denote the wire conducting electrical current. Basically, there is no difference between a cable and a wire other than the fact that the term cable is used for all heavy section insulated conductors, whereas a wire means a thin section insulated/bare conductor.

Wire size

Wires used for electrical purposes may be a single copper rod as in electrical overhead transmission lines of the Indian railways. However, such rods would lack flexibility, so, conducting wires and cables are usually made of several wires of smaller diameter (wire size) stranded together to give both strength and flexibility.

Copper wires are expressed in number of strands twisted together/wire gauges system, for example, 3/22, 3/20, 7/22, 7/20, etc. A 3/22 wire system means a cable has three smaller wires of 22 standard wire gauge (SWG) stranded together. However, when the SWG reaches a value of 0, any higher diameter wire is specified as 00, 000, 000, 000, etc. or 2/0, 3/0, 4/0 up to 7/0 (Table 3.3). A higher wire gauge size means a thinner wire. The smallest wire gauge is no. 50 (or 50 SWG), which has a diameter of 0.0048 inches/0.0254 mm.

Total number of wires stranded together and the wire size depends upon the desired current-carrying capacity of the wire. For example, 40/36 wire (40 strands and 36 SWG, i.e., 0.0076 inches/0.1930 mm diameter of each strand of wire) has a current carrying capacity of 7A, while a 162/36 wire will have a current carrying capacity of 28 A.

Table 3.3 Standard wire gauge table

SWG gauge no	Diameter		SWG gauge no	Diameter	
	Inch	mm		Inch	mm
7/0	0.500	12.7	23	0.024	0.6096
6/0	0.464	11.786	24	0.022	0.5588
5/0	0.432	10.973	25	0.020	0.5080
4/0	0.400	10.160	26	0.018	0.4572
3/0	0.372	9.449	27	0.0164	0.4166
2/0	0.348	8.839	28	0.0148	0.3759
0	0.324	8.230	29	0.0136	0.3454

(Contd)

Table 3.3 (*Contd*)

SWG gauge no	Diameter		SWG gauge no	Diameter	
	Inch	mm		Inch	mm
1	0.300	7.620	30	0.0124	0.3150
2	0.276	7.010	31	0.0116	0.2946
3	0.252	6.401	32	0.0108	0.2743
4	0.232	5.893	33	0.0100	0.2540
5	0.212	5.385	34	0.0092	0.2337
6	0.192	4.877	35	0.0084	0.2134
7	0.176	4.470	36	0.0076	0.1930
8	0.160	4.064	37	0.0068	0.1727
9	0.144	3.658	38	0.0060	0.1524
10	0.128	3.251	39	0.0052	0.1321
11	0.116	2.946	40	0.0048	0.1219
12	0.104	2.642	41	0.0044	0.1118
13	0.092	2.337	42	0.0040	0.1016
14	0.080	2.032	43	0.0036	0.0914
15	0.072	1.8288	44	0.0032	0.0813
16	0.064	1.6256	45	0.0028	0.0711
17	0.056	1.4224	46	0.0024	0.0610
18	0.048	1.2192	47	0.0020	0.0508
19	0.040	1.0160	48	0.0016	0.0406
20	0.036	0.9144	49	0.0012	0.0305
21	0.032	0.8128	50	0.0010	0.0254
22	0.028	0.7112			

Figure 3.2 shows a standard wire gauge for checking the wire size (by inserting the wire to be checked into the appropriate hole).

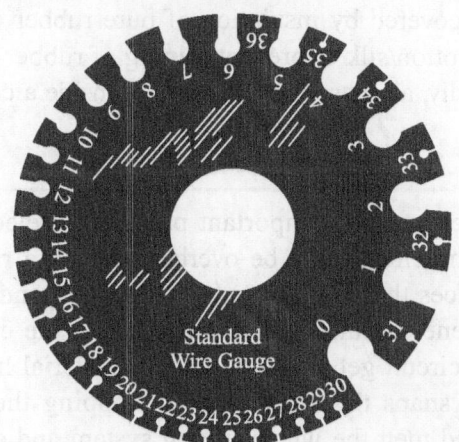

Fig. 3.2 A standard wire gauge for checking the wire sizes

Various types of conductors/cables

There are several kinds of conductors. These are as follows:

Vulcanized indian rubber (VIR) This type of wire is used in casing, capping, and conduit wiring (discussed later in Section 3.4.2). It consists of copper wire covered with a layer of rubber insulation with a protective cotton braid over it. It is almost obsolete nowadays since it absorbs moisture on site. It is still used in hand-held irons, etc., where maximum flexibility is required. But it is also observed that soon the rubber insulation melts and its performance deteriorates.

Lead alloy sheathed wires Ordinary VIR cables and wires are unsuitable for damp conditions. So, in order to use those wires under such conditions, a thin lead covering is made on the VIR. These wires are called lead alloy sheathed wires. These are expensive wires.

Cab type sheathed/tough rubber sheathed (CTS/TRS) These wires are somewhat moisture proof, because instead of vitrified rubber, they are covered with a tough rubber compound which does not deteriorate under damp conditions. These are cheaper than lead sheathed cables. However, they are also obsolete nowadays.

Polyvinyl chloride (PVC) wires In this type of wires, bare conductors are insulated with PVC insulation. They are used in batten wiring, conduit wiring, and cleat wiring in domestic wiring system.

3.3.2 Flexible Cable

These cables are used for giving connections to heating appliances and other portable appliances. Flexibility is of prime consideration here. The flexible cord usually consists of two or three separately stranded conductors. Each of them are covered by insulation of pure rubber or vulcanized rubber and then covered by cotton/silk to prevent joining of rubber insulation of different strands together. Finally, all these are put together inside a common covering layer of cotton or silk.

3.3.3 Fuse

Fuses are very important part of any electrical system. For many reasons, part of the circuit may be overheated with a risk of electrical fire. Fuses are special devices that are inserted in the circuit and consist of wires of low melting point. Whenever current through a part of the circuit becomes very high, the wires in the circuit get heated. The fuse material having low melting point melts quickly and snaps the circuit, thus, stopping the current flow before further heating could melt the whole wiring system and cause a possible fire. Figure 3.3 shows the scheme of a fuse. The circuit continues through the fuse wire. As soon as overheating starts, the fuse wire melts; the circuit continuity breaks and the current flow stops. So, the following may be formally stated.

Fuse is a device which cuts off the circuits when more than the predetermined value of current flows in a circuit. It is the weakest point of the circuit, which breaks when more than normal current flows in the circuit.

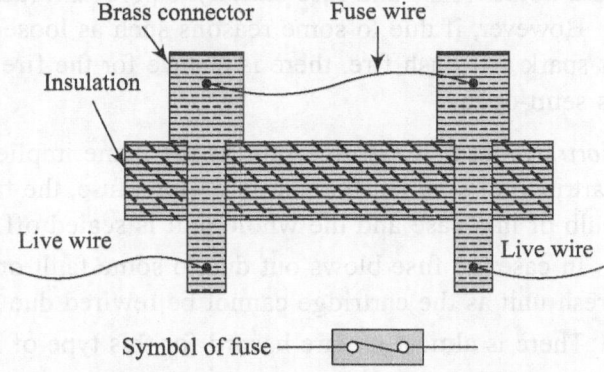

Fig. 3.3   Scheme of fuse and its symbol

The standard fuse sizes are

1. 2 amps (for lamp application)—Consuming not more than 480 W at 240 V
2. 5 amps—Consuming not more than 1200 W
3. 10 amps—Consuming not more than 2400 W
4. 13 amps—Consuming not more than 3120 W

Materials for fuse wires

Fuse wire must be able to carry the normal current without much heating but should effectively melt when normal working current is exceeded. Either copper or lead-tin alloy is mostly used as a material for an ordinary fuse wire. Lead-tin alloy with 37 per cent lead and 67 per cent tin is best suited for the purpose as it is very hard. Minimum length of a fuse wire should be between 65 mm to 100 mm.

Types of fuse holders

There are many types of fuse holders inside which the fuse wires are housed and which are finally placed in the circuit. They are

1. Semi-enclosed fuse or re-wirable fuse or Kit Kat fuse unit
2. Totally enclosed or cartridge fuse
3. High rupturing capacity fuse or HRC type fuse
4. Miniature circuit breaker (MCB)

Semi-enclosed fuse or re-wirable fuse or Kit-Kat fuse unit   It is the most important and common type of fuse unit used for day-to-day work in domestic installations.

It can be rewired without any safety hazard, even if the cut-out terminals are energized. The component of the fuse unit, which holds the fuse wire is called a fuse carrier. It is a separate unit and can be taken out from the base; the damaged wire is replaced by a good fuse wire and can be inserted with care onto the base unit. Incoming and outgoing live wire is permanently connected to the base. The material of the fuse holder (base and fuse carrier) is porcelain (an insulator and heat-resistant).

However, if due to some reasons such as loose contact inside the fuse there is spark and flash fire, there is chance for the fire to come out of the fuse as it is semi-open.

Cartridge fuse   Cartridge fuse as the name implies, has a shape similar to the cartridge of a bullet. In a cartridge-type fuse, the fuse wire is enclosed in a tube bulb or in a case and the whole unit is sealed off.

In case the fuse blows out due to some fault or overload, it is replaced by a fresh unit as the cartridge cannot be rewired due to its sealing.

There is almost no fire hazard for this type of fuse as it is fully enclosed.

High rupturing capacity fuse or HRC type fuse   It is similar in construction to the cartridge type with the exception that the fuse wire can carry short circuit heavy current for a known period of time. During this time, if the fault (may be a short circuit, etc.), which has caused is rectified it does not blow off, otherwise it blows off (melts) and breaks the continuity of the circuit.

Miniature circuit breaker (MCB)   A miniature circuit breaker automatically switches off the current instantly if there is a short circuit or power overload. It thus prevents damage to expensive wiring and the risk of fire. Supply is restored by manually switching 'ON' again after the fault is rectified. There is no fuse to replace or rewire. Automatically, it will switch 'OFF' if overload persists. Many houses, factories, and plants have installed MCBs replacing the earlier rewirable kit-kat fuses.

3.3.4   Switch

Quite often, a particular appliance needs to be put off by stopping power supply (i.e. current supply) to it. This is done by means of a switch placed in the circuit of the appliance, which can be turned 'ON' or 'OFF', simply by pressing a knob or by operating a handle, thereby making or breaking the circuit (Fig. 3.4 a and b).

In a household, the connection lines immediately after entering the premises are provided with a main switch, which when switched OFF disconnects the entire household internal circuit from the supply and it becomes safe to work anywhere in the domestic internal circuit. For each individual equipment and appliance, there will be a dedicated switch. These switches have many types of classifications, such as (a) surface switch, (b) flush switch, (c) tumbler switch,

(d) piano switch, (e) one-way switch, and (f) two-way switch, depending on the purpose of classification.

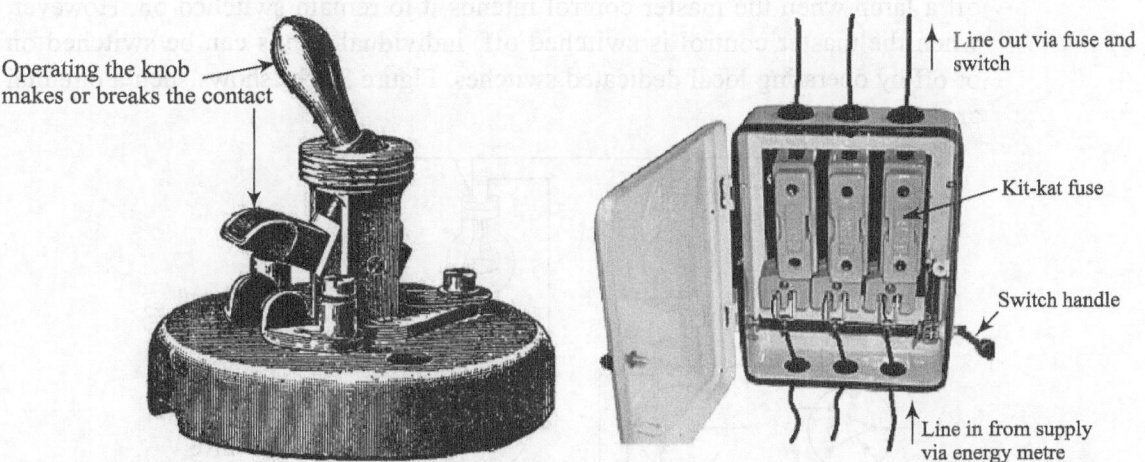

Operating the knob
makes or breaks the contact

Line out via fuse and
switch

Kit-kat fuse

Switch handle

Line in from supply
via energy metre

Fig. 3.4 (a) A common tumbler switch

Fig. 3.4 (b) A view of three-pole main switch integral with fuse

(*Courtesy:* www.fromoldbooks.org)

In many applications, in addition to dedicated switches, a group of appliances may have an additional master switch which includes the following types depending on their functions.

Master off switch It puts off all the lamps irrespective of the positions of individual dedicated switches. This prevents anybody from switching on the individual lamp when the master switch is put off. Figure 3.5(a) shows a typical circuit for such an arrangement.

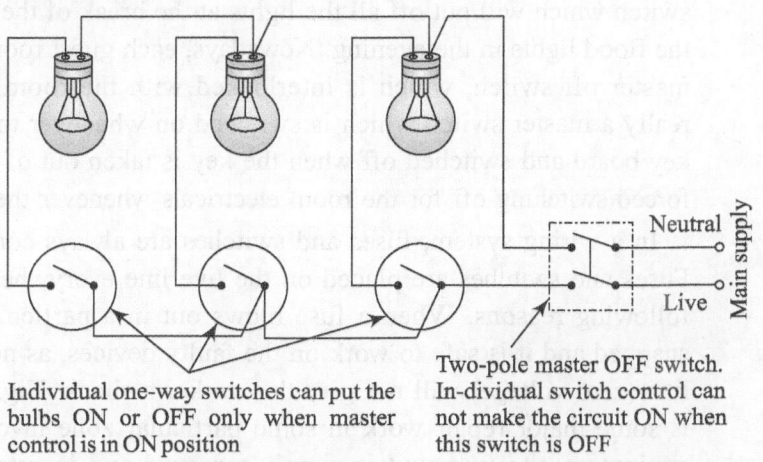

Neutral

Live

Main supply

Individual one-way switches can put the
bulbs ON or OFF only when master
control is in ON position

Two-pole master OFF switch.
In-dividual switch control can
not make the circuit ON when
this switch is OFF

Fig. 3.5 (a) Lamp circuit with 'master off' control switch

Master on switch It puts on all lamps irrespective of the positions of individual dedicated switches. This arrangement prevents any individual from switching off a lamp when the master control intends it to remain switched on. However, when the master control is switched off, individual lamps can be switched on or off by operating local dedicated switches. Figure 3.5(b) shows such a circuital arrangement

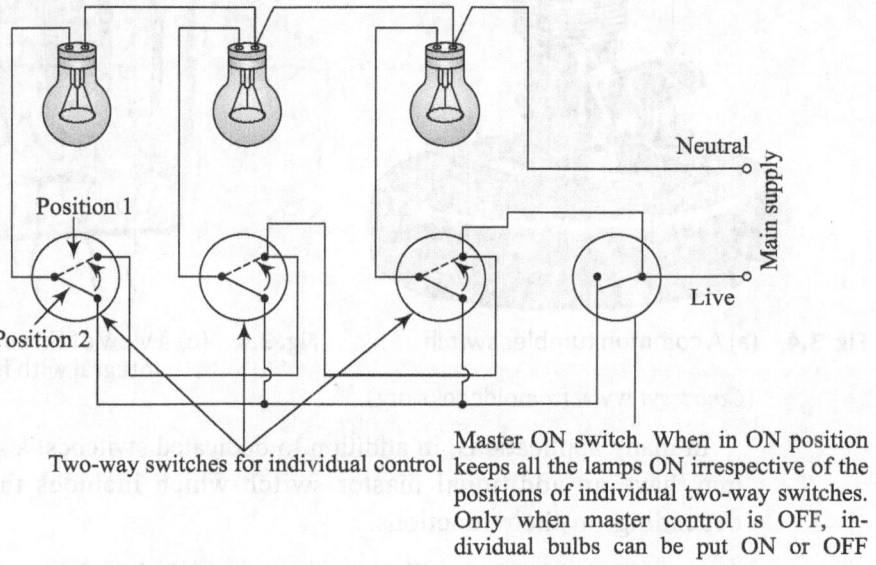

Two-way switches for individual control

Master ON switch. When in ON position keeps all the lamps ON irrespective of the positions of individual two-way switches. Only when master control is OFF, individual bulbs can be put ON or OFF

Fig. 3.5 (b) Lamp circuit with 'master on' control switch

Variable master control switch It controls both switching off or on of the circuit. For example, flood lights in the hotel premises may have a variable master control switch which will put off all the lights at the break of the morning and put on all the flood lights in the evening. Nowadays, each guest room in a hotel has its own master off switch, which is interlocked with the room key. The key board is really a master switch which is switched on whenever the key is inserted in the key board and switched off when the key is taken out of the board. This ensures forced switching off for the room electricals whenever the guest leaves the room.

In a wiring system, fuses and switches are always connected to the live line. Fuses and switches are placed on the live line everywhere. This is done for the following reasons. When a fuse blows out in a particular line, the live line is snapped and it is safe to work on the faulty devices, as now touching the neutral line (zero voltage) will not give the worker a shock (Fig. 3.6a). Further, if there is some major repair work in some particular zone involving several pieces of equipment, the fuse unit is simply removed and inserted only after the repair work is complete. This would not have been possible if the fuse were put on the live line, as removing the fuse would still keep the equipment in a live line and would result in a shock. Putting off only individual switches is not safe for major repair works because somebody may accidentally switch on the circuit causing fatal injury to the worker.

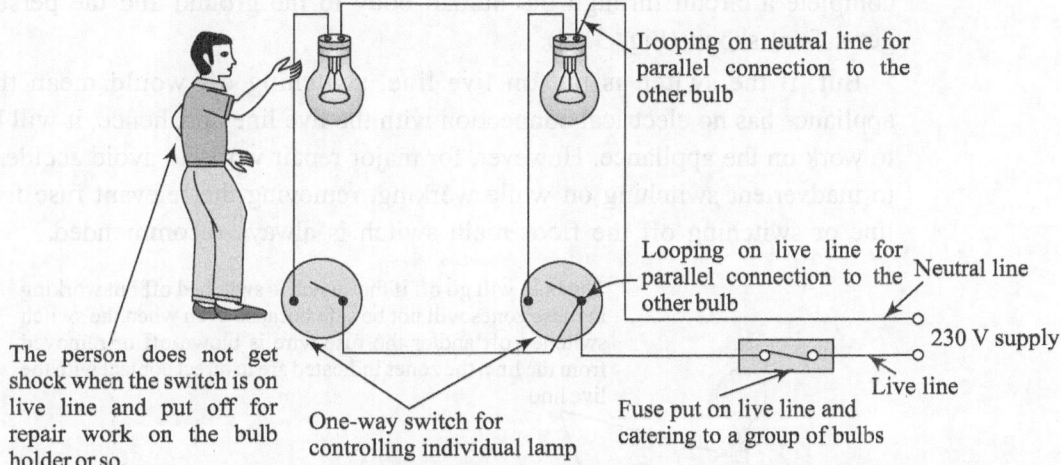

Looping on neutral line for parallel connection to the other bulb

Looping on live line for parallel connection to the other bulb

Neutral line

230 V supply

Live line

Fuse put on live line and catering to a group of bulbs

One-way switch for controlling individual lamp

The person does not get shock when the switch is on live line and put off for repair work on the bulb holder or so.

Fig. 3.6 (a) A simple lamp circuit with placement of switches and fuse on the live line

A switch is, in a similar fashion, always placed in the live line of a circuit (Fig. 3.6 b). This is done because after the switch is put off, the appliance should be cut off from the live line so that people can then work on the appliance, if needed, and even touch the wiring inside, change some parts, etc.

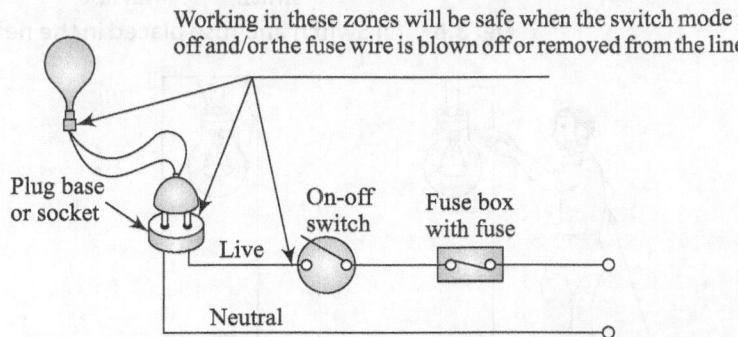

Working in these zones will be safe when the switch mode is off and/or the fuse wire is blown off or removed from the line

Plug base or socket

On-off switch

Fuse box with fuse

Live

Neutral

Fig. 3.6 (b) Switch and fuse placed in the live line

Referring to Fig. 3.6 (a) and (b), again, where switches are put on live lines, it is seen that if the switches are put off, the circuit continuity will be broken and the appliance will cease to work. The same effect could be achieved even if the switches were put on the neutral line (Fig. 3.6 c and d), however, with a difference.

If the switch were placed in the neutral return line, switching off would definitely put off the appliance because the circuit continuity is broken but it is observed that the appliance is still in contact with the live line (230 V) and if now the internals of the appliance are touched, the live line will get ground voltage to complete a circuit through the human body to the ground and the person will get a fatal shock (Fig. 3.6d).

But, if the switch is put on live line, switching off would mean that the appliance has no electrical connection with the live line and, hence, it will be safe to work on the appliance. However, for major repair work, to avoid accidents due to inadvertent switching on while working, removing the relevant fuse from the line or switching off the floor main switch is always recommended.

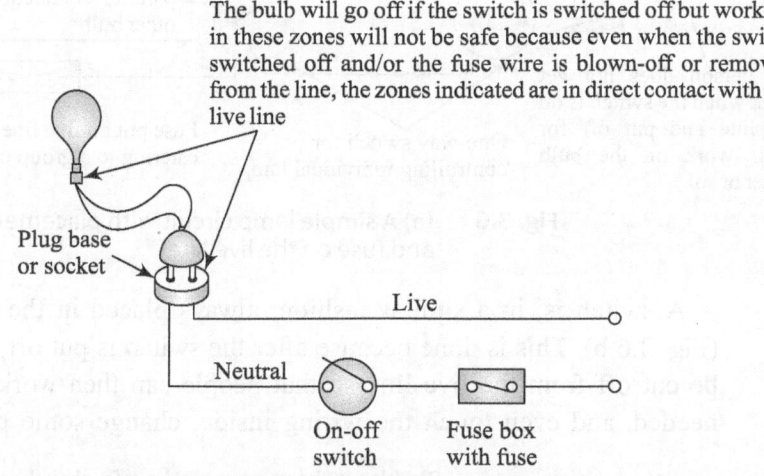

The bulb will go off if the switch is switched off but working in these zones will not be safe because even when the switch switched off and/or the fuse wire is blown-off or removed from the line, the zones indicated are in direct contact with the live line

Plug base or socket

Live

Neutral

On-off switch

Fuse box with fuse

Fig. 3.6 (c) Switch and fuse placed in the neutral line

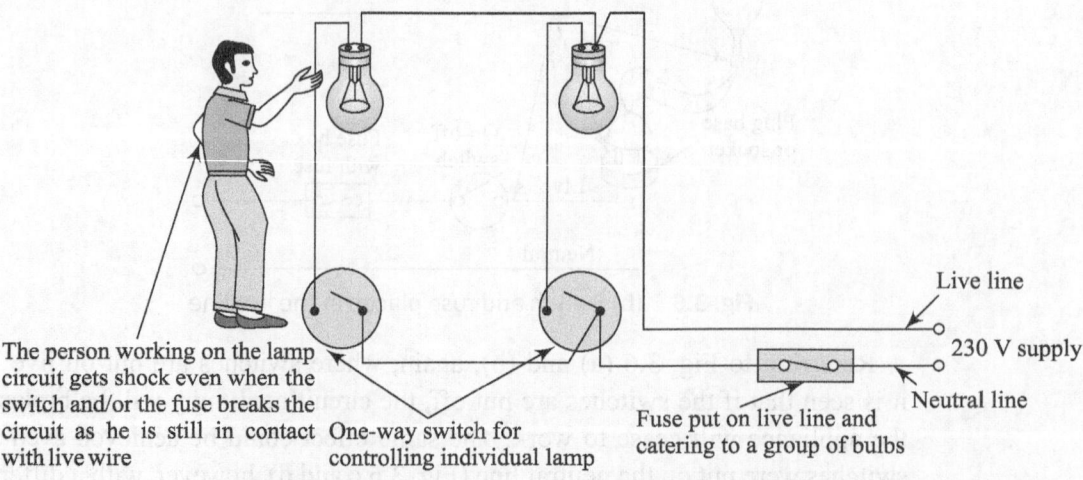

The person working on the lamp circuit gets shock even when the switch and/or the fuse breaks the circuit as he is still in contact with live wire

One-way switch for controlling individual lamp

Fuse put on live line and catering to a group of bulbs

Live line

230 V supply

Neutral line

Fig. 3.6 (d) Switch and fuse placed in the neutral line

So, fuses and switches are placed on a live line to make the corresponding appliance safe to work.

3.3.5 Thermostat

A refrigerator while in the running mode suddenly stops operating for some time and then again its compressor motor starts and the refrigerator starts operating again. This is because when the refrigerator's internal temperature decreases below a certain value, which can be set (Hi-Lo-Med) by the temperature-adjusting knob, the motor stops working. The temperature after some time rises again to some higher value (due to heat released by substances stored inside and heat leakage from outside) and again the motor starts to bring down the temperature. Essentially, there is a temperature-sensing device that senses the temperature inside and makes or breaks a circuit for the motor. Thermostat is such a device which operates the on–off switch by responding to temperature changes. In other words, it is a temperature-sensitive device, which opens or closes electrical contacts depending on temperature.

There are three types of thermostat devices, namely (a) mechanical, (b) electrical, and (c) digital.

Mechanical thermostats

Mechanical thermostats are extensively used in heating, cooling, automobiles, and refrigeration system. In automobiles, a thermostat is used to regulate the flow of coolant to the engine. A mechanical type may be a (a) wax pellet type, or (b) gas bulb type, or (c) bimetal type. Let us discuss only about the operating principles of these mechanical thermostats in some detail.

Wax pellet type This type of thermostat makes use of a wax pellet inside a sealed chamber. The wax is solid at low temperatures but as the engine heats up, the wax melts and expands enough, corresponding to certain set point temperature, and thus regulates the control valve for the flow of coolant, etc. The melting temperatures are different for different compositions of wax.

Gas expansion type or remote bulb type This type of thermostat is sometimes used to regulate temperatures in gas ovens. It consists of a gas-filled bulb connected to the control unit by a long and thin capillary copper tubing. The bulb is usually located at the top of the oven. The tube leads to a chamber sealed by a diaphragm. As the oven heats up the gas inside, the bulb expands and consequently applies pressure on the diaphragm, which actuates a flow control valve for the entry of fuel to the burner. These controls are usually equipped with about 5 ft tubing, but additional lengths may be added.

Bimetal thermostat The length of a material always changes when the material changes its temperature (increase in temperature or decrease in temperature). If there is an increase in temperature, the length increases (thermal expansion) and if there is a decrease in temperature, the length decreases (thermal contraction). The change in length for 1°C change in temperature is called coefficient of thermal expansion (or contraction). Metals have very high values of thermal coefficients.

Bimetal thermostat is a temperature-sensitive device made up of a thin duplex strip of two dissimilar metals having different thermal expansion coefficients (Fig. 3.7). This means that for the same temperature change, the two dissimilar metals change their lengths by different amounts. As the temperature changes, the difference in expansion, therefore, creates a bending action on the strip. It is very easy to observe that the metal strip A having higher coefficient of expansion will tend to stretch more than metal strip B. But since the two metal strips are joined together, the only way this is possible is for the strip to bend the way, as shown in Fig. 3.7, whereby the strip A is placed such that the outer side of the bent

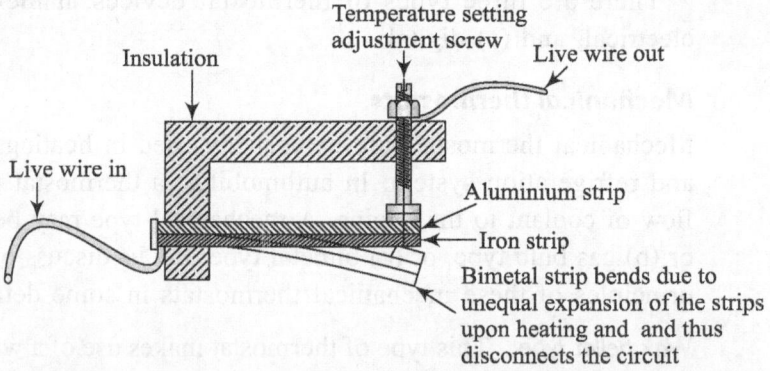

Fig. 3.7 Schematic arrangement of a bimetal thermostat

piece has a longer length than the bottom piece. Now this change of configuration will open the circuit and stop the current. By suitable arrangement, it is also possible to switch a circuit on when the combined strip bends. So when the temperature exceeds a certain value, the motor can be swiched on or off. This is the working principle of a Bimetal thermostat. Such thermostats are used in refrigerators in the domestic set-up, where the compressor quite often switches off when the temperature goes low and then restarts when it rises to a certain set value. The temperature at which this action will be triggered in the driving electric motor can be controlled by a setting of the thermostat bimetallic strip with the help of an adjusting screw. In order to avoid frequent on–off situations, quite often, an arrangement is made when the connection switches off at temperature T_1 and continues till the temperature goes up (in case of a refrigerator) and then again switches on at some higher value T_2. The lower the difference ($T_2 - T_1$), higher is the frequency of start–stop action of the motor. This difference is known as temperature differential for the thermostat.

Adjustable differential thermostat An adjustable differential thermostat is a temperature control device designed so that the operating on–off differential may be adjusted within the range of control. One popular model has served ranges of 0°F to 70°F (–18°C to 21°C) and 160°F to 280°F (71°C to 138°C) with differentials (ON-OFF action) adjustable from 0.5°F to 15°F (0.3°C to 8°C).

3.4 WIRING IN HOTELS—DISTRIBUTION SYSTEMS AND LAYING METHODS

While the wiring for lights, fans, room ACs, etc., follow similar methodology and principles in commercial and industrial set-up, including the hotel industry, the wiring for electric motors and other three-phase machines follow more complicated methodology and principles. In the following sections only the basic schemes of wiring related to domestic wiring and the scheme of distribution of electric lines from supply main for other power loads are discussed.

3.4.1 Distribution Systems

At the electrical post of a locality, three live lines (three phases) and a neutral line reach from the local transformer or from the substation transformer (see Chapter 2) output. At the entry point of a hotel premises, wires are laid from the street electric post directly through overhead lines or through underground cables into the premises. Depending on the particular requirement of the consumer, lines are drawn as either

(a) Single phase-two-wire line—one live (one phase) line and one neutral line from the post as shown in Fig. 3.8.

or

(b) Three-phase-four-wire line—three live lines (three phases) and one neutral line from the post

At the substation, the neutral is earthed and has the earth voltage or zero voltage. Each live line will have an AC supply of 230 V. Normally, for industrial premises and hotels, which require higher voltages to run heavy-duty motors,

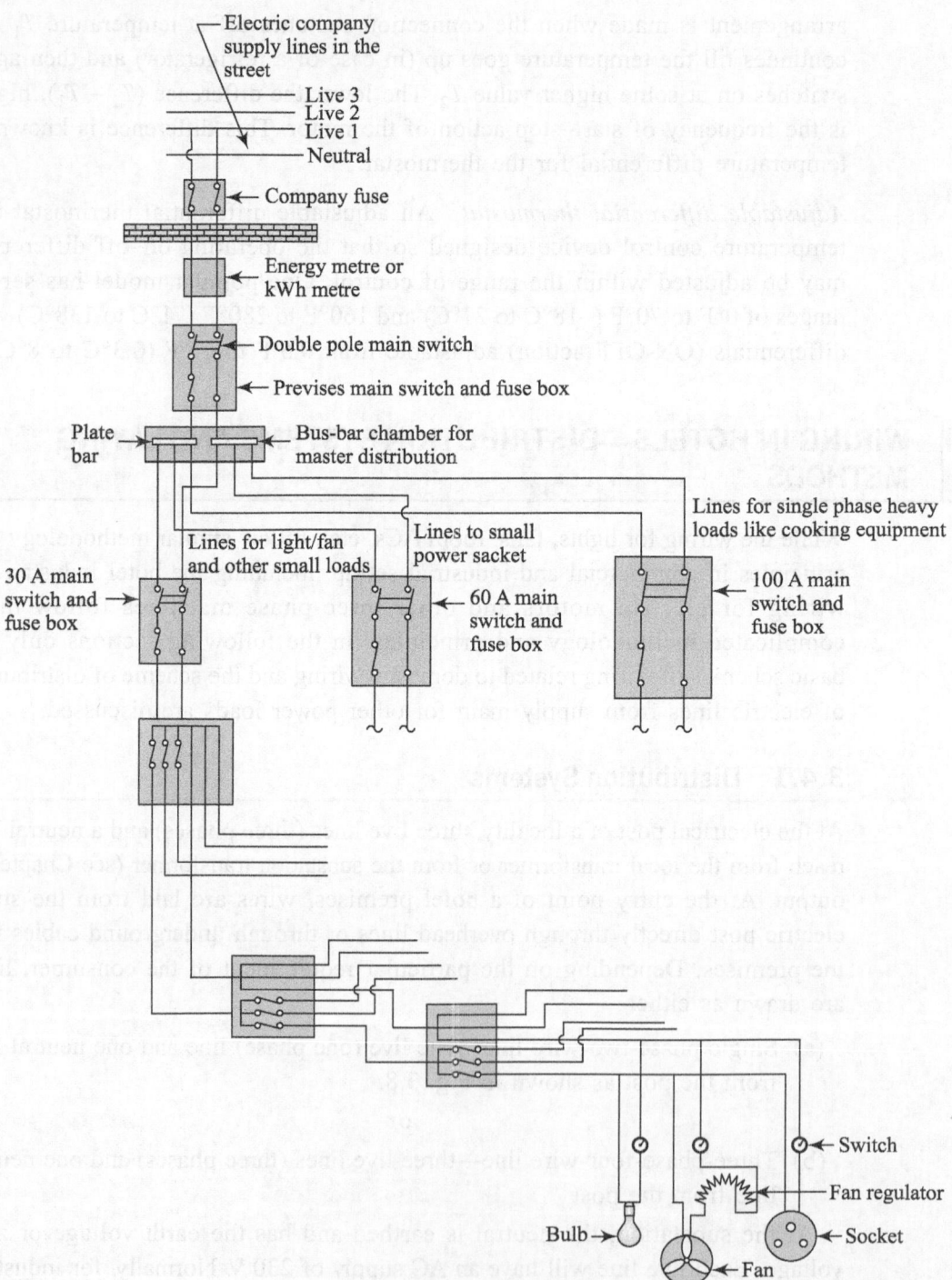

- Electric company supply lines in the street
- Live 3
- Live 2
- Live 1
- Neutral
- Company fuse
- Energy metre or kWh metre
- Double pole main switch
- Previses main switch and fuse box
- Plate bar
- Bus-bar chamber for master distribution
- Lines for single phase heavy loads like cooking equipment
- Lines for light/fan and other small loads
- Lines to small power sacket
- 30 A main switch and fuse box
- 60 A main switch and fuse box
- 100 A main switch and fuse box
- Switch
- Fan regulator
- Bulb
- Socket
- Fan

Fig. 3.8 Electric supply lines into the premises and system of distribution inside for various uses

etc. three-phase-four wire system is used. Single phase-two-wire system can be obtained from the three-phase-four-wire system very easily and is used for wiring serving lighting and other small load equipment. For domestic use, single-phase-two-wire lines are laid into the premises.

We restrict our discussion of wiring methods in hotel premises, which are partly like industrial premises for heavy loads and partly like domestic premises for light loads. After entering the premises, and before the lines are connected to the main switch and energy metres, the electric company provides a fuse called 'board fuse' or 'company fuse'. This fuse is placed to save the building from fire hazard in case the total current in the premises circuit exceeds a certain value that is preset for the particular premises. This maximum value of current is set based on its total maximum electrical load envisaged for the premises (total maximum kilowatt anticipated divided by the supply voltage roughly gives the maximum current for the house passing through the main switch, for example, 150 Amp for a medium size hotel, etc.).

After the company fuse, the lines pass through electrical energy metre (or kWh metre) and then to the bus-bar chamber (Fig. 3.8). Bus-bar chamber has strips of copper bar mounted on a structure where each of the incoming lines is screwed onto a particular strip of copper bar. From the bus-bar chambers, leads are taken out from each of the bus-bars to cater to specific purposes forming a group containing separate main switches and fuses. Groups may form a three-phase system (Fig. 3.8 is for a single-phase system. A three-phase system will be similar, where connections are taken from other two phases running on the street, etc.), while others may form a single-phase system that may serve lighting loads and yet another may form a supply for the heating load. Each of the single-phase connections for single-phase groups is taken from different phases in the bus-bar to make a phase load balance. A number of groups may be created to cater to different floors in the hotel. These groups of cables are called 'sub-mains'. These sub-mains would have a sub-main distribution system to cater to the needs of different parts of a particular floor through what is called the floor distribution board. This system consisting of bus-bar and sub-mains containing main switches and fuses for each sub-main is called 'main switchboard'. These main switches are normally two-pole switches (for single-phase sub-main) as shown in Fig. 3.8 or three-pole switches (for three-phase sub-main). When the main switch is switched, both the neutral and live lines from the supply are cut-off and rest of the circuit corresponding to the sub-main circuit inside the premises is quite safe to work on. Lines come out of the floor distribution boards and form the basic supply input to the wiring in the particular floor.

In an electric circuit electrical loads are put in parallel. Following this principle, wiring is done in such a way so as to provide for parallel load connections. Final wiring from floor distribution box to individual loads can have two schemes as follows.

1. Tree system   2. Distribution board system

In each of the systems, fuses are placed in each individual sub-circuit to protect the particular zone from overload heating.

Tree system

Figure 3.9 shows a tree system of wiring. It is observed that farther it is from the supply point (230 V), higher is the drop in the available voltage, as a result of line resistance loss. So, appliances connected in zone B will have a little less voltage (say, 225 V) compared to what zone A will get (say, 229 V). Additionally, fuses for each sub-circuit are located at different positions. Due to these reasons, the tree system of wiring has almost fully given way to the distribution board system.

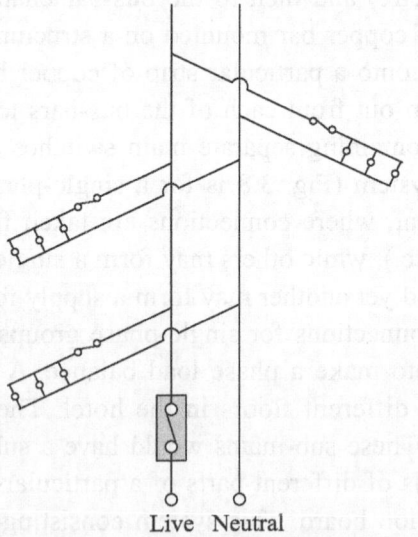

Live   Neutral

Fig. 3.9   Scheme of Tree system of wiring

Distribution board system

Figure 3.10 shows a distribution board system of wiring. It is the most widely used system of domestic wiring. In such a system it is observed that the length of the wires from supply point to each load zone is almost the same, causing almost the same voltage drop due to line loss at each zone. So, voltage available at each zone will be almost the same (230 V minus voltage drop due to line loss) and as a result all the loads connected will operate at nearly same voltage (all 100 W bulbs will glow by the same amount everywhere). In this system, leads (live and neutral) from the main switch are brought to a main distribution board. From there, lines are taken to sub-distribution boards and so on. It is observed that there are fuses in each distribution box protecting parts and sub-parts of the circuit. All the fuses for a particular zone are, thus, located in the same distribution board. Sub-distribution boards are needed as Indian electricity rules stipulate the maximum number of points that can be connected to a particular board at six to eight with a maximum power limit of 500 W. This is the reason why many distribution boards are needed.

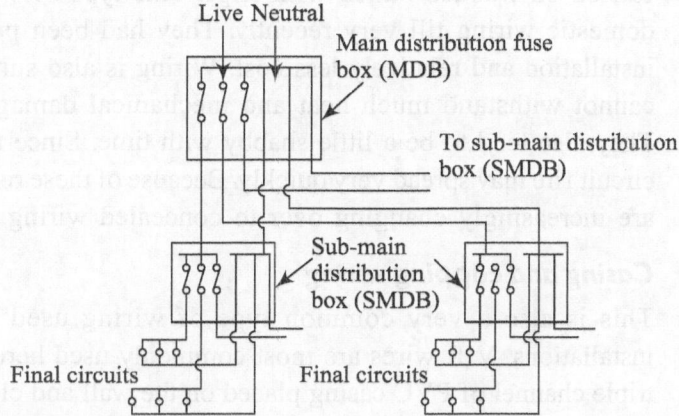

Fig. 3.10 Distribution board (DB) arrangement of wiring

3.4.2 Laying Methods for Building Wiring

The various wiring laying methods are as follows:

1. Cleat wiring
2. Wooden batten wiring
3. Casing and capping wiring
4. Lead sheathed wiring
5. Conduit wiring

Cleat wiring

This type of wiring is suitable for temporary wiring purposes such as in marriage halls and has advantages such as it saves labour and overall costs. The porcelain or wooden cleats are very easy to erect and fixed at a distance of 4 to 5 m apart. VIR or PVC wires are normally used in this system of wiring. The number of grooves in the cleat base depends on the number of wires intended to run side by side. This type of wiring can be dismantled very quickly and the material can be recovered for further use. They are not recommended for permanent installation as although they appear very neat and attractive in the beginning, they become shabby with dirt and discoloration in time. Further, oil and smoke are detrimental for VIR wires. The wires are to be laid very taut between cleats to avoid contact with wall. But with time, they sag and may touch the wall and may pose safety hazards, if the wall is damp.

Wooden batten wiring

In such type of wiring either TRS or PVC wires are used. Here the wires are carried on wooden batten with clips. This type of wiring was very popular in domestic wiring till very recently. They had been popular because of ease of installation and relatively less cost. Wiring is also suitable in damp climate, but cannot withstand much heat and mechanical damage including fire hazards. They also tend to be a little shabby with time. Since the wires are open, a short circuit fire may spread very quickly. Because of these reasons, modern installations are increasingly changing over to concealed wiring system.

Casing and capping wiring

This is also a very common type of wiring used for indoor and domestic installations. VIR wires are most commonly used here and are carried in two or triple channel of PVC casing placed on the wall and closed by PVC covers called capping. Unlike in batten wiring, wires in casing and capping wiring are not visible from outside. They are costlier but at the same time more reliable and pleasing to look at.

Lead sheathed wiring

In lead-sheathed wiring system, conductor cores are separately insulated and then covered with common lead alloy sheath containing 95 per cent lead. The lead-sheathed wires are easily fixed by means of metal clips on wooden batten and form a good surface system. The lead covering provides protection to the wire from mechanical damage. As these wires are very costly, they are not in use nowadays.

Conduit wiring

In this system, VIR or PVC wires are carried through steel or iron pipes. These conduits may be running over walls or concealed. Concealed wiring lends a good appearance to the interior of a room. They also give good protection from mechanical injury or fire risks. For workshops and public buildings, conduit wiring is the best and most desirable system of wiring.

Nowadays, PVC conduit pipes are also available, which do not require any threading. For concealed wiring, the pipes are directly embedded on the walls and roofs and then wires are drawn through them. If they run over the wall, they run on blocks of wooden bases like cleats at intervals of not more than 1 m. These wooden blocks are secured on the walls and the metal conduit held on the blocks by clamps.

Table 3.4 provides a comparison of different types of wiring laying options.

Table 3.4   Comparison of wirings in building installations

S.No.	Basis of comparison	Cleat wiring	Batten wiring	Cap and casing	Lead sheathed	Conduit wiring
1.	Life expectancy	Short	Fair	Long	Long	Very long
2.	Mechanical protection	NIL	Fair	Fair	Good	Very good
3.	Cost	Low	Medium	Medium	High	Costliest
4.	Fire possibilities	High	High	High	Poor	NIL
5.	Protection from dampness	NIL	Slight	Good	Good	Very good
6.	Types of labour required	Semi-skilled	Skilled	Skilled	Skilled	Highly skilled

3.5 | SAFETY COMPONENTS IN WIRING AND PRECAUTIONS IN HANDLING ELECTRICAL SYSTEMS

In this section, first few terms that are important from the electrical safety point of view are discussed and then the 'dos' and don'ts' while handling electrical systems are enumerated.

3.5.1   Safety Components in Wiring

Earthing—Three-pin plug and socket

Connecting outer metal cover of electrical appliances to the general mass of the earth with the help of wire is called 'earthing'. The wires from individual appliances are connected to a metal plate (electrode), which is buried 2.5 m to 3 m deep in the ground. Earthing is done to save human life from the danger of shock while touching any apparatus that has become faulty due to the wires inside coming in contact with the outer metal body. These conditions of danger in handling a faulty appliance are examined and how earthing prevents this danger even though the user handles the faulty apparatus is explained.

Figure 3.11 depicts a series of situations when the equipment internal wiring insulation is (a) good, (b) good and the operator touches the metal body of the appliance, (c) weak and the operator touches the metal body, and (d) completely broken and the operator touches the metal body. In the cases of (c) and (d), the operator is likely to get mild to severe shock. Usually, this happens if some loose metal pieces bridge the gap between bare wire inside and the metal body or if the insulation in between the and the metal body gets deteriorated. Now, when the metal body is not touched, the current normally flows through the equipment and comes back through the neutral line. But if an operator touches the metal body, the current gets the opportunity to flow to the ground through a parallel path via the human body and the resistance of the body being small will have a large current flowing through this parallel path. This is sufficient to give a severe shock to the operator and if the full voltage acts across the operator, it may cause even death. To prevent such situations a piece of wire is connected to the metal body inside the equipment, which comes out of it as the third pin in a three-pin socket fitted with the equipment. Figure 3.11 (a, b, c, and d) explains the events with equivalent circuits.

To obviate this danger, earthing is done where the metal body is connected to the earth (a large sink where electricity can fluently flow in) as shown in Fig. 3.12. To achieve this in practice, a metal wire is connected to the metal body inside the appliance and brought out through the power connection socket of the appliance. For this purpose a three-pin connection is used for such appliances (Fig. 3.13). When a three-pin plug is fitted in it, the connection now continues via the third hole/pin in the plug inserted and goes into the third hole of the wall plug base via the connecting flexible cord as shown in Fig. 3.13. The wire comes out of the wall connection board as a metal wire, runs along the main wiring for the particular board side-by-side (watch for a bare wire running with the main wiring visible in case of batten wiring); all such wires are joined at some point and then led deep inside the ground. This arrangement is called earthing.

If we observe, we find that this arrangement provides another (the third one) parallel path for the current but with much less resistance compared to the path to the ground through human body; so major part of the current will flow through the earthing wire thus saving the human life. Figure 3.12 explains the equivalent electric circuit for the earthing system. Of course, if bare wire inside comes in contact with the metal body, whether or not a human being touches the cover, there will always be a flow of current to the ground through the earth wire and a huge current will flow (because of low resistance in the path leading to the ground), possibly blowing off the fuse. However, if the wire is not bare but the insulation is weak, the person will get a weak shock in absence of the earthing.

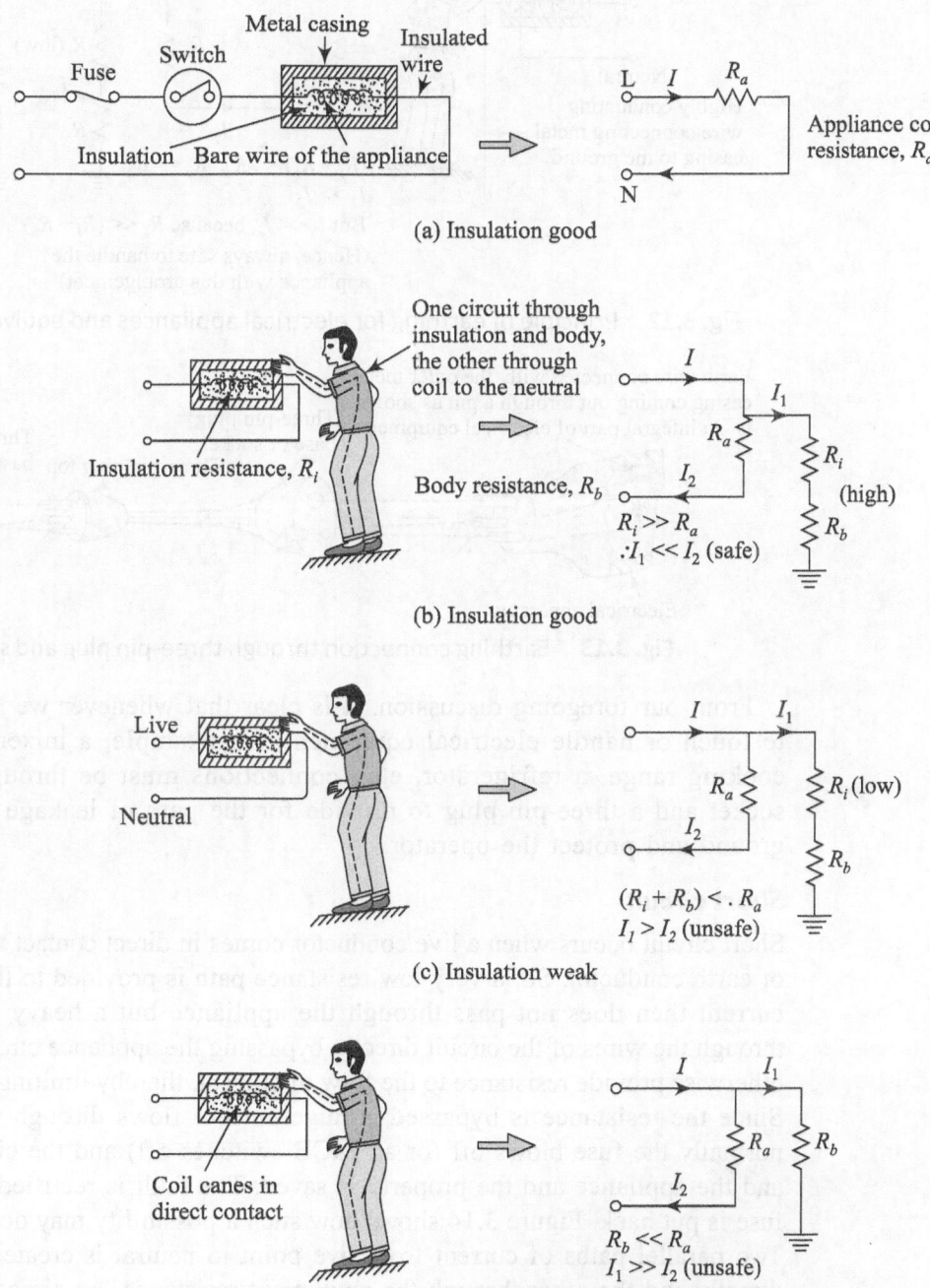

(a) Insulation good

(b) Insulation good

(c) Insulation weak

(d) Insulation broken completely

Fig. 3.11 Different conditions of insulation and coil inside an electrical appliance

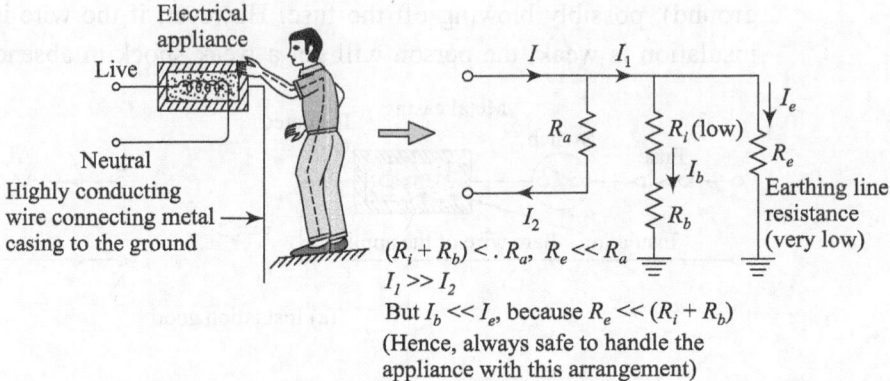

$(R_i + R_b) < . R_a, R_e \ll R_a$
$I_1 \gg I_2$
But $I_b \ll I_e$, because $R_e \ll (R_i + R_b)$
(Hence, always safe to handle the appliance with this arrangement)

Fig. 3.12 Principle of earthing for electrical appliances and equivalent circuit

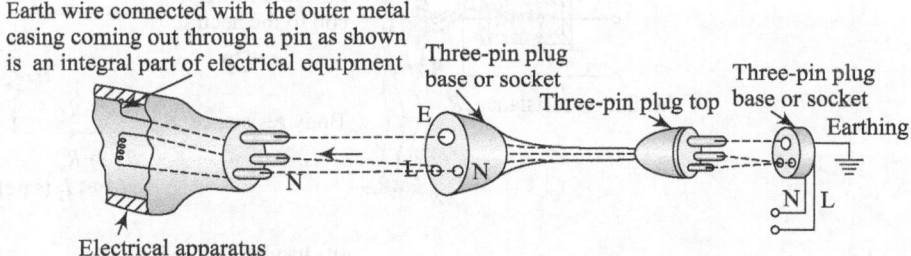

Fig. 3.13 Earthing connection through three-pin plug and socket

From our foregoing discussion, it is clear that whenever we have occasions to touch or handle electrical equipment, for example, a mixer, an electrical cooking range, a refrigerator, etc., connections must be through a three-pin socket and a three-pin plug to provide for the path of leakage current to the ground and protect the operator.

Short circuit

Short circuit occurs when a live conductor comes in direct contact with the neutral or earth conductor. So, a very low resistance path is provided to the current. The current then does not pass through the appliance but a heavy current passes through the wires of the circuit directly bypassing the appliance circuit. This would otherwise provide resistance to the flow of current, thereby limiting its magnitude. Since the resistance is bypassed, a huge current flows through the circuit and normally the fuse blows off (or an MCB switches off) and the circuit is cut off and the appliance and the property is saved. The fault is rectified and then only fuse is put back. Figure 3.14 shows how such a possibility may occur in practice. Two parallel paths of current from live point to neutral is created; one of them directly and the other through the equipment resistance. As almost no resistance is offered in the first path, huge current will flow through it far exceeding the rated current for the circuit, causing the fuse to blow off. Short circuit may also occur within the appliance if the live contact inside comes in direct contact with the earth line inside, leading to earth potential. This path is also a very low resistance parallel path compared to the original circuit and huge current flows. Short circuit may not only occur inside an appliance but also outside in the wiring system where two wires (live and neutral) run side by side separated by insulation coating.

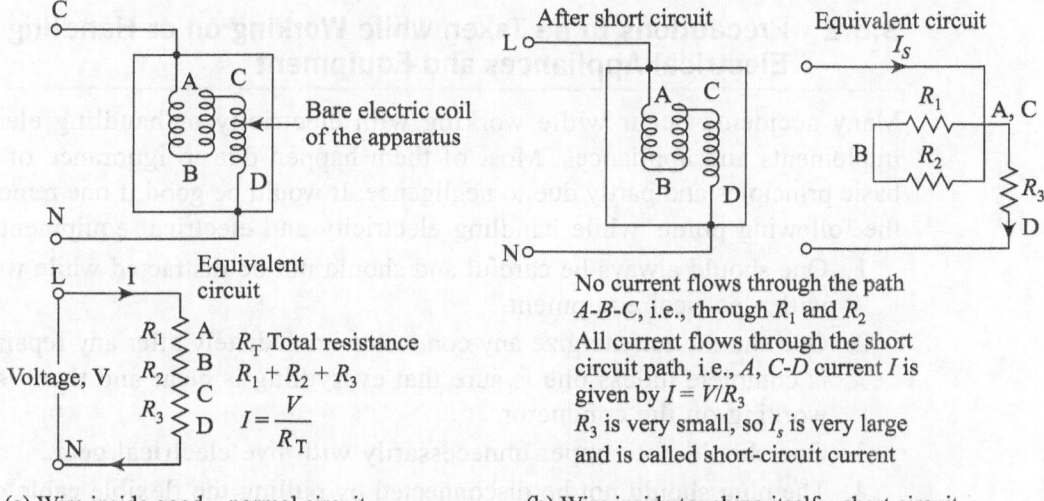

After short circuit

Equivalent circuit

Bare electric coil of the apparatus

Equivalent circuit

R_T Total resistance
$R_1 + R_2 + R_3$

$I = \dfrac{V}{R_T}$

No current flows through the path A-B-C, i.e., through R_1 and R_2

All current flows through the short circuit path, i.e., A, C-D current I is given by $I = V/R_3$

R_3 is very small, so I_s is very large and is called short-circuit current

(a) Wire inside good—normal circuit

(b) Wire inside touching itself—short circuit

Fig. 3.14 Condition of short circuit inside an appliance

If this insulation separating the bare conductor becomes weak, or worn out, short circuit occurs. So, the usual causes for short circuit could be

(a) A worn flexible wire

(b) A plug with its wires loosened or connections that have come together (Fig. 3.15).

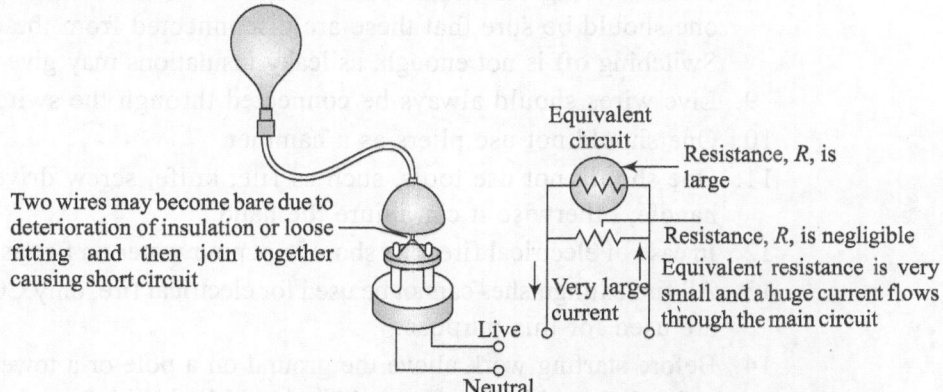

Two wires may become bare due to deterioration of insulation or loose fitting and then join together causing short circuit

Equivalent circuit

Resistance, R, is large

Resistance, R, is negligible

Equivalent resistance is very small and a huge current flows through the main circuit

Very large current

Live

Neutral

Fig. 3.15 Short circuit occurring outside the appliance due to two bare wires coming in contact directly before the circuit is completed through the appliance

No serious damage is likely to occur to the circuit or to human life beyond the blowing of a fuse in case of short circuit if proper fuses/MCBs are put in the appropriate positions and safe wiring practices are strictly followed. However, poorly designed and/or maintained electrical systems have been the culprit for many a devastating fire due to short circuit.

3.5.2 Precautions to be Taken while Working on or Handling Electrical Appliances and Equipment

Many accidents occur while working with electricity or handling electrical implements and appliances. Most of them happen due to ignorance of a few basic principles and partly due to negligence. It would be good if one remembers the following points while handling electricity and electrical equipment.

1. One should always be careful and should not be distracted while working with electrical equipment.
2. One should not energize any conductor immediately after any repair work is complete unless one is sure that everything is clear and there is none working on the conductor.
3. One should not tamper unnecessarily with live electrical gear.
4. The plug should not be disconnected by pulling the flexible cable off the socket.
5. Before doing any work or replacing parts, always remember to put the main switch 'off'.
6. Safety demands good earthing. Hence, earth connection should always be kept in a good condition.
7. Before using portable electrical equipment, one should check that these are well earthed.
8. While moving electrical appliances, such as table fan, iron heaters, etc. one should be sure that these are disconnected from the electric supply. Switching off is not enough, as leaky insulations may give serious shocks.
9. Live wires should always be connected through the switch.
10. One should not use pliers as a hammer.
11. One should not use tools, such as file, knife, screw driver, etc. without handle; otherwise it can injure the hand.
12. In case of electrical fire, one should not pour water on fire as it is dangerous.
13. All fire extinguishes cannot be used for electrical fire; only CO_2 extinguishers are used for this purpose.
14. Before starting work above the ground on a pole or a tower, one must use safety belt and ladder. The ladder should be held by another person on the ground so that one does not slip. This type of work, of course, must be done by a qualified electrician.

15. One should not tie wire with electric pole for hanging and drying clothes, etc.
16. The hands should not be wet while handling electrical appliances.
17. One should wear rubber-sole footwear while handling electrical appliances.
18. Even the slightest feeling of discomfort while touching appliances should be brought to the notice of an electrician or a competent person.
19. Every electric appliance depending on its wattage has to be connected to a proper socket on the wall, for example, a 15 A plug or 10 A plug has to be inserted to the corresponding compatible socket.

Electrical testing tools for safe working

Two commonly used electrical testing tools for safe working are discussed as follows.

Tester It is a very useful tool for testing the presence of electricity in a contact such as a switch point or anything like the metal body of an appliance. The tester completes the circuit through human body but we remain safe because the amount of current flowing through the body is very less due to high resistance of the tester wire itself.

Megger The instrument by which the insulation resistance of a conductor can be measured is known as an insulation testing megger.

3.6 | CONVERSION OF AC TO DC SUPPLY

There are two systems of electric supply for domestic and industrial service, namely direct current DC supply and alternating current AC supply. Because of engineering and economic reasons, almost all supply systems in India have converted to the AC system. However, many electrical equipment and machines used in household and industrial purposes have to be supplied with DC because of technical reasons, for example, DC motors are better for traction purposes such as in tram cars, electric locomotives, etc. DC supply is also used in electrical arc welding for improved quality of welding. So, there is a need to convert AC supply to DC supply in the hotel premises for particular machines. The process of converting AC to DC is known as 'rectification' (Fig. 3.16).

AC is converted to DC by one of the following means:
1. Generator set
2. Rotary converter
3. Solid sate rectifier
4. Mercury arc rectifier

Generator set In a generator set, an AC motor runs on AC supply, which drives a DC generator that produces a DC voltage. This DC voltage is then used by the particular machine.

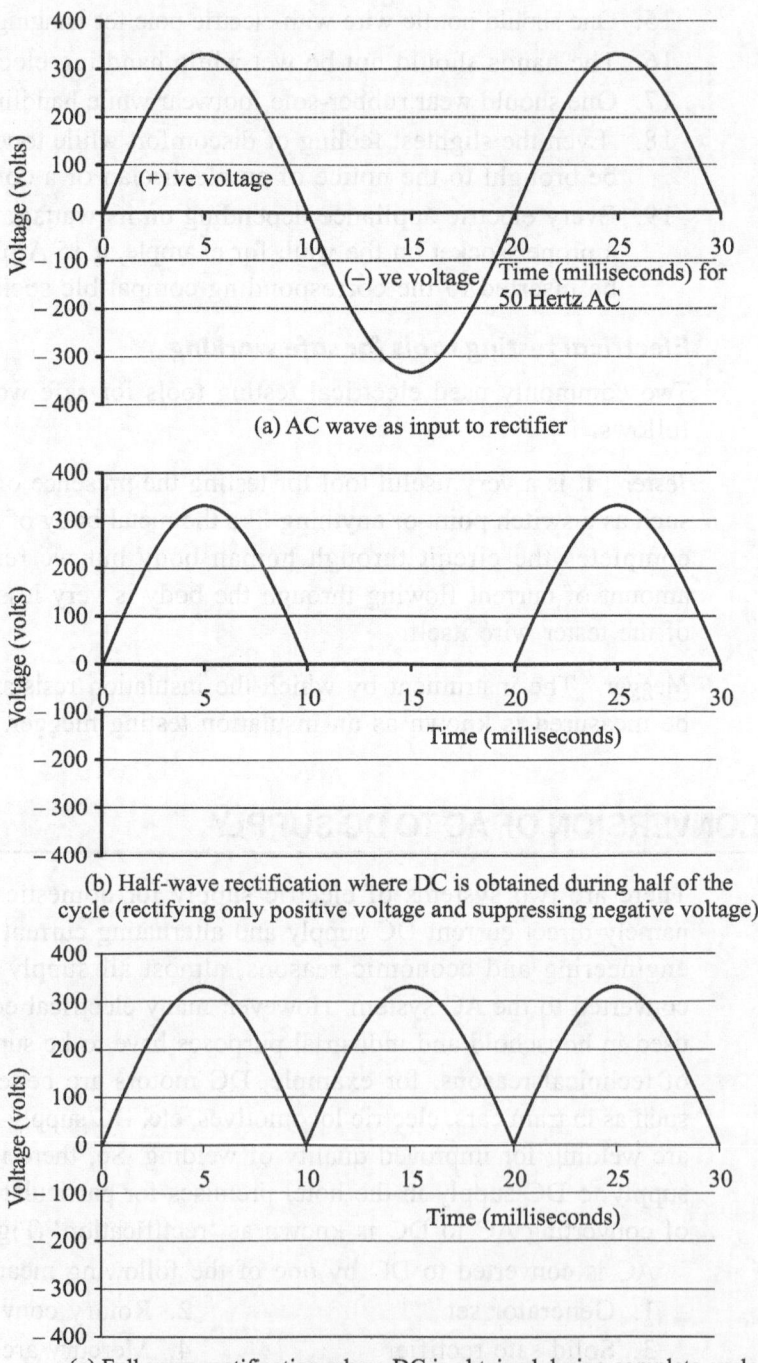

(a) AC wave as input to rectifier

(b) Half-wave rectification where DC is obtained during half of the
cycle (rectifying only positive voltage and suppressing negative voltage)

(c) Full-wave rectification where DC is obtained during complete cycle
(rectify both positive and negative voltages)

Fig. 3.16   (a) AC wave as input to rectifier, (b) Half-wave rectification, and
(c) Full wave rectification

Rotary converter It is an AC motor and DC generator combined into one armature, which receives AC and the output of the same armature is DC voltage. This avoids the energy loss associated in case of a motor-generator set, which receives AC and converts to mechanical rotational energy of the motor, which further drives the DC generator. Thus, a rotary converter is not only energy efficient but also compact and smaller in size. However, a motor-generator set has its own advantages that relate to specialized electrical engineering aspects.

Solid state rectifier It is works on the principle of receiving AC (Fig. 3.16a) at one end, but allows only the positive current to flow through it (half-wave rectification), as shown in Fig. 3.16(b) or convertings the negative half-cycle of current also to positive current (full-wave rectification) as shown in Fig. 3.16(c).

Mercury arc rectifiers Till the mid 1970s, *mercury arc rectifiers* were used for high-voltage DC conversion for high power. Nowadays, they have been replaced by silicon solid state rectifiers.

3.7 | ELECTRIC TARIFF AND ENERGY BILL

Any electrical equipment consumes energy and charges need to be paid for the units of electricity consumed. If one knows the power rating of appliances, one can calculate the energy consumed for running them for some period time by multiplying power with the time for which this power is drawn from supply.

So, if an electric bulb of rating 40 W/220 V has been glowing for two hours, the energy consumed is calculated as follows:

Power = 40 W and time = 2 hours = 7200 sec.

So, electrical energy consumed

= power × time = 40 × 7200 watt-sec = 288000 J

SI unit of energy is joule or watt-second. From the above example, we find that joule or watt-second is a very small unit of electrical energy; so for commercial purposes, energy is measured in kilowatt hour (kWh). This unit is also called board of trade (BOT) unit.

1 BOT unit = 1 kWh = 1000 Wh = 36,000 Watt-s = 36,000 J

In the above example, the bulb thus consumes 288000/36000 = 8 kWh or 8 units.

There are various systems of electric tariffs for charging consumers for electricity. They are

1. One-part system
2. Two-part system
3. Three-part system

Let us first discuss how to calculate electrical expenses based on wattage of loads and their running hours only, with the help of a sample problem.

Find out the bill for the month of April for the following loads in a domestic apartment.

(a) Twelve 100 watt bulbs working for 10 hours a day
(b) Twenty 60 watt ceiling fans working for 15 hours a day
(c) One 2 kW heater working for 5 hours a day
(d) One 3 kW oven working for 10 hours a day
(e) One 2 hp motor for drinking-water pump running for 2 hours a day

There are two electrical energy metres in the hotel. Electricity charge for 1 kWh (1 BOT unit) energy used is Rs 5.00.

Metre rent for the month is Rs 20 per metre.

Solution: Note that the month of April has 30 days (be particular about the number of days of each month in a particular year).

Convert 2 hp motor to its kW equivalent. 1 hp = 746 Watt = 0.746 kW.

Therefore, 2 hp = 2 × 0.746 kW = 1.492 kW.

Next, find the energy consumed by each type of load by noting that

$$\text{Energy} = \text{power} \times \text{time}$$

Or, energy consumed in kWh = power in watt × number of units for the particular type of load × hours run in a day × number of days of the month/1000

If power is given in kW, division by 1000 is not needed.

Following this, proceed as follows:

(a) Energy consumed by 100 watt lamps = 100×12×10×30/1000 = 360 kWh
(b) Energy consumed by 60 watt fans = 60×20×15×30/1000 = – 540 kWh
(c) Energy consumed by 2 kW heaters = 2×1×5×30/1000 = 300 kWh
(d) Energy consumed by 3 kW ovens = 3×2×10×30/1000 = 1800 kWh
(e) Energy consumed by 2 hp (1.492 kW) motor = 1.492×1×2×30 = 89.52 kWh

Total energy consumed by all loads during the month = 3089.52 kWh

So, total electrical units consumed is 3090 BOT units (rounded off).

Rate per unit = Rs 3.00.

Amount of electricity charge = 3089.52 × 3 = Rs 9268.56 = Rs 9269 (rounded off)

Metre rent for 2 metres = Rs 20 × 2 = Rs 40

Therefore, total amount of bill = electricity charge + metre rent for the month

= Rs 9269 + Rs 40 = Rs 9309

The method described above is the way electricity bills are calculated for normal domestic connection. While the principles are essentially the same, the calculations for industrial units are less simple and described later in this chapter.

3.7.1 Electrical Metre Reading

The total energy consumed in a period is recorded by an instrument called the 'electric energy metre'. The unit of electrical energy consumption is 1 kWh (I kilowatt-hour), popularly known as one electrical unit (BOT unit) and hence, this metre is also known as the kWh metre. Figure 3.17 shows the front face of an 'analog kWh metre'.

Fig. 3.17   GE-make analog type kWh metre
(*Source:* www.tdsurplus.com)

The analog kWh metre is essentially an electric motor with a large circulating display disc that rotates as energy is passed through it. The total number of revolutions are transformed into corresponding total units and a series of gears are used to store total units consumed in different recorders by displaying them in digits of tens, hundreds, thousands, etc. of units (much in the same manner as seconds, minutes, and hours are shown in a clock; stop watches are closer comparisons, as they store the total time elapsed) in different display dials. Reading is taken at the beginning of a particular period and then at the end of the period. Difference in the kWh readings between the two gives the electrical energy units (BOT) consumed during the period. Figure 3.18 shows typical display dials and units consumed, shown at the beginning and end of a period.

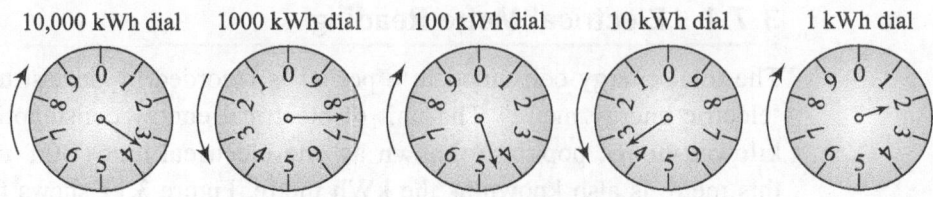

(a) Metre reading at the beginning of the month or bill period

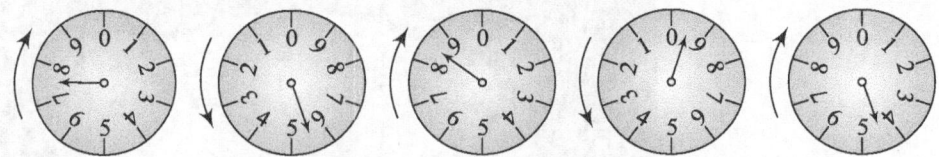

(b) Metre reading at the end of the month or bill period

Fig. 3.18 Status of the dial indicators in kWh metre for calculation of energy units consumed during a period of time

The units are calculated as follows:

From the metre dials, it is observed that at the beginning of the period the reading in the 10000 kWh dial is 2 (if the pointer is in between two marks, take the lower value, 2 in this case). So, corresponding to this dial the consumption is 2×10000 = 20000 units. Similarly, all such values from individual dials are calculated and then added together for the total units. The calculation results are furnished in Table 3.5.

Table 3.5 Calculation results in metre reading

Time of reading	kWh metre dial reading					TOTAL UNITS
	10000 kWh dial	1000 kWh dial	100 kWh dial	10 kWh dial	1 kWh dial	
Beginning of period	3 3 × 10000 units	7 7 × 1000 units	4 4 × 100 units	3 3 × 10 units	2 2 × 1 units	37,432
End of period	7 7 × 10000 units	5 5 × 1000 units	8 8 × 100 units	9 9 × 10 units	4 4 × 1 units	75,894

So, the units consumed during the period is total units recorded at the end of the period–total units recorded at the beginning of the period i.e., 75,894 – 37,432 = 38,462 units. Nowadays, these types of energy metres have become obsolete and digital-display type metres are in common use, which are very user friendly.

Hotels have many functional areas such as accommodation, kitchen, laundry, banquet hall, administration, etc., with some of them working as profit centres. Some of them, such as kitchen, may also have separate tariff systems. Statutory metre reading taken by electricity supply companies is usually on a monthly basis. However, it is always advisable for hotels to have metre readings weekly or fortnightly for internal checks. To keep a tab on energy consumption at different functional areas and in order to calculate profit earned by each profit centre, separate energy metres are installed for the different areas and profit centres.

Energy bill also depends upon the peak power demand, and hence hotels will have a separate peak demand metre which records the maximum demand at any point of time during the period of billing. Electric tariff also depends upon the timing of the usage of power. An additional metre called time of day (TOD) metre, which records the power usage timings in the establishment.

3.7.2 Tariff System

The word 'tariff' means the schedule of rates framed by electricity supply companies for their consumers. The following factors decide the tariff scheme:

1. Type of consumer (domestic, commercial, or industrial)
2. Peak or maximum demand of power
3. Total electrical energy (or BOT units) consumed
4. Time of the day when the energy is used
5. An electrical engineering factor called the power factor (ideal maximum of 1)

There are many different schemes of electric tariff systems followed by different electricity supply companies. They have different schemes and rates for domestic set-up, industrial set-up, railways, etc. According to one such scheme, if an industry takes voltage higher than, say, 13.2 KV, it will be entitled to a rebate as indicated below.

Supply voltage	Rebate
33/66 kV	1 paise/unit of energy consumed
110 kV	2 paise/unit of energy consumed
220 kV	3 paise/unit of energy consumed

Of course, in such cases, the industry has to install its own step-down transformer for getting working voltage output required for the electrical appliances and equipment.

Another scheme called 'time of day' (TOD) tariff system, also called 'off-peak' tariff, is available as an option for the consumer. Following is an example of the scheme:

Time of day of use	Increase + / reduction (–) in energy charges over the normal tariff applicable
22.00 hrs to 06.00 hrs	(–) 50 paise per unit
06.00 hrs to 10.00 hrs	0
10 hrs to 12.00 hrs	+ 40 paise
12.00 hrs to 18.00 hrs	0
18.00 hrs to 22.00 hrs	+ 50 paise

This incentive is given to even out fluctuations of demand of electricity during 24 hours of a day, which makes the power generation plants run efficiently and economically.

3.7.3 Three-part, Two-part, and One-part Tariff Systems

There are many systems of tariff being proposed to different categories of customers and by different electricity supply companies. All these tariff systems have been derived from the following relationship.

$$C = AX_1 + BX_2 + D$$

where C = total charge for a certain billing period (usually, 1 month or 2 months)

X_1 = maximum (as peak) power demand during the period or 80 per cent (say) of total connected load, or installed load (for which the premises is licensed by the electricity supply company which has accordingly designed and is maintaining the power supply system for the premises) whichever is higher.

X_2 = total energy (total unit or BOT) consumed during the period

A = charge per kW maximum demand

B = charge per unit of electricity consumed

D = fixed charge for the billing period

So, it is observed that the total charge is the sum of three components. The first part is purely based on the maximum power (not total units) drawn during the period, averaged over a short period of time. This part is also known as 'demand charge'. The second part is called 'energy charge', which charges according to total units consumed, as calculated in the example, earlier. The third part is a billing period 'fixed charge'.

Demand charge = normal demand charge + time of day charge – incentive

Energy charge = normal energy charge + time of day charge – incentive

Fixed charge is based on the following considerations:

1. The floor area or number of habitable rooms
2. The installed load

Based on different combinations of these three parts, the following tariff systems in vogue.

1. Three-part tariff system consists of all three components of tariff, i.e., demand charge, energy charge, and fixed charge. In general, most electricity supply companies have a three-part tariff system.
2. Two-part tariff system consisting of energy charge and fixed charge.
3. One-part tariff could consist of only the energy charge.

A typical three-part composite tariff scheme may be as follows:

Demand charges for peak demand   Rs 180 /kW of billing demand month

Energy charges

For the first 1 lakh units	360 paise per unit
For the balance units	440 paise per unit

Fixed charges

Metre rent	Rs 100 per month
Fixed charge for installed load	Rs 20 per kW per month

Electrical energy and time of day (ToD) metres are installed in the consumers' premises for recording the corresponding quantities for the purpose of billing for a particular period. To be more precise, demand charges are expressed in terms of rupees per kVA instead of KW to take care of the power factor consideration for the AC supply system, as discussed later.

Note that these charges are in addition to statutory taxes.

For street lighting, sign lighting, signal systems, and irrigation of tube wells, a flat demand rate tariff system is used. This does not mean that they are not charged for energy, but since the total connected load and their duration of operation is known, no metering of consumption is needed for them.

Power calculation in single-phase and three-phase AC appliances and motors—Power factor (PF)

Power calculations for AC systems are a little different from systems using a steady DC system or battery as the voltage and power source, which have been the basis of calculations earlier in the chapter.

Power consumed by single-phase equipment

Power in single-phase AC load is given by

$P = VI \cos \phi$, where V is the phase RMS voltage (230 V in India), I is the RMS current and $\cos \phi$ is a factor called the power factor, abbreviated PF henceforth.

Power factor should be ideally unity. But in practice, it is less than 1. In an AC system, that are different types of loads such as electric motors, fluorescent tubes, electric ovens, electric lamps, etc. Due to the particular electrical nature of some of these loads, especially electric motors and fluorescent tubes in the electrical systems, the current and the supply voltage are not in the same phase but the current lags the voltage (Fig. 3.19) by some value (in fact, ϕ is this angle of lag). Larger the lag, lower is the power factor with a theoretical minimum of 0 and a maximum of 1.0. If both voltage and current are in the same phase, we have the ideal maximum power factor of unity.

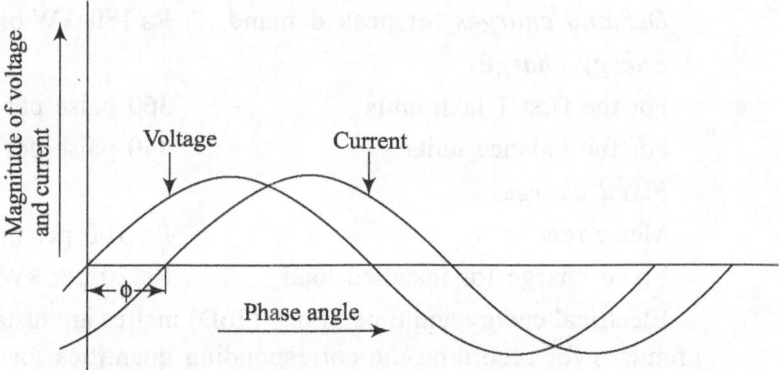

Fig. 3.19 Current lags behind voltage in many AC systems

There are many ways to improve power factor in an electric load system. One such way is the installation of what is called capacitors, among others, in the electrical system. A cost-benefit analysis has to be done in this regard before installing any such device.

Let us consider a fluorescent tube light of rating 230V/40 W connected to 230 V (rms value) AC supply. We all know that a tube light system has a choke coil as one of its components for its running. This choke coil causes the current to lag behind voltage resulting in a power factor appreciably lower than unity. Let the power factor be 0.8. When the tube light is switched on, the power drawn is 40 W.

Now, power, $P = V \times I \times$ power factor

and hence,

current, $I = 40/(230 \times 0.8) = 0.217$ ampere.

If there are two hundred such tube lights in a hotel each running (not necessarily simultaneously) for 10 hours a day, on an average, in a particular month of 30 days, the total energy consumed will be $40 \times 200 \times 300 = 2400000$ W-hr or 2400 kWh or 2400 electrical units (BOT). Electricity authority normally charge for this number of units only. However, we find that if the power factor could be unity, the power consumed would have been VI, i.e., 230×0.217 i.e. 50 Watt. Thus electricity authority can say that they have supplied 230 volt and a current of 0.217 ampere so that the customer could have taken the full 50 W and the electricity company could earn more revenue. Thus, the consumer is less efficient in utilizing energy available. This part of VI in ac power is thus important and is called VI part of power and has unit of volt-ampere or VA, in short, and which in this case is 50 VA. As stated earlier, electricity company charges for the actual units and does not have any system of penalty for domestic units for low power factor (the lower limit is set by the type of appliances used but seldom goes below 0.85) nor any incentive for improved power factor. However, for industrial units (including hotel industry), there is a scheme of tariff wherein electricity company charges for peak VA demand during a specified period (may be averaged over typically a period of half an hour). In our example, let there be a short period in the month when all the tube lights were 'on'. Therefore, although the peak power consumed during the time is $40 \text{ W} \times 200 = 8$ kW, the peak demand charge would be for $50 \times 200 = 10$ kVA. Demand charge is set as Rs per kVA, Detailed working on demand charge computation is carried out later. Obviously, if an establishment improves power factor, the total peak demand charge would come down.

Further, low power factor also attracts financial penalty while an improvement in it attracts incentives in tariff. Typically, if power factor is less than 0.97 (actual figure set by electric companies may vary), the premises owner has to pay a penalty.

Power consumed by three-phase equipment Motors operating in hotels and other industry (e.g., in lifts, pumps, refrigeration plants, compressors, etc.) are connected to a three-phase system, as they offer many advantages such as compactness and lower cost compared to the single-phase motors. Voltage supply in the three-phase system is normally specified in terms of line voltage (Fig. 3.20), e.g., 400 V in India. The scheme of three-phase supply and connection to a three-phase motor is shown in Fig. 3.20. The power calculation for such a system is given by

Power $= \sqrt{3} \times$ line voltage $\times$ line current drawn by the appliance $\times$ power factor (PF)

Or $P = \sqrt{3}\ VI \cos$

The line voltage in India is 400 volts.

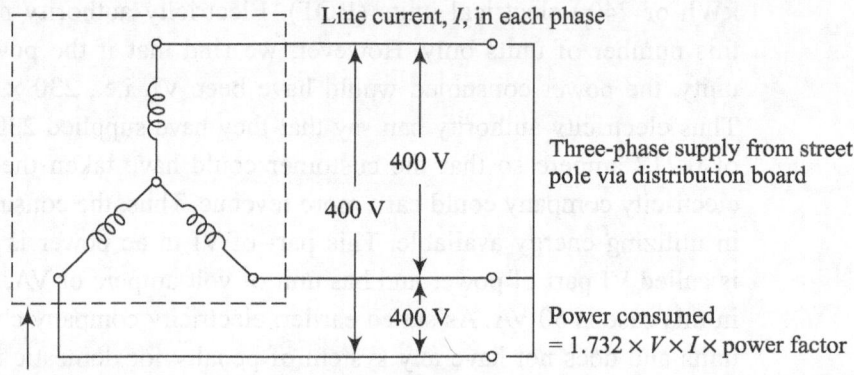

(a) Y connection inside equipment

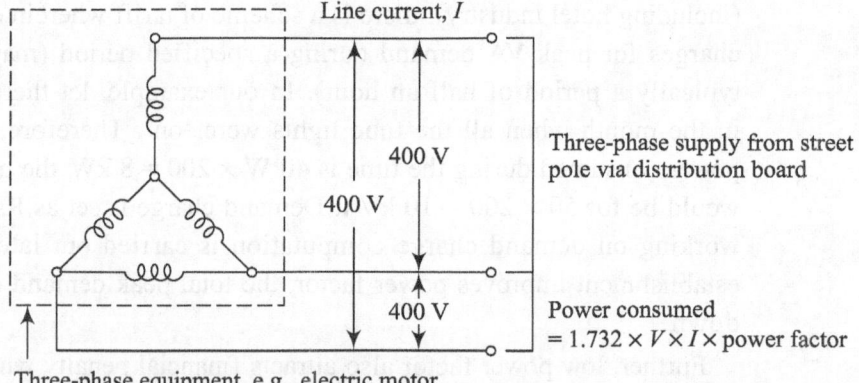

(b) Delta connection inside equipment

Fig. 3.20 Three-phase connection of equipment

Let us solve a few problems related to electricity bills.

- Suppose a hotel has a three-phase supply with a power factor of 0.9. A 3 kW electric motor is supplied with 400 V. Suppose, the mechanical power load (power required to grind 10 kg of substance in a mixer-grinder, say, that the motor is driving is equal to 1.5 kW. To drive this load, the electrical power requirement as the input to the motor will be slightly higher than this. If motor efficiency is 90 per cent, the motor input power will be 1.8/0.9 = 2 kW. This is the power the motor will draw from mains. For this power, the relation is given by

$P = \sqrt{3}\ VI \cos\phi$, where $\cos\phi$ is the power factor (PF).

Remember that for a single-phase AC system, P= $VI \cos\phi$.

(Contd)

(*Contd*)

$\sqrt{3}$ *VI* part (or *VI* for single phase) of the power, or the VI component, or VA is the one that the motor could utilize but, instead, it is utilizing less amount (P), power factor being less than unity.

So, VI component of power is equal to power/power factor.

Thus, for the preceding example,

$(\sqrt{3} \text{ VI}) \times \text{PF}) = 2.0 \text{ kW}$.

Or VI component = 2.0/0.9 kilovolt_ampere (kVA) = 2.22 kVA

It is this component of kVA and not kW which is used by the electricity company for its maximum or peak demand tariff calculation as already explained.

The current drawn by the motor under this condition is

$$I = 2.22 \times 1000/(1.732 \times 400) = 3.2 \text{ A}$$

In the theoretically worst case, when the PF would be zero, the energy consumed would have been zero even if there is voltage supplied to the motor, is drawing current, and the motor does not do any work. Surely, the energy utilized would be zero although the electricity company is supplying voltage and current to the motor and thus energy utilization factor is zero.

Suppose, the motor runs for 10 hours a day. Energy consumption (for calculating energy charge) is 2.0×10 kWh (BOT unit) = 20 units. Unit charge for energy is Rs 4.00. Therefore, energy charge portion of the total bill for the motor per day is 20×4 = Rs 80 i.e., Rs 2400 per month.

- Now, let us solve a total electricity bill problem for a hotel, showing different components of tariff.

Suppose a hotel has consumed 20,000 units during a month. On an average, the power demand is only 50 kW. But the average peak demand over a short period of time during the billing period, when all the ovens, AC machines, refrigerators, and flood lights were put 'ON' simultaneously, the power demand has been 130 kW. Power factor for the unit may be taken as 0.9. The total installed electrical load in the premises is 150 kW. The tariff is calculated as follows.

Demand charges	Rs 180 /kVA of billing demand month
Energy charges	
For the first 1 lakh units	360 paise per unit
For the balance units	440 paise per unit

Fixed charge Rs 20 per kW installation load

Based on these data, the total bill may be calculated as follows.

Maximum demand load (VI component)   = Max. power demand/power factor
= 130/0.9 = 145 kVA (rounded off)

Total connected load (kVA) = Total connected/installed power/power factor
= 150/0.9 = 166.67 kVA.

(*Contd*)

(Contd)

Take 75 per cent (the exact value to be fixed by the electric company) of this total connected load and actual maximum demand load and take the higher of the two for the purpose of determining the demand charge. In this case, $0.75 \times 166.67 = 125$ kVA, which is lower than 145 kVA. So, we take 145 kVA as the maximum demand for further calculation.

Energy charge	: Rs $3.60 \times 20,000$	= Rs 72,000
Demand charge	: Rs 180×145	= Rs 26,100
Fixed charge	: Rs 20×150	= Rs 3,000
Total bill		= Rs 99,100

Note that the demand charge has gone up to Rs 26,100, because over a short period of time, the average power demand went up. If that condition could be avoided by careful planning (by not operating all the loads at a time) and the maximum average power demand could be limited to 50 kW, say, the demand charges would be only Rs 10,000 (Rs $180 \times 50/0.9$).

- Let us solve another problem.

A 5-star hotel has a total connected load of 500 kW. The energy units for a particular billing period is 1,00,000 units. The maximum average power taken during a short period is 450 kW. Maximum demand charge is Rs 200 per kVA. Power factor of the premises is improved to 0.91. Earlier, the power factor was 0.9. The fixed charge for the billing period is Rs 10,000. The energy rate (considered flat) is Rs 5 per unit. The time of use charge for maximum demand is Rs 15,000 while incentive for operating at off-peak time is Rs 9,000. Corresponding values for calculating energy charges are Rs 15,000 and Rs 10,000. Now calculate the bill for the period.

Electricity company offers an incentive of 0.15 per cent of energy charges for each 0.01 unit increase in power factor from 0.9. The maximum demand calculations will be guided by the following stipulations of the electricity company.

Billing demand shall be the recorded maximum demand for the month in kVA or 80 per cent of the contract demand (as per the agreement), whichever is higher.

When the actual maximum demand in a month exceeds the contract demand as per the agreement the excess demand shall be charged at a rate of 150 per cent of the demand charges applicable. A power factor of 0.9 is considered for conversion from kW to kVA or vice versa.

Maximum demand = 450 kW = VI part/power factor.

Therefore, VI part = $450/0.9 = 500$ kVA.

Connected load (VI part) as per contract = power in kW/power factor = $500/0.9 = 556$ kVA (rounded off)

80 per cent (exact value to be specified by electricity authority) of the connected load is $556 \times 0.8 = 445$ kVA. We have to take higher of the two values of 500 kVA and 445 kVA for calculating normal demand charge.

(Contd)

(*Contd*)

Therefore, normal demand charge = Rs 200 per kVA × 500 kVA = Rs 1,00,000

(A) Total demand charge = normal demand charge + time of use charge for maximum demand − incentives

$$= \text{Rs } 1,00,000 + \text{Rs } 15,000 - \text{Rs } 9,000 = \text{Rs } 1,06,000$$

(B) Total energy charges = Rs 5 per unit × 1,00,000 units + Time of use charge-incentives

$$= \text{Rs } 5,00,000 + \text{Rs } 15,000 - \text{Rs } 10,000 = \text{Rs } 5,05,000$$

(C) Incentive due to improvement of power factor = 0.91− 0.9 = 0.1.

For each improvement of 0.01, the saving is 0.15 per cent of the energy bill. Hence total saving for 0.1 improvement in PF will be 1.5 per cent of energy bill i.e., 0.015 × 505000 = Rs 7575

(D) So, total energy bill = (B) − (C) Rs 5,05,000 − Rs 7575 = Rs 4,97,425,

(E) Fixed charge during the period = Rs 10,000

Therefore, total bill for the period is (A) + (D) + (E) = Rs 1,06,000 + Rs 4,97,425 + Rs 10,000 = Rs 6,13,425

3.8 | CONCLUSION

This chapter provided an introduction to simple electrical wiring systems, common electrical safety devices, and electrical tariff calculations. While the detailed technicalities are avoided in the discussion and complex industrial electrical safety systems are not included, the discussion in this chapter should indeed explain in some detail the various schemes of domestic electrical wiring so that the systems in use are comprehensible and useful to the students of hotel management, as also to other readers. The electrical power economy and tariff systems are also discussed at some length and it shows that by judicious planning and careful use of electrical power, a lot of economy can be achieved. This is very important for the students of hotel engineering who are to become future hotel managers.

KEY TERMS

BOT	It is the electrical commercial unit of energy and expressed in kWh, which is equal to 36,000 J.
Cable	It is a wire with higher size, usually having much higher current-carrying capacity.
Conductors	Conductors are materials through which charges can move, i.e., electricity can pass easily without much resistance. Materials

such as copper, aluminum, iron, etc. come under this category. Silver is the best conductor of all.

Demand charge	This is the tariff levied by electricity supply companies for maximum demand of power during the bill period.
Earthing	Earthing is a safety arrangement that connects the metal part of an electric equipment to the general mass of the earth, thereby protecting the operator in case the metal body comes in contact with the live wire inside.
Electric tariff	It is the energy charging system and rate for consumption of electric units by the consumer.
Energy charge	This is the tariff levied by electricity supply companies for the total units consumed by the customer during the bill period.
Fuse	Fuses are like safety valves in an electrical circuit. In case there is more than rated current in the particular circuit, the fuse wire melts and breaks the circuit
Insulators	Insulators are materials which offer tremendous resistance to movement of charge through them so that they allow practically no electricity to flow through them. Materials such as rubber, asbestos, Bakelite, mica, ebonite, etc. fall in this category.
Power factor	It is a factor in AC system that reduces the efficiency of the consumer's electrical system in extracting maximum power available from the board supply. Its value is between 1 and 0. The consumer is charged for lower power factor.
Rectifier	It is a device which converts AC supply to DC.
Short circuit	It is a phenomenon of electrical accident when a live line comes in direct contact with the neutral or the earthing line by-passing a load resistance—huge current flows and the fuse blows off.
SWG	It is the acronym for standard wire gauge. It is a numbering system which specifies the diameter of the wire.
Switch	It is an electromechanical device, which makes or breaks a circuit. It is put in the live line to provide safety to workers.
Thermostat	A thermostat is a device for regulating the temperature of an environment so that its temperature is maintained near a desired temperature. The temperature may be set within a range of temperature values. The thermostat switches on or off the electric circuit that provides power to the heating or cooling devices as needed to maintain the correct temperature that is set.
Three-part tariff system	This is an electrical tariff system composed of demand charge and energy charge and a fixed charge.
Two-part tariff system	This is an electrical tariff system composed of both demand charge and energy charge.
Wire	It is a metal conductor used for conveying and distributing electricity. Electrical wire is usually made up of copper and aluminium.

REVIEW QUESTIONS

3.1 Define electric resistivity of a material.

3.2 Physically explain electrical conductivity of a material.

3.3 Explain what do you mean by a 'good electrical conductor' and name a few conductors.

3.4 List a few good insulators and state why we need insulators in an electrical system.

3.5 What is an electric wire and what is a cable?

3.6 Explain how wires are designated.

3.7 Explain what is meant by wires of specification (a) 5/0 and (b) 3/20.

3.8 Write, in short, different types of electrical wires used in domestic wiring.

3.9 List different types of laying methods in domestic wiring, highlighting their relative merits and demerits.

3.10 Explain the function of fuse. Why is it put in the live line?

3.11 List and describe various types of fuses used in domestic wiring installation

3.12 What is the function of main switch in a domestic circuit? Explain why switches are put in the live line?

3.13 Draw a distribution board system of wiring installation and explain its advantage over a tree system.

3.14 Explain why earthing is necessary for electrical appliances.

3.15 What is a 'testing megger'?

3.16 Describe the precautions to be taken while handling electricity and electrical equipment.

3.17 What is rectification? Why is it necessary?

3.18 Explain BOT.

3.19 Explain, in brief, the working of an electrical energy metre.

3.20 Explain a three-part tariff system.

3.21 What are the considerations for assessing fixed charge component in a three-part tariff system?

3.22 What is power factor and its maximum possible value in an AC system?

REVIEW PROBLEMS

3.1 A 60 W, 230 V tungsten filament bulb has a coiled coil wire of diameter 0.015 mm. Specific resistance for tungsten may be taken as 0.53×10^{-7} ohm-m. What will be the corresponding straight length of the coil when unwound?   [2.94 m]

3.2 Following data relate to a one-star hotel in respect of its electrical load in the business for the month of February 2008.

(a) Fifty 100 watt bulbs working for 12 hours a day

(b) Forty 60 watt ceiling fans working for 15 hours a day

(c) Two 2 kW heaters working for 4 hours a day

(d) Four 3 kW ovens working for 10 hours a day

(e) Two 2 hp motors for a drinking-water pump, running for 4 hours a day

(f) Four 1 kW refrigerators working for 14 hours a day

(g) Four 2 kW A.C. machines working for 6 hours a day

Electricity charge for 1 kWh (1 BOT unit) energy used is Rs 3.50.

Metre rent for the month is Rs 50.

Calculate the energy bill for the month, considering power factor to be unity.

[Rs 35,121.90]

3.3 A five-star hotel has consumed 60,000 units in a particular month. The peak demand for power during the period has been recorded as 200 kVA. Calculate the total bill to be paid if energy charge is Rs 3.50 per unit and demand charge is Rs 180 per kVA. Take total installed load is 250 kVA . Demand charge would be based on higher of the maximum demand and 80 per cent of the installed load. Fixed charge for the month is Rs 10,000. Taxes may be taken as 12 per cent of total electricity charges.

[Rs 3,15,840]

4 Chapter

Elementary Illumination Science and Lighting Systems

4.1 INTRODUCTION

It can be observed that a football stadium, a bedroom, a hotel banquet, or a restaurant, each has its own unique requirement for lighting arrangement, both in terms of lighting devices, as well as in terms of layout and design. Specialist illumination engineers and architects are required to decide about this aspect in various such application areas. This is especially true for a hotel whose business is to offer a comfortable and peaceful stay for guests. Productivity of people and sense of comfort depends critically on the level, type, and layout of lighting. Illumination science and engineering deal with this aspect of technology. Science of illumination deals with the measurement of light energy emitted by a light source and the illumination produced at the surfaces.

The science of illumination is very important in our working life as different types of works demand different forms and levels of illumination. It is obvious that a study table, an operation theatre in a hospital, or a hotel lobby will all need different types and levels of illumination.

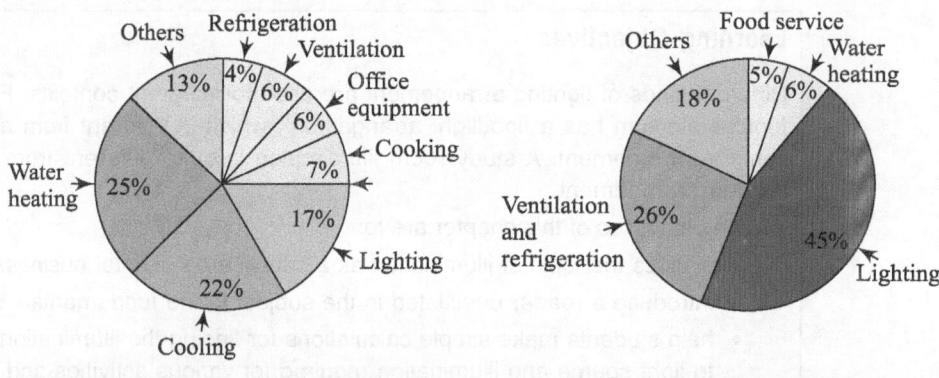

(a) Typical energy consumption according to end-use in a big hotel without space heating.

(b) Typical electrical energy distribution by end-use

Fig. 4.1 Pie-chart distribution of end-use functions

The science of illumination deals with both the light source and the recipient surfaces. Illumination engineering deals with design and manufacture of lighting devices and control, assessment of lighting requirement for each of the areas of lighting, and choosing the most optimum devices and their layout. As regards the hotel industry, proper illumination and lighting system form a critical functional area involving guest comfort, safety and security, and pleasant wor king environment. It assumes additional significance when we note that lighting system accounts typically for about 12–20 per cent of total energy cost in a modern hotel of moderate size. Figure 4.1(a) and (b) shows pie-chart distribution of various end-use functions in a typical hotel. Actual figures will vary somewhat depending upon the object conditions prevailing for a particular property.

4.2 | PLANE ANGLE AND SOLID ANGLE

Lets us define two different types of geometric angles for an understanding of the elementary aspects of photometry and illumination.

4.2.1 Plane Angle

We all are familiar with the term angle. When two lines meet at some point, they form an angle. This angle is called plane angle (Fig. 4.2). The common unit of angle is degrees (°). If two lines are opposite to each other and meet at a point, the angle between them is said to be 180°. An arc of circle makes an angle at the centre of the circle and the unit of angle used in all scientific relations is defined with respect to this configuration (Fig. 4.3). This unit of angle is not measured in degrees but radians.

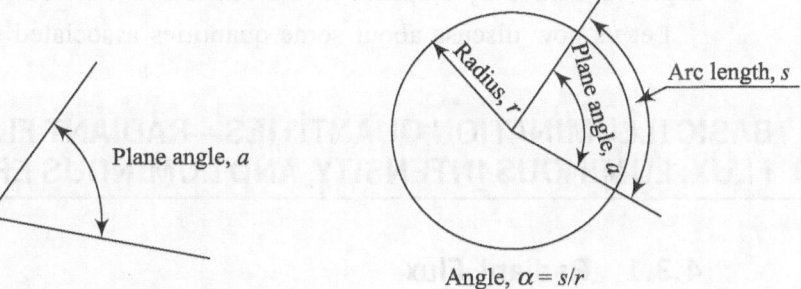

Plane angle, *a*

Radius, *r*

Plane angle, *a*

Arc length, *s*

Angle, $\alpha = s/r$

Fig. 4.2 Plane angle   **Fig. 4.3** Arc of a circle

Plane angle (α), subtended by an arc of length *s* at the centre of a circle of radius, *r*, is given as

$$\alpha = s/r \qquad (4.1)$$

Therefore, when the radius of the circle is unity and the length of the arc is also unity, the angle is said to be 1 radian. A semicircle makes an angle of 3.1415 radian at the centre, i.e., 3.1415 radian is 180°. This number 3.1415 is abbreviated as π(pi). Therefore, $\pi^c = 180°$. The superscript 'c' denotes radian.

4.2.2 Solid Angle

Taking a cue from plane angle, think of an angle subtended at the centre of a sphere by a stretch of the area, *A*, on the surface of the sphere, as shown in Fig. 4.4. This angle is called 'solid angle' and is defined as

$$\varpi = A/r^2 \qquad (4.2)$$

Compare Eqn (4.2) with Eqn (4.1).

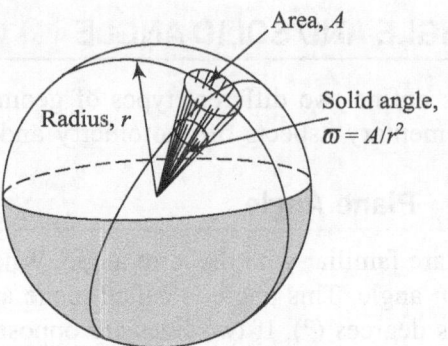

Fig. 4.4 Solid angle

The unit of solid angle is steradian. So, it can be observed that if, $A = 1$ and $r = 1$, then the solid angle subtended is 1 steradian. It can be proved that total solid angle subtended by a sphere at the centre is 4π steradian.

Let us now discuss about some quantities associated with a light source.

4.3 | BASIC ILLUMINATION QUANTITIES—RADIANT FLUX, LUMINOUS FLUX, LUMINOUS INTENSITY, AND LUMINOUS EFFICACY

4.3.1 Radiant Flux

A light source emits light, which is a form of energy and is transmitted as an electromagnetic wave such as heat wave, radio wave, etc. However, a light source not only emits energy that is visible as light but also other forms of electromagnetic energy such as heat and other invisible radiations. It can be observed that if an electric filament type bulb glows, the room gets heated as well. The total energy radiated by a light source per second is called radiant flux. Its unit is, therefore, joule/sec or watt.

4.3.2 Luminous Flux (*F*)

It has already been state that a light source emits all sorts of electromagnetic waves including visible light. Now, different types of electromagnetic waves (infrared, thermal, visible light, microwave, radio wave, etc.) have different wavelengths. Usually, wavelength is expressed in unit of angstroms (Å), which is 10^{-8} m. Only those radiations that have a wavelength between 4000 and 7800 Å will produce visual sensation. So, out of the total radiation energy flux, only a fraction will be radiated as visible light and the corresponding energy rate is called luminous flux and denoted by F.

So, it can be stated that luminous flux (F) is the photometric energy emitted per second by a light source. However, this energy flux does not give an indication of the corresponding sensation to the eyes. For example, suppose the total radiation flux by a light source is 1 W. Let the corresponding luminous flux be 0.6 W. Now let this source emit more of green-yellow light. Let another light source also have a luminous flux of 0.6 W for the same radiation flux of 1 W. But let this source emit more reddish light. So, according to the definition, both have the same luminous flux. But it can be observed that the green-yellow rich light appears more sensitive to eyes than the reddish one. Therefore, the luminous value of a source is measured not merely by the total light energy flux but also by the effect it produces on the eyes. Figure 4.5 is drawn to show this effect graphically.

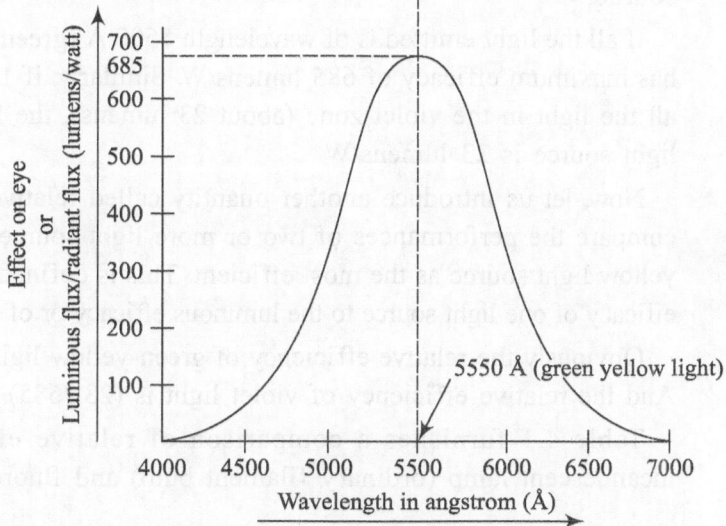

Fig. 4.5 Luminous value of a source of light

It can be observed from the figure that the effect on the eye per unit watt of radiant flux is the most for wavelength of 5500 Å (green-yellow light) and very low for both sides of the curve (reddish and violet) for the same conversion efficiency. If a 1 watt lamp produces all green yellow light (5500 Å), a maximum value of 685 is assigned for this maximum effect on eye and so, the luminous flux of this light said to be 685 lumens/W. If the same amount of energy is radiated by visible light of wavelength 4500 Å (violet) per watt of radiant energy, it is seen that the effect is equivalent to about 23 lumens/W only.

So, formally, the unit of lumen is assigned to measure luminous flux and it is defined as follows.

The luminous flux of a light source is said to be 685 lumens, if the radiant flux is 1 W and the wavelength of the all the emitted visible light is 5500 Å.

Simply stated, more lumens imply more light. The maximum sensation to the eyes does not mean the best effect on the eyes. As a matter of fact, the green-yellow combination is not suitable for many situations and is also not soothing for eyes. People usually reject this colour rendition as it is very uncomfortable.

4.3.3 Luminous Efficacy of Electric Light Source

More watts of a lamp does not necessarily mean more light, but lumens per watt is an efficacy measure of the bulb.

The luminous efficacy, η, is defined as the ratio of the luminous flux, F, (in lumens) emitted by the source to the electric power input, P, (in watt) to the light source.

If all the light emitted is of wavelength 5500 Å (green-yellow light), the source has maximum efficacy of 685 lumens/W. Similarly, if 1 W of light source emits all the light in the violet zone (about 23 lumens), the luminous efficacy of the light source is 23 lumens/W.

Now, let us introduce another quantity called relative luminous efficiency to compare the performances of two or more light sources considering the green-yellow light source as the most efficient. This is defined as the ratio of luminous efficacy of one light source to the luminous efficacy or of green-yellow light source.

Obviously the relative efficiency of green-yellow light source is 100 per cent. And the relative efficiency of violet light is $(23/685) \times 100 = 3.35$ per cent.

Table 4.1 furnishes a comparison of relative efficiencies of tungsten incandescent lamp (ordinary filament bulb) and fluorescent tube light.

Table 4.1 Illumination performance of tungsten filament lamp and fluorescent tube light

Tungsten lamps (GLS)				Fluorescent tube light (FTL) (standard length for each category)			
Power input, watt (A)	Luminous flux, lumens (approx.) (B)	Efficacy, lumens/ watt B/A (C)	Relative efficiency C/685 × 100 (%)	Power input, watt (D)	Luminous flux, lumens (approx.) (E)	Efficacy, lumens/ watt (D/E) (F)	Relative efficiency F/685 × 100 (%)
40	460	11.5	1.7	8	325	40.53	5.9
60	840	14.0	2.0	20	950	47.5	6.9
100	1630	16.3	2.4	30	1500	50.9	7.4
200	3660	18.3	2.7	40	2300	57.5	8.4
500	9950	19.9	2.9	100	4400	44.0	6.4

It is observed that fluorescent light is much more energy-efficient. This is due to two reasons. The filament incandescent type of bulb works on the principle of light energy emitted from a very hot source. Consequently, much of the energy supplied is emitted in the region of wavelength corresponding to thermal radiation, which does not contribute to the visual effect. Secondly, the light wavelength emitted lies away from the green-yellow zone, so effectively lumen/W decreases. In case of fluorescent lamps, visible light wave is produced by electric discharge (discussed later), and hence the temperature of the light source is quite low. So, much of the electrical energy supplied is converted to radiation falling in the visible light zone and, additionally, the wavelengths of different colours of light radiation in such tube lights lie closer to 5500 compared to an incandescent bulb. These factors result in higher efficiencies for them. However, a look at the relative efficiencies, in both the cases, indicates that they are far away from the maximum achievable of 100 per cent (all light emitted is the green-yellow type) although, as already pointed out, the particular combination of light colours corresponding to the 100 per cent emission may not be acceptable in practical applications.

4.3.4 Luminous Intensity (*I*)

We observe that in Fig. 4.6(a) and (b) we place the eyes at the same distance from the aperture. Obviously, when light is received in the eyes, more energy is falling on the eyes and, hence, there is more sensation of brightness in the case where it is coming out through a small aperture. So, it is observed that as far as the recipient of light energy is concerned, there will be some physical quantity of the source that will determine this effect on the recipient. This physical quantity or property of the light source is called the luminous intensity.

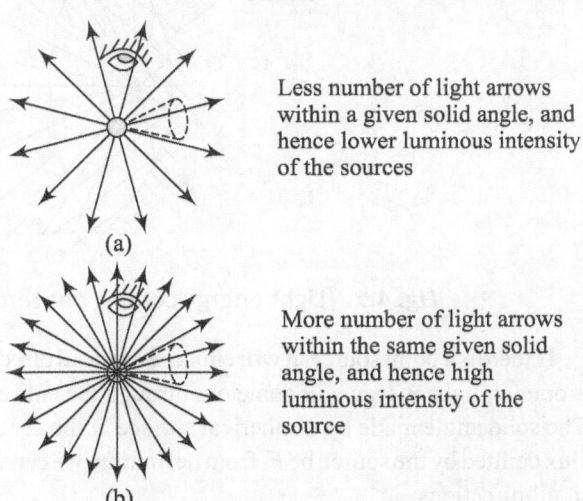

(a)

Less number of light arrows within a given solid angle, and hence lower luminous intensity of the sources

(b)

More number of light arrows within the same given solid angle, and hence high luminous intensity of the source

Fig. 4.6   Two different light sources

Thus, luminous intensity (I) of a light source in a particular direction is defined as the luminous flux emitted by the source per unit solid angle in that direction.

Or
$$I = F/\varpi$$

Unit of luminous intensity is lumen per steradian or candela (cd) or candle power (CP). 1 lumen per steradian is called 1 candela.

It is easy to see that in case of a large aperture, the solid angle is more than that in case of small aperture having the same source. So, for the same amount of luminous flux, following the definition of luminous intensity, the source of light having all the flux coming out through a small aperture has more luminous intensity than when it is coming out through a larger aperture.

Let us work out an idealized problem.

A light source has a luminous flux of 200 lumens. The whole energy comes out through a spherical aperture (in fact, energy will be radiated in all directions, not just only through the aperture; assume that the energy emitted in other directions is fully reflected back by some reflector inside and the whole of the energy comes out through the aperture as shown in Fig. 4.7. For example, car head lights have very efficient reflectors behind the bulb.). Let the solid angle made by the area of opening be 2 steradians. The intensity of light is, therefore, 200/2 = 100 lumens/steradian or 100 candela. If the same energy comes out through an enlarged aperture of, say, 3-steradian solid angle, the intensity will be 66.66 candela.

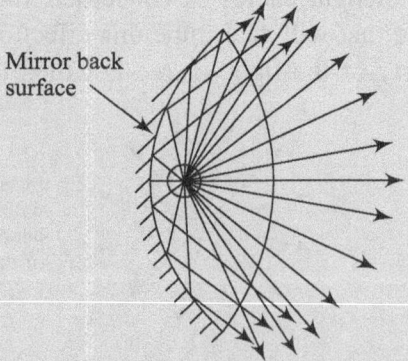

Mirror back surface

Fig. 4.7 Light energy coming out through the aperture

If there is a point source, it will emit energy in all directions, i.e., if one imagines a sphere around it, light energy will come out through the entire surface of the imaginary sphere. The solid angle made by a spherical surface at the centre is 4π. So, if the total luminous flux emitted by the source be F, from definition, we can write, assuming uniform emission in all directions,

(Contd)

(Contd)

$$I = F/4\pi \qquad (4.3)$$

If, for a source, luminous intensity, *I*, is specified, the total luminous flux emitted can be written as

$$F = 4\pi I$$

4.3.5 Illuminance or Intensity of Illumination

Now that the terms and specifications associated with a light source have been established, the attention is turned to the processes at the surface that receive the energy. Figure 4.8 finds that the total luminous flux received by the two receiving surfaces is the same which is *F*, for assumption. However, it is clear that surface *A* has more intense illumination than surface *B* because of same energy being received over a small area.

Let us now define a quantity called illuminance or intensity of illumination (*E*) to quantify the intensiveness of illumination, which is the luminous flux received per unit area of the receiving surface. Intensity of illumination must be distinguished from luminous intensity, which is an exclusive property of a light source.

So, from the given definition, we can write

$$E = F/A$$

From definition, unit of illuminance becomes lumen/m² or lux. In FPS system, the unit of luminous intensity is foot-candle which is lumen/ft².

This may be a requirement of particular space that has to receive light for some specific purpose; for example, a study table may require 150 lumens of light flux per square metre of area. Now, choose a proper lighting device that would be able to provide this light energy on the table. So, from this specification of requirement, let us now look at the source (Fig. 4.8).

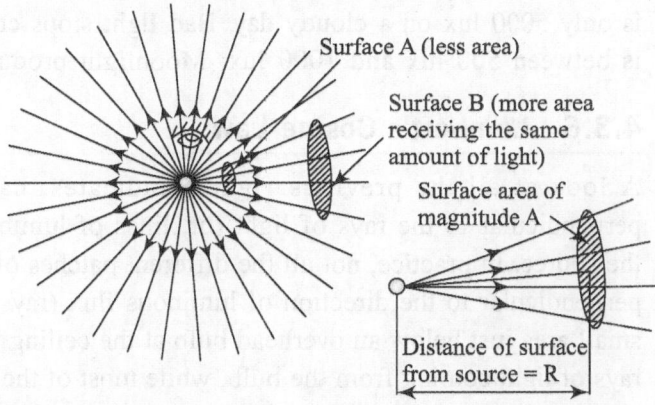

Surface A (less area)

Surface B (more area receiving the same amount of light)

Surface area of magnitude A

Distance of surface from source = R

Fig. 4.8 Intensity of illumination for a source-receiving surface combination

In terms of I, the luminous intensity of a source, which will be usually given for a light source, we can write

$$F = Ix\varpi = Ix\, A/R^2$$

or

$$F/A = I/R^2,$$

or

$$E = I/R^2 \qquad\qquad (4.4)$$

Suppose, we have a 60 W incandescent bulb hanging right over the table, the total luminous flux (Table 4.1) is 840 lumens and hence the intensity of illumination, I is $F/4\pi = 840/(4 \times 3.1415) = 66.847$ candela. If the lamp is held 2 m (about 6.6 ft) away from the table, Eqn (4.4) tells us that the illuminance or intensity of illumination as experienced by the table will be only 16.71 lux. Bringing it to 1 m height will increase the quantity to 66.85. So, it can noted that reducing the distance by half increases illuminance four times.

A 40 W fluorescent lamp at 2 m height will have 45.75 lux of illuminance. All these arrangements are found to be inadequate for the study table, which should have a lighting level, or illuminance of 150 lux. But, one finds such arrangements only common in practice and should try to provide the level of illumination as specified in standard codes of practice, e.g., Indian standard code of practice for interior illumination, IS 3646 (Part I).

So, it is found that the illuminance of a given area due to light from a given source of intensity, I, decreases very rapidly (with the square of the distance from the source of light) as the distance from the source increases. A two times increase in distance will result in four-fold decrease in illuminance. Equation 4.4 also tells us that the illuminance, E, at a place on the surface of a sphere of unit radius due to a point source of unit intensity kept at the centre of the sphere will be 1 lux (if unit radius is 1 m) or 1 foot candle (if the unit radius is 1 foot).

$$1 \text{ foot candle } \approx 10.764 \text{ lux}$$

Just to have a realistic idea about the unit of illuminance, it may be noted that the level of natural light on a bright sunny day is about 50,000 lux, while this is only 5000 lux on a cloudy day. Bad light stops cricket game when this level is between 500 lux and 1000 lux. Moonlight produces only 0.2 lux.

4.3.6   Lambert's Cosine Law

A look at all the previous figures indicates that everywhere the area is perpendicular to the rays of light (or, lines of luminous flux) coming out from the source. In practice, not all the different patches of areas that receive light are perpendicular to the direction of luminous flux (rays of light). For example, the small area just below an overhead bulb at the ceiling will be perpendicular to the rays of light coming from the bulb, while most of the areas on the side walls will not be so. So, when Eqn (4.4) is used to find the illuminance of a surface due to a given luminous flux intercepted by the surface, it is implied that the surface is perpendicular to the rays of light. Therefore, a general relationship between *E, I,* and *r* is needed which takes into account this inclination of the surface with the direction of flux. Here, take a specific case where the light rays are coming from a distant source so that all the rays are parallel to each other when they reach the surface (Fig. 4.9).

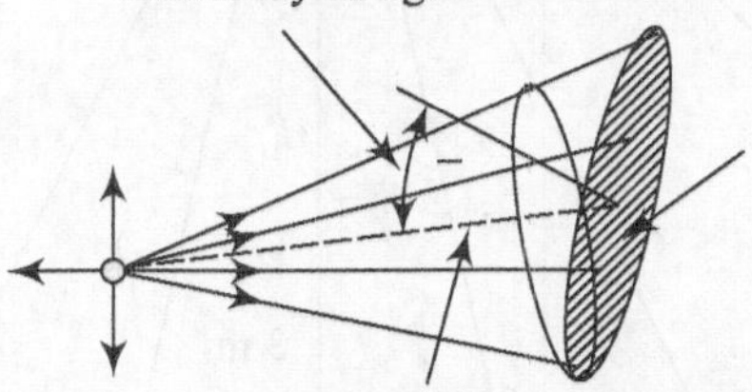

θ, the angle between the normal to the area and the direction of the central ray of light

The area is oblique and not normal to the direction of the central ray of light

r, the distance of the oblique area from the source of light

Fig. 4.9 Rays parallel to each other

The relation is now given, with reference to Fig. 4.9, as

$$E = Ix \cos \upsilon / r^2 \tag{4.5}$$

where υ is the angle between the direction of flux and the direction of the normal to the area. This is Lambert's cosine law. For a special case of $\upsilon = 0°$, this equation boils down to Eqn (4.4).

A student with preliminary knowledge of trigonometry can find out the value of $\cos \upsilon$, if the value of υ is known. Otherwise, any scientific calculator will give this value with its costise function key.

However, for distant sources like the sun or the moon, whatever be the orientation of the area (horizontal or vertical), the rays coming out of these distant sources being parallel in whatever orientation we look at them, they will always be perpendicular to any surface which views them and hence will have same intensity anywhere on the surfaces.

Let us now solve a problem.

A lamp is fixed at a height of 3 m from the floor (Fig. 4.10). It has a mean luminous intensity (sometimes called the mean spherical candle power [MSCP]) of 1000 candela. Determine the (a) total luminous flux of light, *F*, (b) illumination at a point on the floor, just below the lamp, and (c) illumination at a point on the floor at a position 2 m away on the floor. Consider the bulb as a point source of light.

(Contd)

(Contd)

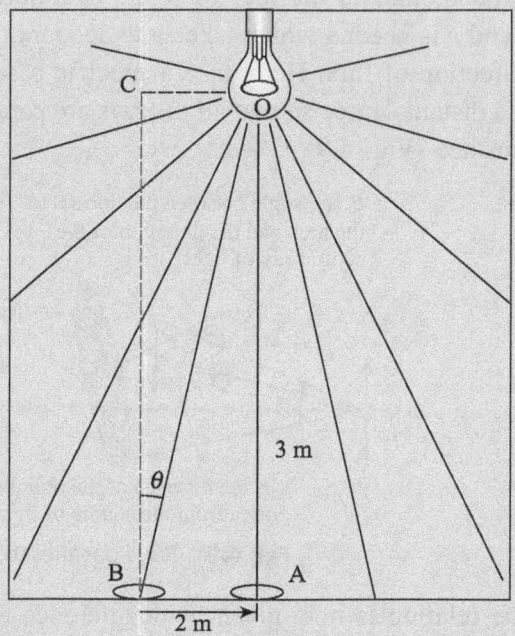

Fig. 4.10 Lamp fixed at a height of 3 m from the floor

Given, $I = 1000$ candela.

(a) We have from Eqn 4.3 total luminous flux F,

$$= 4\pi I = 4 \times 3.1415 \times 1000 = 12566 \text{ lumens}$$

(b) By comparing Fig. 4.9 and Fig. 4.10, we have $\theta = 0°$ for the point A, directly under the lamp.

Now, $\cos 0° = 1$, $r = 3$, and hence,

$$E_A = I \times \cos \theta / r^2 = 1000 \times 1/3^2 = 111.11 \text{ lux}$$

(c) For the point B, from the basic definition of cosine (cos) function,

$$\cos \theta = BC/OB = OA/OB$$

$OA = 3$ m and for a right angled triangle (OAB), Pythagoras' theorem gives us

$$OB^2 = OA^2 + AB^2 = 3^2 + 2^2 = 13 \text{ m. So, } OB = 3.6 \text{ m}$$

Therefore, $\cos \theta = 3 / 3.6 = 0.833$

Now, illuminance at B can be calculated as

$$E_B = I \times \cos \theta / r^2 = 1000 \times 0.833 / OB^2$$
$$= 1000 \times 0.833 / 3.6^2 = 64.27 \text{ lux}$$

Comparing E_A and E_B, one finds a decrease of illuminance at B compared to that at A by about 50 per cent by simply moving away a little bit from the direct line of sight of the light rays.

If this stage, inverse calculations can be early done for finding out the lamp intensity requirements and their locations when illumination requirements at different points on the floor (based on specific purposes) are given. Table 4.2 furnishes average illumination requirements for some specific jobs and functions, in particular relation to the hotel industry.

4.3.7 Brightness or Luminance

Now, let us define one more quantity to establish the complete cycle of interrelationship comprising (a) a source of light, e.g., a lamp, (b) a surface receiving the light, e.g., a book, (c) a sensor, e.g., a human eye receiving the light reflected back from the illuminated surface (the book).

For a particular surface to be visible to human eyes, it is not enough that the surface receives sufficient light. The light after being received by the surface must be reflected to the eyes so that the surface can be seen clearly. Here comes the question of the quality of the surface. Consider two surfaces of same illuminance, (the same intensity of energy reception, i.e., they are kept at the same distance away, at the same inclination, from a common light source emitting light energy uniformly in all directions). However, suppose one surface is dirty, while the other is clean. It is a common experience that the clean surface would appear brighter. Similarly, the colour of the surface also influences the degree of brightness, e.g., a black surface would appear darkest even if it has a high degree of illuminance but a white surface would appear very bright. Therefore, it can be concluded that the appearance of brightness of surface not only depends upon illuminance, but also on the nature (colour and surface condition) of the surface of the object which is being viewed.

This property of visibility of a surface, which receives luminous flux from one or many sources of light and then reflects back to the eyes for its visibility, is called brightness or luminance.

4.4 | LIGHTING REQUIREMENTS IN HOTEL INDUSTRY

Proper design of lighting is one of the key factors for the successful functioning of the hotel industry. While design should have an eye on the energy economy as a whole, proper levels and duration of lighting must be maintained throughout. The principal objectives may be listed as follows:

- To provide good working conditions to prevent strain and fatigue to the personnel
- To create a proper ambience
- To help people know the directions and destinations by use of proper lighting signs

- To help maintain safety standards
- To enhance security in the premises
- To selectively attract people to some focal points

Table 4.2 provides typical lighting requirements at different functional areas in a hotel. Let us recall that the level of natural light on a bright sunny day is about 50,000 lux, while this is only 5000 lux on a cloudy day. Moonlight produces only 0.2 lux.

Table 4.2 Lighting levels at different functional areas of a hotel

Functional areas	Illuminance (lux)
Entrances and reception areas	200–300
Bedroom—general	100
Bedroom—bed-head and mirror	150
Corridor and passage	100
Kitchen and office	500
For work requiring detailed and minute observations	1000–1500
Restaurants, bars, etc.	50–150

In connection with the end use, we define a few terms the specification of which would help selecting bulbs for specific requirements.

4.4.1 Colour Temperature

Often terms such as 'warm,' 'soft-white,' 'cool,' and 'daylight' are used to specify the heat feeling of a light. Colour temperature is a quantitative measure of this feeling. Colour temperature is generally represented in Kelvin, referring to absolute temperature (a temperature expressed in degrees Celsius plus 273 gives the value of the corresponding absolute temperature. For example a temperature of 27 degrees Celsius is 300 Kelvin or 300 K. So, a temperature of 400 K is accordingly 127 degrees Celsius). Table 4.3 furnishes colour temperature for different warmth effects of light sources.

Table 4.3 Colour temperature for different types of lights

Effect of light	Colour temperature (Kelvin)	Use
Warm/soft white	≤ 3200 K	Daily living, bedroom, recreation room
Cool/bright white	3200–4000 K	Work place, garage, kitchen, bathroom, etc.
Daylight	≥ 4000 K	Areas requiring precision activity; reading or where accurate colour rendition is wanted

4.4.2 Colour Rendering Index

True colour of any object refers to the perception of colour of the object to an average man when the object is kept in open daylight condition. However, lights never reproduce the conditions of open daylight particularly in absence of any daylight. We introduce a concept of colour rendition, which indicates the efficacy of a light to reproduce daylight colour condition of the object and express it by means of a quantity called colour rendering index (CRI). Colour rendering index is, thus, a quantitative, not subjective, measure of the ability of a light source to reproduce the colour of various objects faithfully as in natural light. Colour rendering index refers to the spectrum of light, or it is a measure of how well a light supports the full range of colours visible to the human eye, with daylight being given a value of 100 at a temperature of about 5000–6000K. One has to consider temperature too because CRIs are really comparable across the same temperature range. Soft-white bulbs will make things look more yellow even if they have a high CRI. Colour rendering index should be considered if accuracy is important to bulb selection decision.

The CRI of incandescent and light emitting diode (LED) bulbs (discussed later) are quite close, while compact fluorescent lamps (CFLs) (also discussed later) are a little less in the blues. For good colour quality, CFL may have a value over 80 and as close to 100. But a high CRI only does not imply good colour rendition. However, without going into the complexities of the subject, it can be said that a CRI above 80 is necessary for better colour rendering of people and things in general.

4.5 | LIGHT SOURCES USED FOR DOMESTIC AND COMMERCIAL PURPOSES

Different types of common light sources such as common electric bulbs, fluorescent tube lights, mercury vapour lights, neon signs, sodium vapour lamps, etc are all familiar terms. Each of them has different nature of light emission as far as its effect on eyes and on the brightness of the surroundings is concerned. Blue-green colours are considered cool, while yellow-red colours are considered warm. Cool light produces higher contrast and is considered better for visual tasks. Warm light is preferred for living spaces because it is considered more soothing to skin tones and clothing. However, people's acceptance to blue-green colours is very limited, in general, particularly in public meeting places and living places. All of these light sources can be broadly classified into two basic types depending on the physical principles which they employ for emission of light. These are

1. Incandescent lamps or general lighting service (GLS) lamps
2. Electric discharge lamps
3. Light emitting diodes (LED)

4.5.1 Incandescent Lamps or GLS Lamps

Incandescent lamps, colloquially known as electric bulbs, are general purpose lighting sources and are a synonym for general lighting service lamps because of their versatility of uses. There are quite a few developments of incandescent bulbs in recent times for better efficacy and diversity of uses. They are described as follows.

Conventional incandescent lamp

Incandescent lamps work on the principle that when an object is heated to high temperature, it radiates visible light. It is our common experience that when coal burns, it gets red hot and if it burns in an otherwise dark region, the region not only gets heat but is also illuminated to some extent. In incandescent lamps, current is passed through a fine metallic wire (a high resistance) by connecting it to a voltage supply line. The current, while passing through the wire, heats it and raises the temperature and as a consequence, energy is radiated. At low temperatures, most of the energy radiated will have wave-lengths corresponding to thermal radiation only and we can feel the heat only. As the temperature increases, visible light fraction increases and at 3000°C (usual for tungsten filament bulbs), there is bright yellowish light from them.

To prevent the wire from oxidation at a very high temperature, it is sealed inside an evacuated (free from atmospheric oxygen) bulb with two lead points coming out of the bulb for connecting to the supply lines (Fig. 4.11). However, completely evacuated condition leads to evaporation of metal filament coil and the internal surface blackens. To overcome this difficulty, the internal air is first drawn out (evacuated) and then the evacuated space is filled in with some inert gases (which do not react with filament metal) like a mixture of argon gas and nitrogen gas, which again raises the internal pressure, thus reducing evaporation loss. This arrangement prevents oxidation as well as evaporation of the coil element. Low wattage lamps, however, may have evacuated bulb.

The different types of materials that can be used for making the filament wire are carbon, tantalum, and tungsten. Tungsten is the most commonly used material. Its advantages are as follows.

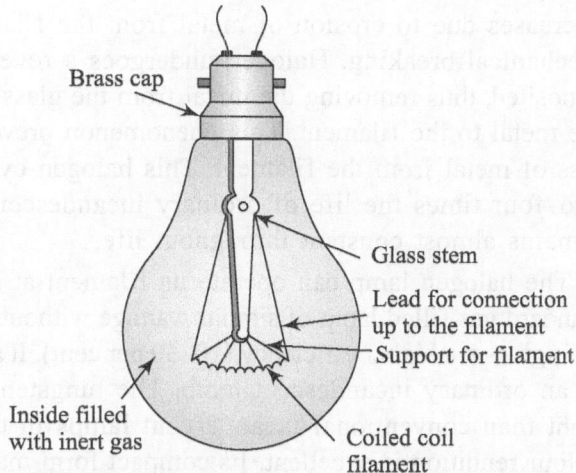

Brass cap

Glass stem

Lead for connection
up to the filament

Support for filament

Inside filled
with inert gas

Coiled coil
filament

Fig. 4.11 Components of an incandescent lamp

(*Source:* calgary.rasc.ca)

- Its melting point is high, of the order of 3400°C. Although carbon has a still higher melting point (3500°C), it cannot work at high temperatures as it starts disintegrating:

- It is mechanically strong and can be mechanically processed into thin wire.

- Its length does not change much with temperature (so the resistance remains fairly constant even at higher temperature).

- Its vapour pressure is low so evaporation loss of material is very low.

The hotter the filament, higher will be the efficacy but lower will be the lifetime of the bulb, which is determined by evaporation of the tungsten filament. So, an optimization is made between higher lumen output per watt and life of the bulb. The life span, usually, ranges from a few hundred hours to 2000 hours. Figure 4.12 shows a view of a typical incandescent lamp.

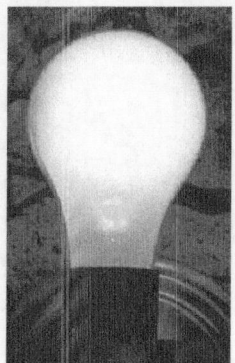

Fig. 4.12 Incandescent lamp

(*Source:* calgary.rasc.ca)

Tungsten-halogen (TH) lamp

In this lamp, the glass is filled with an inert gas, plus a small amount of halogen such as iodine or bromine gas. In ordinary incandescent bulb tungsten metal of the filament gets evaporated due to high temperature and slowly gets deposited on the inside of the glass bulb surface, thus blackening it. Due to this, brightness decreases with time. Burning life (total illumination hours before failure) also decreases due to erosion of metal from the filament, making it weak against mechanical breaking. Halogen undergoes a reversible reaction with the metal deposited, thus removing the metal from the glass surface and then re-depositing the metal to the filament. This phenomenon prevents bulb darkening as well as loss of metal from the filament. This halogen cycle, thus, enhances life (about two–four times the life of ordinary incandescent bulbs) and the light output remains almost constant throughout life.

The halogen lamp can operate its filament at a higher temperature than in a standard gas filled lamp of similar wattage without any reduction of operating life. This gives it a higher efficiency (10–30 per cent). It also emits cooler light compared to an ordinary incandescent lamp. The tungsten halogen lamp emits a whiter light than conventional incandescent lamps in the range of warm white. The colour rendition is excellent. Its compact form makes the tungsten halogen lamp an ideal point light source. However, it is definitely much costlier than its non-halogen counterpart. In general, as a rough guide, this lamp should be used in place of ordinary incandescent lamp when single-lamp wattage requirement is more than 200 W due to decreasing efficiency of the later kind of bulb beyond the stated voltage. Figure 4.13 shows a typical halogen lamp.

Fig. 4.13 View of an outdoor Halogen lamp for security purpose

(*Source:* calgary.rasc.ca)

Merits and demerits of incandescent lamp

Incandescent lamps are cheaper, reliable, very easy to install and provide good quality bright light. They are rich in warm colours like red, orange, and yellow. Cooked food items, especially reddish in colour, for example, cooked meat, red wine, etc., look very bright under this lamp. People also look much brighter in this lighting environment and, as such, these bulbs have a wide acceptance. However, they are not energy-efficient and high power lamps emit a lot of heat to the surroundings causing discomfort during summer period and increasing AC load in an AC room. Their life is also less compared to a fluorescent light.

4.5.2 Electric Discharge Lamps

In any gas, there are three types of particles. One is neutral atoms (with electrons and protons bonding stable), loose positively charged nucleus (consisting, primarily of, positively charged protons and neutral neutron particles) and negatively charged electrons. These lamps work on the principle that when high voltage is applied across a stream of particles of gas, the negatively charged loose electrons move with high velocity towards the positively charged end of the voltage source of the bulb and positively charged protons move towards the negatively charged end. Some of the high-energy electrons when colliding with neutral atoms of the gas impart their energy to the atoms to 'excite' their electrons to move to outer orbits (electrons move around nucleus in orbits like our planets in a solar system according to Bohr's model) with higher energy. But this state is unstable and these electrons stay there for only 10^{-8} seconds and fall back to some lower or the original orbit (Fig. 4.14). While doing this they emit electromagnetic radiation mostly in the visible range and light is emitted. In this section, (a) sodium vapour discharge lamp, (b) mercury vapour discharge lamp, and (c) metal halide lamp are discussed.

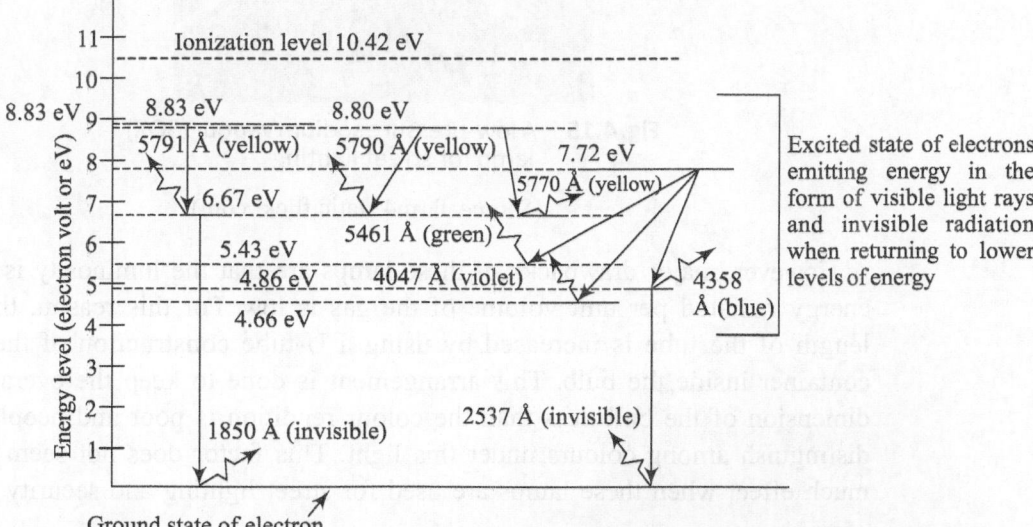

Fig. 4.14 Excited electrons emitting light while falling back to the original orbit

Sodium vapour discharge lamp

This type of lamp uses sodium vapour inside the bulb. The radiations resulting from excitation of sodium atoms can be theoretically shown to be consisting of mainly radiations in the visible zone only, i.e. the energy radiated is mostly light energy. It is the only lamp having this property.

There are two types of sodium vapour lamps, namely (a) low-pressure sodium (LPS) vapour lamps or SOX lamps and (b) high-pressure sodium (HPS) vapour lamps or SON lamps.

LPS Vapour (or SOX) lamps   In LPS, most of the radiation is in the bright yellow zone (about 5890A). This fact results in a high value of efficacy of about 200 lumen/W at the highest wattage rating level. These bulbs also maintain their lumen output fairly constant with age. Their high efficacy results in low power cost and is suitable where long duration of operation is imperative. Because of the attractive power economy coupled with its bright yellow radiation, these bulbs are used widely in outdoor lighting. Wattage normally ranges from 10 to 160 W. Figure 4.15 shows a SOX lamp for street lighting.

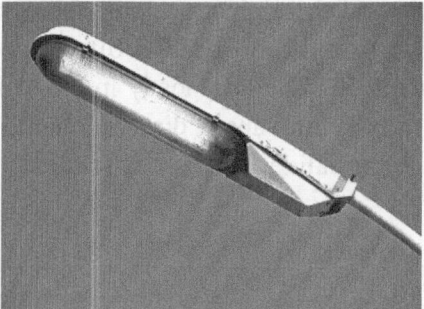

Fig. 4.15   A low-pressure sodium vapour (SOX) lamp for street lighting

(*Source:* farm4.static.flickr.com)

However, major drawbacks of these lamps are that the luminosity is low and energy radiated per unit volume of the gas is low. For this reason, the actual length of the tube is increased by using a U-tube construction of the vapour container inside the bulb. This arrangement is done to keep the overall length dimension of the bulb less than the colour rendition is poor and people cannot distinguish among colours under this light. This factor does not seem to be of much effect when these lamps are used for street lighting and security lighting.

HPS Vapour (or SON) lamps   The disadvantages of LPS are greatly reduced by employing high-pressure sodium (HPS) vapour lamps, also called SON lamps. These lamps operate at high pressure and a little mercury vapour is added to them. Because of these factors, the colour rendition becomes pinkish or orange red and hence colour distinction of colours of objects under this light is possible. Illumination efficacy is also quite high—of the order of 100 lumens/W and can go up to 150 lumens/W. They are favourably smaller in size than the LPS counterpart and more easily started. They are widely used for street lighting, security lighting, and parking lot lighting. The warm colour of SON frequently improves the appearance of stone and brickwork. Its long life coupled with good lumen reliability over the length of life makes it very suitable for lighting where access for the lighting arrangement is difficult as in lofty industrial interiors such as steelworks, power plant turbine house, aircraft hangers, etc. They are available in brands broadly called SON-T and SON-E, where SON-T is tubular and SON-E is elliptical in the structure of the bulb as illustrated in Fig. 4.16(a) and (b). Sodium vapour lamps are not suited for interior lighting, as such. Figure 4.17 shows street lighting by HFS lamps

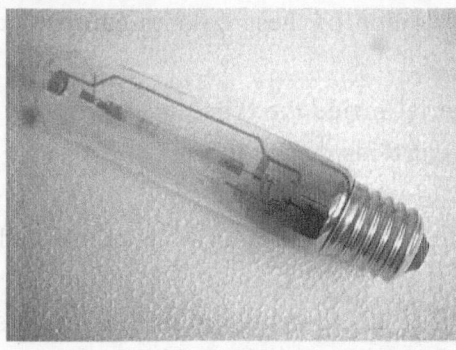 

(a) European GE Lucalox 150w LU150/100/T/40 (SON-T) High Pressure Sodium Lamp
(*Source:* www.freewebs.com)

(b) Osram Vialox 50w NAV-E SON-E lamp
(*Source:* www.freewebs.com)

Fig. 4.16 Types of SON lamps

(a) View of high-pressure sodium vapour lamps on a street

Fig. 4.17 Street lighting by HPS lamps
(*Source:* www.novabrite.com.my)

Mercury vapour discharge

There are low-pressure mercury vapour discharge and high-pressure mercury vapour discharge lamps that are used for lighting purposes.

Low-pressure mercury discharge lamps and fluorescent tube lamps (FTL)   The mercury discharge is quite complex as its electron shell structure is very elaborate. Theoretical analysis as well as experimental evidences indicate that the colours radiated are yellow, green, blue, and violet. So, it is found that this mixture of colours is objectionable for common use. In addition, low-pressure bulbs emit ultraviolet rays, which are injurious to health. They also, like sodium vapour discharge lamps, have to be made longer in size for energy efficien*cy.* Deluxe white mercury lamps are available in the market, which have much improved colour rendition qualities.

Suitable colours have to be added to the basic colour in this type of lamps to make the light acceptable. Addition of these colours can be accomplished by one of the following methods:

- Using fluorescent material inside the tube
- Increasing vapour pressure inside the tube
- Using mixed vapour

Let us now discuss mercury fluorescent lamps, commonly called fluorescent tubes lamps (FTL).

All bodies on which radiations fall absorb a part of it, transmit some of it, and reflect some part. A part of this reflected light can be seen as coming from the object and sense it as the colour of the object. There are some materials that reflect energy, which is different in wavelength from the incident rays, the reflected wavelength usually being more than the wavelength of the incident ray. This phenomenon is known as luminescence. Luminescence is different from normal reflection in that this is really a re-emission process where atoms of the luminescent materials are excited by the incident radiation. Luminescence can further be classified into two categories:

- Fluorescence where energy is re-emitted during the period of excitation only and
- Phosphorescence where the re-emission persists even after the incident excitation is removed.

Fluorescent powders used in low-pressure mercury lamps are usually called phosphors. They are excited by mercury radiation. Table 4.4 provides a list of few materials used as fluorescent materials to produce different colours upon excitation.

Table 4.4 List of phosphor materials and their fluorescence colours

Phosphor materials	Colour of fluorescence
Magnesium tungstate	Blue-white
Calcium borate	Pink
Zinc silicate	Green
Calcium tungstate	Blue
Zinc beryllium silicate	Yellow-white

For commercial use, phosphor materials are mixed with some heavy metals which serve as 'impurities' to provide a choice of suitable colours. These impurities are also called 'activators'. Table 4.5 gives a list of such 'activators'.

Table 4.5 List of activators and their fluorescence colours

Activators	Colour of flourescence
Gold	Blue-white
Silver	Blue
Copper	Green
Manganese	Deep yellow
Copper plus silver	Greenish-white or bluish depending upon composition
Cadmium silicate	Yellow-pink

The low-pressure mercury fluorescent lamp is essentially a long tube, which is coated with a phosphor material. The tube is filled with a small quantity of mercury vapour and a small quantity of argon. The phosphor material additionally converts ultraviolet (UV) radiation into visible radiation, thus reducing health hazard. Figure 4.18 shows a typical circuit and the components for such a lamp.

Mercury fluorescent lamps emit soothing light and have quite high energy efficiency compared to incandescent lamps. But they are costly, require a number of components, and need skill in fitting a new set. However, worldwide, 80 per cent of lighting arrangement is provided by fluorescent lamps. Table 4.1 furnishes a comparative study of lumen output for conventional incandescent lamps and fluorescent lamps.

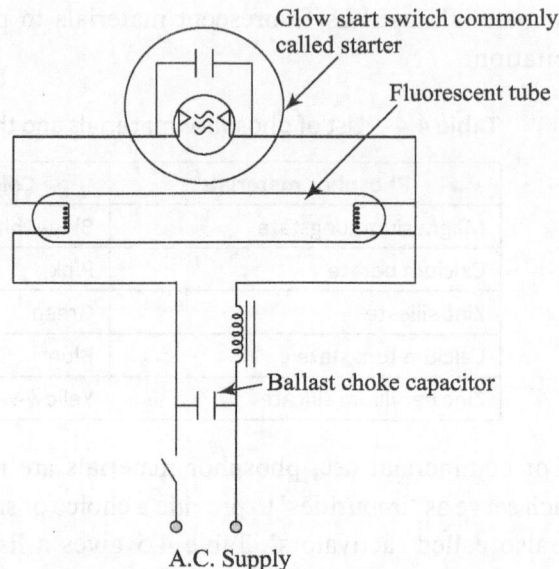

Glow start switch commonly called starter

Fluorescent tube

Ballast choke capacitor

A.C. Supply

Fig. 4.18 Components and internal circuit for a low-pressure mercury fluorescent lamp

Compact fluorescent lamp bulb A recent development is the introduction of energy-saving fluorescent lamp, which is fast replacing filament lamps. These are fluorescent tubes with much smaller diameter and folded several times to reduce the length and make it compact. It is then put inside a glass cover and directly fitted into a conventional filament lamp holder (Fig. 4.19). This lamp is popularly known as compact fluorescent lamp (CFL). Although this bulb is more expensive, its life is five times more than filament lamps and uses about 25 per cent of the energy. It yields about 48 lumens for each watt of electrical power, whereas a 40 W filament lamp produces about 10–12 lumens per watt.

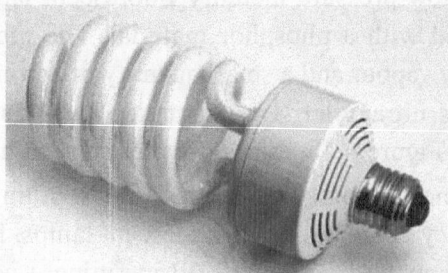

Fig. 4.19 View of a typical CFL lamp

(*Source:* www.crispyneurons.com)

Some CFL manufacturers label their bulbs with a three-digit code to specify their CRI and colour temperature. The first digit represents the CRI measured in tens of percent, while the last two digits represent the colour temperature measured in hundreds of kelvins. For example, a CFL with a CRI of 92 and a colour temperature of 2900 K would be given a code of 929.

The early CFLs had a low CRI value, but today CFLs come with higher CRI values, even in the daylight range. Now, one can have energy conserving CFLs and accurate colour rendition too.

All fluorescent lamps, including CFLs, contain mercury and, therefore, the disposal of fluorescent lamps always poses an environmental problem.

High-pressure mercury vapour lamp Some of the objectionable colours and UV radiation can be eliminated by using high-pressure mercury discharge lamps. They are energy efficient and are used for street lighting and decoration purposes. A high-pressure mercury vapour discharge operates in a quartz arc tube. The internal surface of the elliptic bulb is coated with phosphor material to convert the ultraviolet (UV) radiation emitted from electric discharge inside into visible light radiation.

Metal halide lamps

These lamps (Fig. 4.20) are essentially variants of high-pressure lamps with high intensity and belong to the broad category of mercury vapour lamps. However, they use various other elements in an amalgam with the mercury like sodium iodide and scandium iodide. These lamps can produce much better quality light without resorting to phosphors. They produce much white light than mercury vapour lamp without the use of phosphor coating and are used as stadium flood-light. Metal halide lamps are now available that can have a colour rendition ranging from very yellow to very blue. They are available at a wide range of wattages but at higher ceiling heights, 250 W and 400 W sizes are common.

Fig. 4.20   Metal halide lamp showing the internal arrangement

(*Source:* diytrade.com)

4.5.3 Light-emitting Diode Lamps or LED Lamps

In the past, for quite some time, electronic display boards were used in railway stations, airport, shops, and many other places. These are essentially light emitting diodes, which are nothing but just tiny light bulbs that fit easily into an electrical circuit. Recently, this technology has been developed to manufacture LED bulbs for illumination purpose that is seen by many as the light of the twenty-first century. However, the working principle of these bulbs is different from those of either incandescent or electric discharge bulbs. They are illuminated solely by the movement of electrons in a semiconductor material, working much in similar fashion as a diode transistor that allows electric current to flow in one direction but restricts in the other and they last just as long as a standard transistor. This is the reason why LED lamps are called solid state lamps (SSL).

Light emitting diode lamps are the latest in illumination engineering. They are powered by low-voltage DC supply. These bulbs are very efficient, durable, low cost, and reliable with dimming feature. These bulbs are highly environment friendly and are called 'green bulbs'. Many hotel industries are fast replacing fluorescent and incandescent bulbs with these bulbs. They may even replace CFLs in near future. Figure 4.21 shows LED ceiling light in a room.

Fig. 4.21 LED ceiling light

(*Source:* en. wikipedia.org)

4.6 | CHOICE AND SELECTION OF LAMPS AND LIGHTING ARRANGEMENTS IN HOTEL INDUSTRY

The task of choice and selection of lighting arrangement in different areas of a hotel is a highly specialized area of illumination engineering. The type of colour spectrum (range of wavelength and hence colours emitted by a particular light source) of the light source and the natural daylight colour of the objects intended to be illuminated by the light source are of utmost importance for colour vision of the objects under the situation. For example, a filament lamp gives a very rich white light and hence will enhance yellow, orange, and red bands of colours in the object, which will also become brighter. But care must be exercised while selecting fluorescent lamps as the colour rendering can be quite different. A room, which is decorated in red or pink, will appear dull if daylight fluorescent lamp is used, while it will appear bright if the lamp corresponds to natural light type. Restaurants and food displays should be illuminated by lights which bring out red and green tones. A hotel must be visible from at a distance to attract and impress customers. Neon signs and decorative designs of incandescent lamps on the premises could be very good options. Halogen/metal halide lamps are used for security flood lighting, while soft fluorescent or mercury vapour lamps could be used in parking lot and for landscape illumination.

Just as the move to compact fluorescent bulbs has gained momentum in the mainstream use, hoteliers are switching to an even more energy-efficient lighting source called light-emitting diodes, commonly known as LED. Many hotels already have adopted LEDs for small-scale applications, like exit and other signs. Various types of lighting arrangements such as direct, semidirect, and diffuse lighting are installed at different areas depending on the particular requirements. The hotel management must together with a professionally qualified architect and illumination engineer finalize this important aspect of hotel design keeping in mind the initial cost, running energy cost, replacement cost, and functional efficacy of each type of alternative lighting arrangements.

Table 4.6 furnishes a comparative statement of performance of different types of light sources to help select the right kind of lighting for specific purposes.

Table 4.6 Comparative features of different light sources used in various applications

Light source	Average efficacy (lumens/W)	Advantages	Disadvantages	Applications
Ordinary tungsten incandescent lamp	100 W lamp gives 16.3 lumens/W (see Table 4.1)	• Lowest unit cost • Very good bright colour rendition	• Warm colour • Low life, usually ranging from 750 hrs to 2000 hrs	Living room, study room, general-purpose lamp

(Contd)

Table 4.6 (*Contd*)

Light source	Average efficacy (lumens/W)	Advantages	Disadvantages	Applications
		• Colour rendition is acceptable by people • Bulb replacement is very easy • Gives light almost immediately on applying voltage	• Much of the energy (90 per cent) is emitted as heat instead of light) • Increased load on AC machine • People have feeling of hot environment	
Halogen lamp	100 W lamp gives 16.7 lumens/W	• Emits a whiter light than conventional incandescent lamps • The colour rendition is excellent • Its compact form makes it an ideal point light source • Working life is two to four times that of conventional incandescent lamps	• Much costlier than conventional incandescent lamps	Used widely as street lamp, desk top lamp and car headlight
Fluorescent tubes	100 W lamp gives 4400 lumens/W (see Table 4.1)	• Smooth, cool, white light • Excellent colour rendition • Working life within 10,000 hours	• Higher initial cost; total set costs about 15 times more than a conventional incandescent lamp, while only the tube is about three times costlier • Changing tube accessories requires skilled electrician • Requires more space • Disposal of damaged bulb poses environment problem as it contains mercury	Living room, study room, general-purpose lighting
CFL lamp	20 W lamp gives about 50 lumens/W	• 'Soft white' CFLs are qualitatively similar in colour to conventional incandescent lamps • More durable; much higher rated life ranging from 6000 hrs to 15000 hrs • Very high efficacy • Bulb replacement is very easy • Gives light almost immediately on applying voltage	• Cost is typically four–ten times that of an incandescent lamp for comparable duty • Disposal of damaged bulb poses environment problem as it contains mercury • High wattage lamps are yet to come in the market • CFL may not fit well in existing light fixtures • Gets dimmer like all fluorescent lamps with age	Corridor light, bed room light, general purpose light requiring low wattage
LED	Very high efficacy of 100 lumens	• Long life span, no filament, to break • Radiation-free	• Higher initial cost than CFL • At present, high wattage lighting is not available for	Offices, factories, marketplace, schools, and houses

(*Contd*)

Table 4.6 (*Contd*)

Light source	Average efficacy (lumens/W)	Advantages	Disadvantages	Applications
LED	Very high efficacy of 100 lumens from a 1-W device	• Long life span, no filament, to break • Radiation-free • Stroboflash-free • Easy to install • Environment friendly • Noise-free • Electricity saving can be up to 50% of ordinary fluorescent lamps • Life span: 50,000 to 60,000 hrs • Rigid and safe • Very small size; can be arranged in rows and clusters • Allows brightness control (dimming)	• Higher initial cost than CFL • At present, high wattage lighting is not available for room lighting • Individual LEDs are powered by low voltage DC supply and hence needs rectification from AC to DC or solar cell power can be conveniently used	Offices, factories, marketplace, schools, and houses and so on indoor spaces such as recessed downlight, kitchen under cabinet lighting, portable desk/task lighting
Sodium vapour lamp (low pressure)	Very high value of efficacy of about 200 lumen/W	• One of the highest in luminous efficacy • Life expectancy is very high with modern design going up to 18.000 hrs	• Luminosity is low and energy radiated per unit volume of the gas is low • Colour rendition is poor and people cannot distinguish among colours under this light.	Outdoor lighting
Sodium vapour lamp (high pressure)	Quite high; of the order of 100 lumens/W	• Distinction of colours of objects under this light is possible. • Smaller in size than the LPS counterpart • More easily started • Average lamp life exceeds 20,000 hrs	• Fairly high initial cost • Presence of mercury makes its disposal a problem	Widely used for street lighting and security lighting
Mercury vapour lamp	About 40–65 lumens/W; lowest lumen output of all discharge-typebulb	• Low system cost • Very long rated average; lives exceeding 24,000 hrs • Colour rendition made better with phosphor coating and may be better than sodium vapour	• High initial cost compared to normal incandescent lamp • Low lumen output • Normal colour rendition is blue-green and is objectionable • Deterioration of lumen output with age	Still popular for outdoor lighting
		• High intensity light • Quite cheap	• Ultraviolet radiation is a negative factor • Disposal is a problem	

(*Contd*)

Table 4.6 (*Contd*)

Light source	Average efficacy (lumens/W)	Advantages	Disadvantages	Applications
Metal halide	Quite high ranging from 65 lumens/W to 115 lumens/W	• Produce light much closer to sunlight compared to other high-intensity discharge (HID) bulbs such as high-pressure mercury vapour, sodium vapour, etc. • They burn much cooler compared to normal incandescent lamps • Five times more efficient than a normal incandescent light • Small size, powerful, and efficient light source • Because of smaller size, light can be focused in space • A high life of about 15,000 hrs–20,000 hrs is common	• High initial cost, however, depending on the application, can present a lower installed cost due to fewer fixtures compared to fluorescent lamp • Requires special fixtures to operate • Disposal is a problem	Very widely used in stadiums and other sports facilities and industrial, commercial, retail, and municipal spaces. It is increasingly found in supermarkets, big box retail, offices, and lobbies.

4.7 | CONCLUSION

Illumination science and engineering has come a long way since Thomas Alva Edison implemented the first commercially practical electric incandescent light in 1879. Modern research is oriented towards energy efficiency as well as lumen efficiency of lighting systems. There have been designs for almost all specific needs to cater to. It is evident that hotel industry has its very specific and varied requirements for its lighting arrangements and illumination. Proper illumination is vital not only for immediate satisfaction of guests and good working condition of the employees but also critical from safety and security point of view. From the foregoing discussion, it is apparent that illumination science and engineering is quite complex and is a very specialized branch of science and engineering. Therefore, services of expert and experienced illumination engineers must be deployed for selection and design of illumination systems for such industries. In this regard, every effort should be made to ensure energy saving mechanisms, keeping in mind the huge energy consumed by lighting systems in a modern large hotel, which accounts for about 15 per cent to 20 per cent of total energy cost and about 40 per cent to 50 per cent of the total electrical energy consumed in a hotel.

KEY TERMS

Brightness or Luminance	It is the amount of light reflected from a surface reaching our eyes to give visual sensation about the surface.
Colour rendering index (CRI)	CRI is a quantitative measure of the ability of a light source to reproduce the colour of various objects faithfully in natural light.
Colour temperature	It is a temperature rating of a bulb in degrees Kelvin representing the warmness/coolness effect of the bulb.
Fluorescence	Fluorescence is a phenomenon where energy is re-emitted by a surface during the period of excitation only.
Illuminance	It is the amount of luminous flux received by a surface per unit surface area. Its unit is lux.
Incandescent lamp	It is a type of lamp that has a coil filament inside a glass bulb and radiates light when heated to a high temperature
Luminescence	It is a phenomenon where a surface re-emits incident radiation having wavelengths different from incident radiation.
Luminous flux	It is the light energy part of the radiant flux expressed in some form of effect to the eye. Its unit is lumen/watt.
Luminous intensity	It is the luminous flux radiated by a source per unit solid angle. Its unit is candela.
Phosphorescence	Phosphorescence is a phenomenon where the re-emission persists even after the incident excitation is removed.
Point source	It is source of light which is assumed to be concentrated at a point and radiates light equally in all directions.
Radiant flux	It is the total electromagnetic radiation emitted by a body per second. Its unit is watt.
Solid angle	It is the angle subtended by a part of the spherical surface at the centre of the sphere. Its unit is steradian.
Vapour discharge lamp	It is a type of lamp which emits light based on the principle of atomic excitation of gas particles
Wavelength	It is the distance between successive similar points in a wave. Its unit is (10^{-8} m).

REVIEW QUESTIONS

4.1. Define the following terms mentioning their units.
   (a) Solid angle
   (b) Luminous intensity
   (c) Luminance
   (d) Degree of illumination
   (e) Luminous flux
   (f) Colour Temperature

4.2. Explain why the subject of illumination is important in hospitality industry.

4.3. Using a neat sketch, write Lambert's cosine law.

4.4. Explain how the condition of a surface receiving light is important for its appearance and vision to an observer.

4.5. List the objectives of proper lighting in hotel industry.

4.6. Discuss about the lighting levels needed at different functional areas of hotel industry.

4.7. Describe the principles of incandescent and vapour discharge lamps.

4.8. Explain why tungsten is widely used as filament material in incandescent lamps.

4.9. Explain, in general terms, the working principle of a halogen lamp.

4.10. Write down the merits and demerits of incandescent lamps.

4.11. Write down the merits and demerits of mercury fluorescent lamps.

4.12. Explain, in brief, the working principles of vapour discharge lamps.

4.13. Draw an electric circuit for fluorescent lamp.

4.14. Discuss about choice of proper lighting in the context of colour rendition.

4.15. Compare conventional incandescent lamp, fluorescent lamp, and metal halide lamps.

4.16. State the merits and demerits of high-pressure sodium vapour discharge lamps.

4.17. Discuss, in brief, about LED lamps.

4.18. Discuss, in brief, the choice and selection of lighting arrangement in hotel industry.

REVIEW PROBLEMS

4.1. The uniform luminous intensity (earlier called mean spherical candle power) of an electric bulb is 90 candela. What is the total luminous flux emitted by the bulb? Also determine its luminous efficacy if the power of the bulb is 100 W. Take $\pi = 3.14$.

[1130.4, 113.04 lumen/W]

4.2. A surface is situated at a distance of 1.5 m from a light source of luminous intensity 300 candela. Assuming the area to be everywhere perpendicular to the incident rays of light (Fig. 4.22), calculate the illuminance of the surface. Also find out what will be the luminous flux intercepted by the surface if its area be 0.3 m^2.

[133.33 lux, 40 lumen]

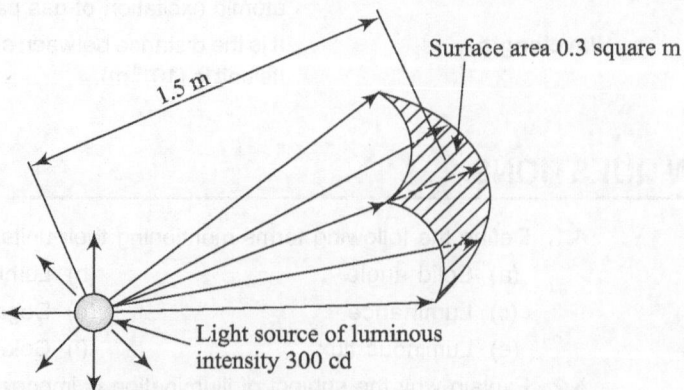

Fig. 4.22 Light source of luminous intensity

4.3. A lamp of 300 candela is hanging from the ceiling of a room as in Fig. 4.23. The bulb is at a height of 4 m from the floor. One side of the room is at a distance of 2 m from the bulb. Calculate Illuminances at points A, B, and C.

[18.75 lux, 26.52 lux, 75 lux]

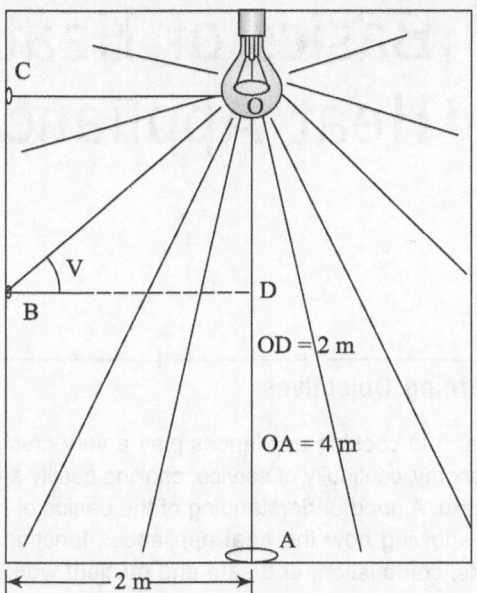

Fig. 4.23   Illuminances at different points (A, B, and C)

5 Basics of Heat, Fuel, and Heat Appliances

Chapter

Learning Objectives

Heat and cooking appliances play a very crucial role in the hospitality industry. Energy economy, continuity of service, cooking quality and safety are very important aspects in this regard. A good understanding of the basics of heat, fuels, and combustion are essential for knowing how the heat appliances function. The fundamental concepts of heating, fuels, combustion, and safe and efficient working of heat appliances become essential for the students in catering technology, chefs, and practicing engineers in the field.

The objectives of this chapter are to:

- introduce the fundamentals of heat, fuel, and combustion, as needed in understanding related applications in a hotel
- discuss different types of fuels used in the hotel industry and their comparative study
- give a preliminary idea about some important heat appliances, used in the hotel industry

After reading the chapter, the reader should be able to comprehend the physics of heat and heat transfer, gain knowledge of combustion mechanism as needed for the proper selection of fuel and heat appliances and associated energy economy.

5.1 INTRODUCTION

Hotel and catering industry deal with many heat equipment and appliances. These include boilers for producing steam, room heating systems, and gas and electric ovens. It is obvious that safe and efficient working of these sets of equipment is vital for ensuring the basic functions of providing food and comfortable accommodation in the hotel industry. A systemic study of the basics of thermal science, including combustion is, therefore, essential. In this chapter, we deal with the fundamentals of heat and heat transfer, fuels, and combustion and then discuss safe and efficient working of a few important equipment such as gas ovens and microwave ovens.

5.2 | TEMPERATURE AND HEAT

Before we go into the fundamentals of thermal science related to heat appliances used in hotel and catering industry, we must have the idea about the two most related fundamental physical quantities, namely temperature and heat. We shall discuss different units of measurement for them and also various forms of heat.

5.2.1 Temperature

Temperature is defined as the degree of hotness or coldness of a body. It also represents a thermal condition of the body from which we know whether the body will absorb heat from another body or will give away the heat to another body.

Units of temperature

Three different units of temperature are used now. These are

1. Celsius
2. Fahrenheit
3. Absolute scale or kelvin

We now form a basis of definition of these different scales. The temperature of an ice–water mixture is taken to be one reference temperature, that of steam–water mixture under normal atmospheric pressure conditions as another reference temperature, and all other temperatures are defined with respect to these temperatures. The first (ice–water mixture) reference temperature is given a value of 0°Celsius (0°C) in the Celsius scale, 32° Fahrenheit (32°F) in the Fahrenheit scale, and 273 kelvin (273 K) in the absolute scale. The second (steam–water mixture) reference temperature is given a value of 100°C in the Celsius scale, 212°F in the Fahrenheit scale, and 373 K in the absolute scale.

Thermometers, thermocouples, and pyrometers are used for the measurement of temperature. A thermometer has a glass tube bulb blown at one end in the form of a bulb and is filled with a liquid, e.g., mercury, alcohol, etc. It works on the principle of thermal expansion of the liquid. A thermocouple consists essentially of two dissimilar metal wires joined together at one end (Fig. 5.1). This end is placed on the part whose temperature has to be measured. According to the physicist Thomas Johann Seebeck, the resultant voltage will be induced between the free ends of the wires. This voltage, called the Seebeck voltage, can be measured by a voltage measuring instrument like voltmeter and related to the temperature of the object. As the temperature changes, the voltage changes and can be correlated with the temperature of the body to be measured. Pyrometers and infrared radiation-sensing devices are used to measure high temperatures such as in a furnace from a distance. They are, therefore, termed as contactless thermometers.

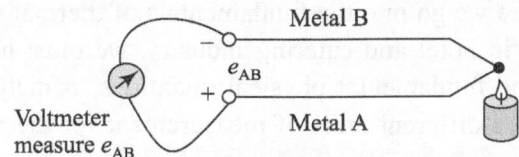

Fig. 5.1   Schematic of a thermocouple

Relationship between different temperature scales

If we know the temperature in degrees Celsius, we can determine the corresponding figure in Fahrenheit scale by the following relation:

$$C/5 = (F - 32)/9 \tag{5.1}$$

The relation between the absolute scale reading and the Celsius scale reading is given by

$$C = K - 273 \tag{5.2}$$

A temperature of 50°C will, therefore, be 122°F in Fahrenheit scale and 323 K in the absolute scale.

5.2.2   Heat

Heat may be regarded as a form of energy whose absorption either makes a body hot or changes its state and whose rejection either makes a body cold or changes its state. When change of state occurs, there is no change of temperature of the body during the period, even though transfer of heat takes place. This will be clear later in Fig. 5.2(a), which represents the heating and cooling curve of water. We observe that whenever water undergoes a change of state (from solid to liquid, called melting or fusion, or from liquid to vapour called evaporation or vapour to liquid called condensation or from liquid to solid called solidification or freezing), there is no change in temperature during the process till complete change of state takes place, i.e., till water completely changes to steam or ice completely melts to water.

Units of heat

We define units of heat in terms of heat required for a definite rise or fall of temperature of a body of given mass.

Since heat is also a form of energy, the SI unit of heat is also joule. However, there are still many old units of heat still in use. They are as follows:

Calorie   It is centimetre-gram-second (CGS) unit of heat. It is defined as the heat required to raise the temperature of 1 gm of water through 1°C (more precisely from 14.5°C to 15.5°C at standard atmospheric pressure of 101.325 kPa).

Kilocalorie   It is a larger unit of heat. It is defined as the heat required to raise the temperature of 1 kg of water through 1°C. Food calorie is often expressed in this unit but without the prefix of kilo:

$$1 \text{ kcal} = 100°C$$

British thermal unit (BTU)   It is the foot-pound-second (FPS) unit of heat and is defined as the heat required to raise the temperature of 1 lb of water through 1°F. This unit is widely used in the USA and abbreviated as BTU.

Therm   It is the commercial unit of heat, in prevalent use in England. The consumers of natural gas are charged according to this unit. The term may be defined as the heat required to raise the temperature of 100,000 lbs of water through 1°F.

$$1 \text{thm} = 100,000 \text{ BTU}$$

Conversion of different units of heat

$$1 \text{ BTU} = 252 \text{ cal}$$

$$1 \text{ cal} = 0.0039 \text{ BTU}$$

$$1 \text{ cal} = 4.186 \text{ J}$$

$$1 \text{ thm} = 100,000 \text{ BTU} = 2.52 \times 10^7 \text{ cal}$$

$$1 \text{ kWh} = 3412 \text{ BTU}$$

A 1 kW electric heater consuming electricity for one hour would produce 3412 BTU/860 kcal/3600 kJ of heat. Table 5.1 shows unit conversion data for heat.

Table 5.1   Unit conversion table for heat

Unit of heat	Quantity of water	Rise in temperature	Relation with calorie
Calorie	1 gm	1°C	1 cal
Kilocalorie	1 kg	1°C	1000 cal
BTU	1 lb	1°F	252 cal
Therm	10^5 lbs	1°F	2.52×10^7 cal
Joule	0.2389 gm	1°C	0.2389 cal

5.2.3 Sensible Heat and Latent Heat

We have seen from the definition of heat that heat absorbed in a body can either increase the temperature of the body or can cause a change of state of the body. When we heat water at room temperature (say 30°C), it slowly gets heated up and we can measure the increase in temperature with a thermometer. This heat, which is used to raise or lower the temperature of a substance (commonly called heating up or cooling down), is called sensible heat. If we continue heating, the temperature of water will reach 100°C. If we still heat the water, no further rise in temperature will occur. Instead, water will start evaporating under normal atmospheric conditions. The temperature will remain constant till all the water has evaporated to steam. If we can collect the steam thus formed, and heat it further, the temperature of the steam will increase and form what is known as superheated steam. We again come to the domain of sensible heat. So, we find that we can deliver sensible heat to water till it reached 100°C (also called the boiling point of water) and then further heating only changes the state of the water from liquid to gas (steam), without any further increase of temperature. This heat required to change the state is called latent heat of evaporation. The amount of heat needed to evaporate water at 100°C to steam at 100°C is the same as the latent heat needed to be extracted from steam at 100°C to condense it to water at 100°C. We now extend the example a little further. Let us start heating ice at a very low temperature, i.e., at less than 0°C. We follow the heating process with time and plot the temperature rise with time as shown in Fig. 5.2(a). The ice will first start heating up (still remaining solid ice) with its temperature increasing till it comes to 0°C (part A–B). This heat supplied is sensible heat, as it changes the temperature. Then, further supply of heat will start melting the ice to water and this is a change of state (from solid to liquid). Temperature of the ice–water mixture will remain the same till all the ice has melted to water (part B–C). This heat needed to cause the change of state is called latent heat. If we continue to heat water further, the temperature of water temperature will start going up till it reaches 100°C (part C–D) and this heat is again sensible heat. Further heating will cause the water to start evaporating to steam, which is a change of state (from liquid to gas) and as before the temperature will not change till all the water has converted to steam at 100°C. This heat is, therefore, latent heat. If we still continue to heat the steam so formed further, the temperature of the steam will now start increasing (part D–E) and we get what is known as superheated steam. This heat is again sensible heat.

A major component of almost all food items is water (refer Table 5.2). Therefore, the heating/cooling curve of food items also involves similar processes, as described earlier. However, there is one important difference between them. Referring to Fig. 5.2(b), which is a typical cooling curve for a common food item, we observe that the freezing/melting zone B–C does not have a constant temperature of freezing and in most cases, freezing starts not at 0°C but at a somewhat lower temperature, i.e., the food item starts freezing at sub-zero temperature. This is due to the fact that water in the presence of other material in it starts freezing at a lower temperature. Further, the distribution of water, and hence the distribution of other matter in water in the bulk of the food item are not same everywhere and different zones of the food item freeze at different temperatures. This explains the nature of the freezing zone B–C. In many cases, an average freezing temperature is assumed. Table 5.2 provides freezing point and latent heat data besides specific heat for the composite bulk of the food item and we need not consider water and solid food component separately for heat calculations.

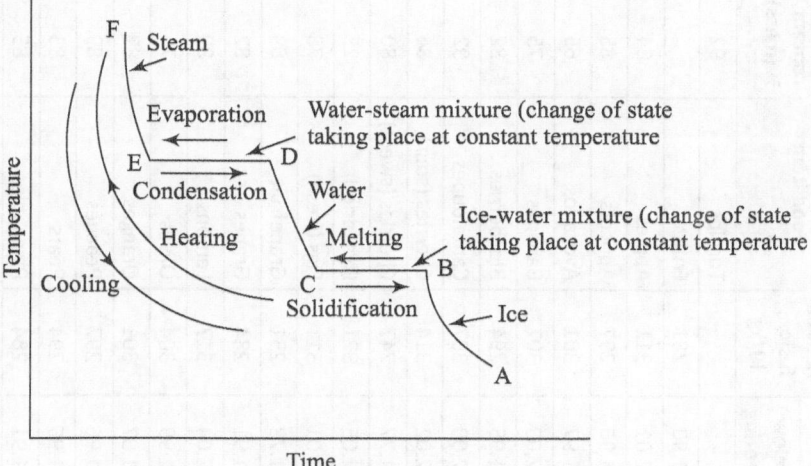

Fig. 5.2 (a) Heating and cooling curve of water

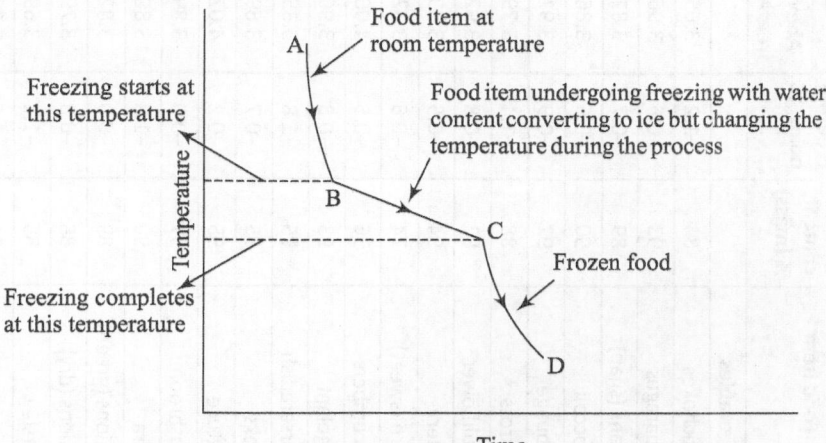

Fig. 5.2 (b) Cooling curve for a typical food item

Table 5.2 Freezing point and latent heat of fusion/freezing of some common food items

Food item	Water content, % (mass)	Freezing point, °C	Specific heat kJ/kg-°C Above freezing	Specific heat kJ/kg-°C Below freezing	Latent heat of fusion, kJ/kg
Vegetables					
Artichokes	84	−1.2	3.65	1.90	281
Asparagus	93	−0.6	3.96	2.01	311
Beans (snap)	89	−0.7	3.82	1.96	297
Broccoli	90	−0.6	3.86	1.97	301
Cabbage	92	−0.9	3.92	2.00	307
Carrots	88	−1.4	3.79	1.95	294
Cauliflower	92	−0.8	3.92	2.00	307
Celery	94	−0.5	3.99	2.02	314
Corn (sweet)	74	−0.6	3.32	1.77	247
Cucumbers	96	−0.5	4.06	2.05	321
Eggplant	93	−0.8	3.96	2.01	311
Horseradish	75	−1.8	3.35	1.78	251
Leeks	85	−0.7	3.69	1.91	284
Lettuce	95	−0.2	4.02	2.04	317
Mushrooms	91	−0.9	3.89	1.99	304
Okra	90	−1.8	3.86	1.97	301
Onions (green)	89	−0.9	3.82	1.96	297
Onions (dry)	88	−0.8	3.79	1.95	294
Parsley	85	−1.1	3.69	1.91	284
Peas (green)	74	−0.6	3.32	1.77	247
Peppers (sweet)	92	−0.7	3.92	2.00	307
Potatoes	78	−0.6	3.45	1.82	261
Pumpkins	91	−0.8	3.89	1.99	304
Spinach	93	−0.3	3.96	2.01	311
Tomato (ripe)	94	−0.5	3.99	2.02	314

Food item	Water content, % (mass)	Freezing point, °C	Specific heat kJ/kg-°C Above freezing	Specific heat kJ/kg-°C Below freezing	Latent heat of fusion, kJ/kg
Turnips	92	−1.1	3.92	2.00	307
Fruits					
Apples	84	−1.1	3.65	1.90	281
Apricots	85	−1.1	3.69	1.91	284
Avocados	65	−0.3	3.02	1.66	217
Bananas	75	−0.8	3.35	1.78	251
Blueberries	82	−1.6	3.59	1.87	274
Cantaloupes	92	−1.2	3.92	2.00	307
Cherries (sour)	84	−1.7	3.65	1.90	281
Cherries (sweet)	80	−1.8	3.52	1.85	267
Figs (dried)	23	—		1.13	77
Figs (fresh)	78	−2.4	3.45	1.82	261
Grapefruit	89	−1.1	3.82	1.96	297
Grapes	82	—	3.59	1.87	274
Lemons	89	−1.4	3.82	1.96	297
Olives	75	−1.4	3.35	1.78	251
Oranges	87	−0.8	3.15	1.94	291
Peaches	89	−0.9	3.82	1.96	297
Pears	83	−1.6	3.62	1.89	277
Pineapples	85	−1.0	3.69	1.91	284
Plums	86	−0.8	3.72	1.92	287
Quinces	85	−2.0	3.69	1.91	284
Raisins	18	—	—	1.07	60
Strawberries	90	−0.8	3.86	1.91	301
Tangerines	87	−1.1	3.75	1.94	291
Watermelon	93	−0.4	3.96	2.01	311

(Contd)

Table 5.2 (*Contd*)

Food item	Water content, % (mass)	Freezing point, °C	Specific heat kJ/kg·°C Above freezing	Below freezing	Latent heat of fusion, kJ/kg
Fish/seafood					
Cod (whole)	78	−2.2	3.45	1.82	261
Halibut (whole)	75	−2.2	3.35	1.78	251
Lobster	79	−2.2	3.49	1.84	264
Mackerel	57	−2.2	2.75	1.56	190
Salmon (whole)	64	−2.2	2.98	1.65	214
Shrimp	83	−2.2	3.62	1.89	277
Meats					
Beef carcass	49	−1.7	2.48	1.46	164
Liver	70	−1.7	3.18	1.72	234
Round (beef)	67		3.08	1.68	224
Sirloin (beef)	56	–	2,12	1.55	187
Chicken	74	−2.8	3.32	1.77	247
Lamb leg	65		3.02	1.66	217
Pork carcass	37		2.08	1.31	124
Ham	56	−1.7	2.72	1.55	187
Pork sausage	38	–	2.11	1.32	127
Turkey	64	–	2.98	1.65	214
Others					
Almonds	5	–		0.89	17
Butter	16	–	–	1.04	53
Cheese (cheddar)	37	−12.9	2.08	1.31	124
Cheese (Swiss)	39	−10.0	2.15	1.33	130
Chocolate (milk)	1	–	–	0.85	3
Eggs (whole)	74	−0.6	3.32	1.77	247
Honey	17	–	–	1.05	57
Ice cream	63	−5.6	2.95	1.63	210
Milk (whole)	88	−0.6	3.79	1.95	294
Peanuts	6	–	–	0.92	20
Peanuts (roasted)	2	–	–	0.87	7
Pecans	3	–	–	0.87	1,0
Walnuts	4	–	–	0.88	13

(*Source*: Water content and freezing-point data are from ASHRAE, *Handbook of Fundamentals*, SI version, Atlanta, GA, American Society of Heating, Refrigerating and Air-Conditioning Engineers, Inc., 1993).

Freezing point is the temperature at which freezing starts for fruits and vegetables, and the average freezing temperature for other foods.

Specific heat data are based on the specific heat values of water and ice at O°C and are determined from Siebel's formula: C (fresh) = 3.35 × (Water content) + 0.84, above freezing point, and C (frozen) = 1.26 × (Water content) + 0.84, below freezing point.

The latent heat of freezing is determined by multiplying the heat of freezing of water (334 kJ/kg) by the water content fraction of the food.

In the light of the foregoing discussion, we have two types of heat in relation to the body which receives or rejects it. They are

1. Sensible heat
2. Latent heat

Sensible heat

Heat which causes a change of temperature in a body is called sensible heat. It can be measured by a thermometer.

In order to find the sensible heat required to be added to a given quantity of a substance for a given rise of temperature (or sensible heat needed to be extracted from a substance to produce a given drop in temperature), there is a term called 'specific heat capacity', which is a property of a substance and varies from substance to substance. They are to be determined experimentally.

Specific heat capacity (C)

- Specific heat capacity, C, also known simply as specific heat, is the measure of the heat energy required to increase (or decrease) the temperature of a unit quantity (say, 1 kg) of a substance by unit temperature interval (say, 1°C change of temperature).

- More heat energy is required to increase the temperature of a substance with high specific heat capacity than one with low specific heat capacity.

- Water (at 25°C) has specific heat capacity of 4186 J (1 kcal) per kg per Celsius degree. This means 1 kg of water would require a heat of 4186 Joule or 1 kcal to raise its temperature by 1°C. Air, at typical room conditions and under constant normal atmospheric pressure, has a specific heat of 1012 J (0.242 kcal) per kg per Celsius degree. It shows that air takes much less heat than water to heat up.

- The definition of specific heat allows us to calculate sensible heat for a substance of a given mass through some temperature change. We now express this in a mathematical equation. The heat needed to raise the temperature of a substance of mass, m, and specific heat, C, from T_1 to T_2 is given by

$$H = m \times C \times (T_2 - T_1) \qquad (5.3)$$

- So, if we know the material of the substance, we know its property of specific heat as available in standard tables. And then, we can calculate the heat requirement for changing the temperature of some mass of the substance.

Table 5.3 provides the values of specific heat for some common materials of interest, while Table 5.4 provides the values for a few common food items.

Table 5.3 Thermal properties of a few common materials

Material	Specific heat, C, at 27°C (J/kg°C)	Thermal conductivity, k, at 27°C (watt/(m°C)	Thermal diffusivity, α, at 27°C (m²/s)
Plain carbon steel	434	60.5	17.7×10^{-6}
Stainless steel (304 grade)	480	14.9	3.95
Copper (pure)	385	401	117×10^{-6}
Aluminium (pure)	903	237	97×10^{-6}
Wood	1260	0.16	1.76×10^{-7}
Brick (common)		0.72	
Brick (fire clay)	790	0.9	5.93×10^{-7}
Mineral wool with asbestos		0.046	
Glass (window)	750	0.7	3.3476×10^{-7}
Felt wool		0.07	
Asbestos (loosely packed)		0.15	

Table 5.4 Some important physical properties of a few food items

Food	Water content, % (mass)	Temperature, °C	Density, kg/m³	Thermal conductivity, k kW/m°C	Specific heat, C_p kJ/kg°C
Fruits/vegetables					
Apple juice	87	20	1000	0.559	3.86
Apples	85	8	840	0.418	3.81
Apples, dried	41.6	23	856	0.219	2.72
Apricots, dried	43.6	23	1320	0.375	2.77
Bananas, fresh	76	27	980	0.481	3.59
Broccoli	90	−6	560	0.385	1.97
Cherries, fresh	92	0–30	1050	0.545	3.99
Figs	40.4	23	1241	0.310	2.69
Grape juice	89	20	1000	0.567	3.91
Peaches	89	2–32	960	0.526	3.91
Plums	—	−16	610	0.247	—

(Contd)

Table 5.4 (*Contd*)

Food	Water content, % (mass)	Temperature, °C	Density, kg/m³	Thermal conductivity, k kW/m°C	Specific heat, C_p kJ/kg°C
Potatoes	78	0–70	1055	0.498	3.64
Raisins	32	23	1380	0.376	2.48
Meats					
Beef, ground	67	6	950	0.406	3.36
Beef, lean	74	3	1090	0.471	3.54
Beef, fat	0	35	810	0.190	—
Beef, liver	72	35	—	0.448	3.49
Cat food	39.7	23	1140	0.326	2.68
Chicken breast	75	0	1050	0.476	3.56
Dog food	30.6	23	1240	0.319	2.45
Fish, cod	81	3	1180	0.534	3.71
Fish, salmon	67	3	—	0.531	3.36
Ham	71.8	20	1030	0.480	3.48
Lamb	72	20	1030	0.456	3.49
Pork, lean	72	4	1030	0.456	3.49
Turkey breast	74	3	1050	0.496	3.54
Veal	75	20	1060	0.470	3.56
Others					
Butter	16	4	—	0.197	2.08
Chocolate cake	31.9	23	340	0.106	2.48
Margarine	16	5	1000	0.233	2.08
Milk, skimmed	91	20	—	0.566	3.96
Milk, whole	88	28	—	0.580	3.89
Olive oil	0	32	910	0.168	
Peanut oil	0	4	920	0.168	
Water	100	0	1000	0.569	4.217
Water	100	30	995	0.618	4.178
White cake	32.3	23	450	0.082	2.49

(*Source:* Data are mainly from ASHRAE, *Handbook of Fundamentals,* SI version, Atlanta, GA, American Society of Heating, Refrigerating and Air-conditioning Engineers, Inc., 1993.)

Most specific heats in the table are calculated from the relation, $C_p = 1.6 + 2.51 \times$ water content, is valid in the temperature range of 3–30°C with reasonable accuracy.

- 5 kg of lamb steak at room temperature of 30°C is to be heated to 80°C for the preparation of kebab. Let us calculate the heat needed to be given to the steak. We assume that the kebab is heated uniformly so that all parts of the kebab have the same temperature. The specific heat of lamb is 3.49 kcal/kg°C.

(*Contd*)

(*Contd*)

From Eqn 5.3, $H = 5 \times 3.49 \times (80 - 30) = 872.5$ kcal; therefore, the heat needed for the purpose is 872.5 kcal.

- 2 kg of water is heated in an electric kettle for making tea. Water is kept at 30°C and is to be heated to 90°C. The water is heated by a 1.2 kW immersion heater. Mass of the kettle is 250 g. The material that makes up the kettle is stainless steel, whose specific heat is 500 J/kg-K. We calculate the time needed to heat water.

We assume that there is no heat loss from the kettle to the surrounding and the temperature of the kettle rises by the same temperature difference. Let us first find the heat needed to heat the water. We do the calculation using the SI unit of heat (i.e., joule), as the power of heater is given in kW or kJ/s.

Heat needed = Heat needed to raise the temperature of kettle + heat needed to raise the temperature of the water $= 0.250 \times 500 \times (90 - 30) + 2 \times 4181.3 \times (90 - 30) = (7500 + 501756) = 509.256$ kJ.

Considering 100 per cent efficiency of the kettle in converting electrical energy to thermal energy, the electrical units consumed $= 509.256/3600$ BOT $= 0.141$ units.

Now, the heater coil is supplying 1.2 kJ of heat in 1 second (power 1.2 kW). So the time taken to heat the water is, $509.256/1.2 = 424.38$ seconds $= 7$ min 4s.

As a matter of fact, the kettle will lose heat to the surrounding atmosphere by convection and radiation (discussed later) and, hence will take somewhat longer time to heat.

Taking the cost of electrical energy to be Rs 4 per unit (1 kWh = 3600 kJ), the energy bill for the heating is Rs $4 \times 0.141 = $ Rs 0.57. The actual bill value will be higher than this due to energy to be supplied for the loss of heat to the surroundings.

- We now solve a simple problem of cooling of apples.

Let 10 kg of apples at a room temperature of 30°C be kept in a refrigerator whose temperature is maintained at 3°C. The specific heat of apple may be taken as is 3800 J/kg-K. Determine the heat to be extracted from it by the refrigerator to cool it down. Take freezing point of apple as -1.1°C.

We refer to Fig. 5.2(b) and see that the condition corresponds to zone A–B, i.e., sensible cooling zone.

We know, $H = m \times C \times (T_2 - T_1) = 10 \times 3800 \times (30 - 3) = 1,026,000$ J $= 1026$ kJ $= 245.11$ kcal.

So, the amount of heat to be extracted is 1026 kJ or 245.11 kcal.

Consider cost of one unit of electricity (1 kWh = 3600 kJ) to be Rs 4. So, $4 \times 1026/3600 = $ Rs 1.14 will have to be spent for this cooling. However, since fruits like apple always release heat, which has to be continuously extracted from it, actual energy bill will be substantially higher than this value depending upon how long and in what quantity the food items are kept in the refrigerator.

Latent heat

Heat, which causes change of state of a body without any change of temperature, is called latent heat.

There are two types of latent heat namely,

1. Latent heat of fusion/solidification/freezing (L_f) and
2. Latent heat of vaporization/condensation (L_{vap})

Latent heat of fusion/solidification (L_f) It is the heat required to either melt (heat is needed to be given) or freeze (heat is needed to be extracted) a product at a given temperature. It occurs at freezing/melting point. In both the cases, the heat requirement is the same. For example, latent heat needed to be given to convert 1 g of ice at 0°C to 1 g of water at 0°C is the same as the heat to be extracted from 1 g of water at 0°C to freeze to 1 g of ice at 0°C.

Latent heat of vaporization/condensation (L_{vap}) It is the heat required to either vaporize (heat to be added) or condense a product (heat to be extracted) at a given temperature. It occurs at the boiling/condensation temperature. Both the quantities for a given substance are same. For example, latent heat needed to be given to convert 1 g of water at 100°C to 1 g of steam at 100°C is the same as the heat to be extracted from 1 g of steam at 100°C to condense to 1 g of water at 100°C.

The heat needed to melt 1 g of ice to 1 g of water is called specific latent heat of melting of ice and is equal to 80 cal/g or 336 kJ/kg. This is also the specific latent heat of the corresponding freezing.

Similarly, specific latent heat of evaporation of water under atmospheric pressure condition is 540 cal/g or 2268 kJ/kg. This is also the latent heat of the corresponding condensation.

Just to see how much this heat is, we note that heat required to raise the temperature of 1 g of water at 0°C to water at 100°C, we need only 100 cal of heat, whereas the latent heat needed to convert this 1 g of water at 100°C to steam at 100°C, we need 540 cal of heat.

From the foregoing definitions, we can say,

$$\text{Total latent heat, } L = \text{mass} \times \text{specific latent heat} \qquad (5.4)$$

1500 g of water at 30°C is kept in the deep freezer whose temperature is –3°C to make tiny ice cubes. We are to determine the heat to be extracted from water to achieve this condition. Specific heat of water and ice may be taken as 4100 J/kg-K and 2000 J/kg-K, respectively. Specific latent heat of freezing of water (L) is 336 kJ/kg.

We refer to Fig. 5.2(a) and by comparing with the present cooling process, we observe that the process is composed of three parts. One is sensible cooling from some intermediate point at 30°C in the region D–C up to C and then a change of state from water to ice at constant temperature (latent heat of freezing of water) corresponding to the curve C–B and,

(Contd)

(Contd)

finally, again sensible cooling of ice at 0°C to ice at –3°C corresponding to the zone B–A from B up to some point at –3°C.

The total heat to be extracted is the sum of the heat needed to be extracted at three different zones of cooling. Let them be H_1 in zone D–C, H_2 in zone C–B, and H_3 in zone B–A.

We have, $m = 200$ g $= 0.2$ kg, $C_{water} = 4100$ J/kg-K, $C_{ice} = 2000$ J/kg-K, $T_3 = 30°C$, $T_2 = 0°C$ and $T_3 = -3°C$, $L = 336$ kJ/kg.

So the total heat extracted from water during cooling, $H = H_1 + H_2 + H_3$.

Sensible heat extracted from water during cooling of water, $H_1 = m \times C_{water} \times (T_3 - T_2)$ $= 1.5 \times 4100 \times (30 - 0) = 184{,}500$ J $= 184.5$ kJ. After extraction of this much of heat, water will be cooled to 0°C. Any further extraction of heat will change water to ice and temperature will not fall till all the water is converted to ice after necessary amount of heat (latent heat) is extracted from water.

$$H_2 \text{ (latent heat)} = m \times L = 1.5 \times 336 = 504 \text{ kJ}$$

So, after extraction of $(H_1 + H_2)$ amount of heat, i.e., 688.5 kJ of heat from water, all water at 30°C will convert to ice at 0°C. Any further extraction of heat will lower the temperature of solid ice (sensible cooling).

Sensible heat extracted from ice during cooling $= m \times C_{ice} \times (T_2 - T_1) = 1.5 \times 2000 \times [0 - (-3)] = 9000$ Joule $= 9$ kJ.

So total heat to be extracted by the refrigerator to make ice cubes

$$H = (184.5 + 504 + 9) \text{ kJ} = 697.5 \text{ kJ}.$$

We note that the latent heat part is substantially more than the sensible heat part in the cooling process of water.

- 5 kg of chicken breast is to be heated from 4°C to 70°C. The specific heat for chicken breast is 3560 J/kg-k. During the process of heating, some of the water content is vaporized and the weight of chicken breast after heating becomes 4.2 kg. The average latent heat of evaporation of water may be taken as 2323 kJ/kg (555 kcal/kg). Determine the heat required for this.

  We observe that there are two modes by which the chicken breast has absorbed heat. One is through sensible heat, which has increased the temperature of the breast and the other is through the latent heat absorbed by water during evaporation. We assume that water has also been heated up to 70°C and then it evaporates although, in reality, water starts evaporating earlier, the rate of evaporation increasing with temperature.

  Mass of water evaporated is (5 kg – 4.2 kg) = 0.8 kg.

  Sensible heat = mass × specific heat × temperature difference = 5 × 3560 × 66 J = 1174. 8 kJ.

  Latent heat = mass of water evaporated × latent heat of evaporation = 0.8 × 2323 = 1858.4 kJ.

  Therefore, total heat needed to heat the breast is (1174. 8 kJ + 1858.4 kJ) = 3033.2 kJ.

(Contd)

(*Contd*)

- 20 kg of salmon at 25°C is kept in a freezer at – 5°C. With the data available to us, we can determine the amount of heat to be extracted from the salmon by the freezer.

  From Table 5.2, we obtain freezing temperature of salmon as –2.2°C, specific heats before freezing and after as 2.98 kJ/kg°C and 1.65 kJ/kg°C, respectively, and latent heat of freezing as 214 kJ/kg.

  Therefore, sensible heat extracted from the salmon before freezing, $H_1 = m \times C \times (T_{initial} - T_{freezing}) = 20 \times 2.98 \times (25 - (-2.2)) = 20 \times 2.98 \times 27.2 = 1621.12$ kJ

  Latent heat to be extracted for freezing, $H_2 = m \times L = 20 \times 214$ kJ $= 4280$ kJ

  Sensible heat extracted from the salmon after freezing, $H_3 = m \times C \times (T_{freezing} - T_{final}) = 20 \times 1.65 \times \{-2.2 - (-5)\} = 20 \times 1.65 \times 2.8$ kJ $= 92.4$ kJ

  Therefore, total heat to be extracted by the freezer becomes

  $$H = H_1 + H_2 + H_3 = 5993.52 \text{ kJ}$$

5.3 | HEAT TRANSFER—DIFFERENT MODES AND COMPARISON

We had discussed the total heat requirement for a substance to be heated/cooled through certain temperature difference with or without a change of state. This heat has to come from a heat-generation source or go into a heat-absorbing sink. A very important factor in this connection is how fast the body is heated/cooled or how the temperature is distributed inside. The way heat is transmitted through the body will dictate these phenomena. Heat transfer is of vital importance in heat appliances because the food items and other substances receive heat from heat sources and in cooling apparatus, substances reject heat to heat sinks. The efficiency of heating and cooling and the way heat is transferred affect not only the energy efficiency of the equipment but also the quality of food being prepared. We discuss various modes of heat transfer and then make a comparison of them to understand the mechanisms better.

5.3.1 Different Modes of Heat Transfer

Heat may be transmitted from one place to another by the following three ways.

1. Conduction
2. Convection
3. Radiation

Conduction

The process by which heat is transmitted from hotter part of a body to its colder part without any visible movement of the material of the body is called conduction.

A solid body or an absolutely static liquid/gas transfers heat from one end or point to the other end or point within the body by conduction. In case of liquid/gas, however, heat transfer takes place within the body also through convection, in general. The rate of heat transfer through the body is governed by a property of the body called thermal conductivity. Higher the thermal conductivity higher will be rate at which heat is transported by conduction from one point to another under a steady-state condition. This is the reason why a glass of hot water enclosed completely by a glass made of copper will lose heat at a much faster rate than when the glass and cover are made of porcelain having a much lower conductivity than copper.

We introduce another important physical property thermal diffusivity, α. It is the ratio of thermal conductivity and the product of specific heat and density, i.e., $k/(\alpha C)$. If the ratio is higher, heat will be transmitted through the material at a faster rate. An example will make it clear. When a metal rod is heated at one end, say, by a flame, the front portion receives heat. A portion of heat is absorbed by the material there (higher the specific heat, higher will be the portion absorbed) and the rest of the heat will be transmitted through it by the property of conductivity. This way the other end will eventually be heated up. Higher the diffusivity, faster the other end will be heated up till steady-state condition is reached. Pure copper has a thermal diffusivity of 117×10^{-6} m^2/s, while ordinary steel has a thermal diffusivity of 17.7×10^{-6} m^2/s. This explains why our hands get heated up very soon when we hold a copper rod touching a burning candle than when we hold a steel rod. This property is, therefore, very important for solid food items, which are to be heated for cooking and where, the heat is transmitted through the body by conduction mainly like baking of cake or roasting of chicken in an oven.

Heat conductors   The substances that can conduct heat easily are called good conductors, while those that cannot do so are called bad conductors or thermal insulators. Metals are good conductors, while wood, paper, polyurethane foam, etc. are bad conductors of heat. Diamond is the best conductor of heat. Then come, in order of decreasing conductivities, silver, copper, gold, and aluminium. Heat conductors have much higher thermal conductivity than heat insulators. Heat will flow through good conductors more easily. If a hotel building wall is made of bad conductor, the rate of heat loss from within the building to outside during winter will be less and one will save on heat required to keep the building warm. Similarly, during summer, heat entry from outside will be less and load on AC system will be less and we save on expenses on energy in the hotel. This is one of the reasons that nowadays, hollow bricks are used with the hollow space being filled with trapped air having a very low thermal conductivity. This effectively reduces the rate of heat entry into the house in summer and rate of heat loss from the house during winter, thus providing economy to heating/cooling expenses.

We refer to Fig. 5.3 and write the basic equation of heat conduction (Fourier's law), which can be simplified to yield:

$$Q = A \times k \times (T_2 - T_1)/L \qquad (5.5)$$

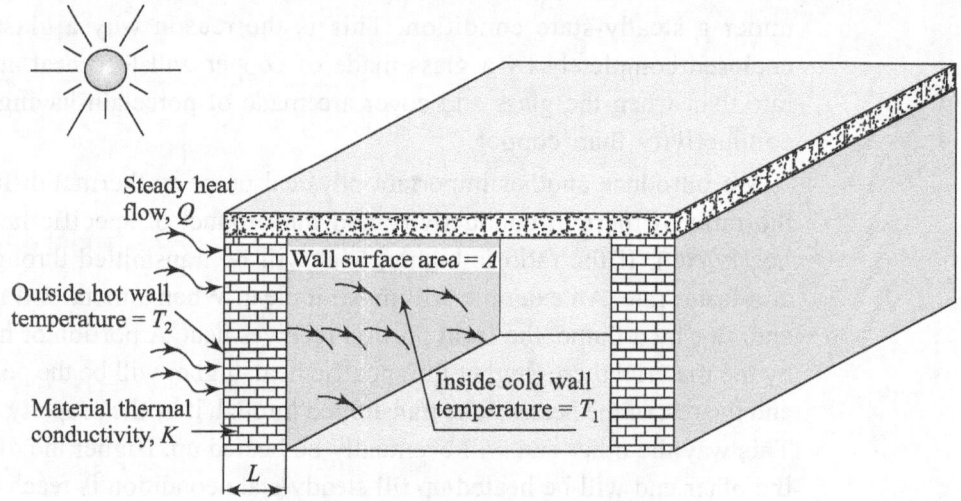

Fig. 5.3 Model for law of conduction

In Eqn 5.5, Q is the rate of heat flow in joule/s watt or kcal/s through the wall, k is the thermal conductivity of the wall material, A is the surface area of the wall on either side, L is the thickness of the wall, T_2 is the temperature of the room or outside air whichever greater and T_1 is the temperature (obviously the lower one) of the other side. We do not have any control on outside temperature and we would also like to maintain room temperature at some level. The surface area of the wall should also be fixed by architectural considerations. So we can change only L, the thickness of the wall or k, the material of the wall. The relation shows clearly that if we increase the thickness, L or decrease thermal conductivity k, then Q, the heat loss/gain from or of the room will be less and thus we can possibly save on the energy bill. From the above equation, we observe that the unit of k is W/°c/m. When surface area, $A = 1$, thickness of the slab, $L = 1$, and temperature difference between the hot side and cold side, $(T_2 - T_1)$ is 1, we find that $Q = k$, which means that thermal conductivity is the rate of heat flow through unit surface area with unit thickness when the temperature difference is 1°C across the slab and denotes how easily heat can flow through the material. We have also seen that the unit of thermal conductivity is J/s/m/°c or W/(m/°c). Therefore, when we say that thermal conductivity of brick is 1.3 W/(m/°c), it means that if we have a brickwork of surface area 1 m² while its two opposite sides are having a temperature difference of 1°C and the thicknesses of the brickwork are 1 m, the heat that will flow through it is 1.3 W or 1.3 J in 1 s or 0.31057 calorie in 1 second.

When we heat a cooking utensil by placing it on top of a hot plate or in an oven, the internal side of the utensil is heated by conduction of heat from its bottom side. So, if the thickness of the plate is small and the conductivity of the utensil is high (thus, making the thermal diffusivity high), the cooking side will get heated up very quickly. When a solid item is cooked, similarly heat is transmitted into bulk of the material by conduction. As such, the rate of heating the material and the rise in temperature will be guided by the property of thermal diffusivity, as discussed.

Materials having conductivity in the range of 0.035 W/(m-°C) to 0.16 W/(m-°C) are called thermal insulators.

If there are more than one layer of materials having same surface area, A, the equivalent conductivity, $k_{eq.}$, is calculated from

$$(L_1 + L_2 + L_3 + ...)/k_{eq} = L_1/k_1 + L_2/k_2 + L_3/k_3 + ...$$

and heat flow, Q, is determined from

$$Q = A \times k_{eq.} \times (T_2 - T_1)/(L_1 + L_2 + L_3 + ...)$$

By adding layers of very low conductivity over a metal sheet of high conductivity, the overall conductivity reduces and this forms the principle of thermal insulation.

Let us solve a problem.

A wall in a guest room in a hotel is made of 250 mm brick work with cement mortar plaster, 15 mm thick, on both sides. The height and width of the wall is 3 m and 5 m respectively. The effective or equivalent conductivity of the composite wall is 0.8 W/(m/°C). During summer the outside temperature is 41°C and the room temperature is maintained at 21°C by running an AC machine. Assuming no other form of heat loss than the one by conduction, we can find out the amount of heat flowing through the wall into the room from outside hot environment.

Area of the wall is 3 × 5 = 15 m². Thickness of the wall is 250 mm and the temperature difference across the wall is (41 − 21)°C or 20°C. Using basic conduction equation, we get

$$Q = 15 \times 0.8 \times 20 = 240 \text{ Watt } (57.336 \text{ cal/s})$$

This means that during summer, when the room is intended to be kept at a temperature of 21°C, heat will flow from outside into the room at a steady rate of 57.336 calorie per second and unless heat is removed from the room at this rate, the room will heat up and eventually rise to the outside temperature. Air conditioner is employed to extract this amount of heat

(Contd)

(Contd)

constantly from the room to maintain the room temperature at 21°C. Putting additional layer of insulating material will lower the effective conductivity and the heat rate can be reduced.

Convection

The process by which heat is transmitted from the hotter part of a body to its colder part by the actual movement of the material of the body is called convection. Therefore, convection is possible only in liquid or gas and not in solid.

Heat is generally transmitted in liquids and gases by convection. Heat transfer rate by convection depends upon mainly on the type of motion of the fluid. When we heat a kettle of water, the water particles close to the bottom of the kettle gets heated by conduction from the hot kettle bottom. As soon as they get heated, they become lighter and hence move up and displace cold and heavier fluid near the top, which come down and gets heated. This cycle continues. During the upward passage of hot fluid, they also transfer some heat to the colder particles. Thus, there is a current of particle flow being set up and heat transfer is enhanced. This forms the essence of convection. In many cases, fluid particles are set in motion during this heating/cooling process externally by mechanical means (like a pump or fan) and the convection current thus generated is called forced convection, whereas the earlier convection type, as occurring in a kettle, is called free or natural convection.

We heat a whole chicken in a *tandoor*, as shown in Fig. 5.4. The air inside the *tandoor* gets heated up due to heat released by the burning wood charcoal or coke and becomes lighter and goes up and heats the chicken piece hanging up. Heavy cold outside air passes through the opening at the bottom of the *tandoor* to fill in the space left by hot oven air rising up. This cold air, thus, continuously provides oxygen for burning as well as maintains the constant circulation of gas through the oven. Thus heat is continuously transferred from the coke bed to the roast by this movement of cold air–hot gas flow system. This kind of air-gas circulation obtained due to rise of lighter hot gas/liquid and descending of heavier cold air/liquid resulting in heat transfer is known as free convection or natural convection, as also explained earlier. However, the speed of air flow over the surface of the streak being roasted is low and hence the roasting rate will be low. In many installations, a fan may be provided at the base to assist this circulation. This arrangement increases heat transfer and the piece is roasted faster. This convection mechanism is known as forced convection. However, it is to be borne in mind that ultimately the hot air particle on coming in contact with the chicken piece transfers heat at the skin level purely by heat conduction.

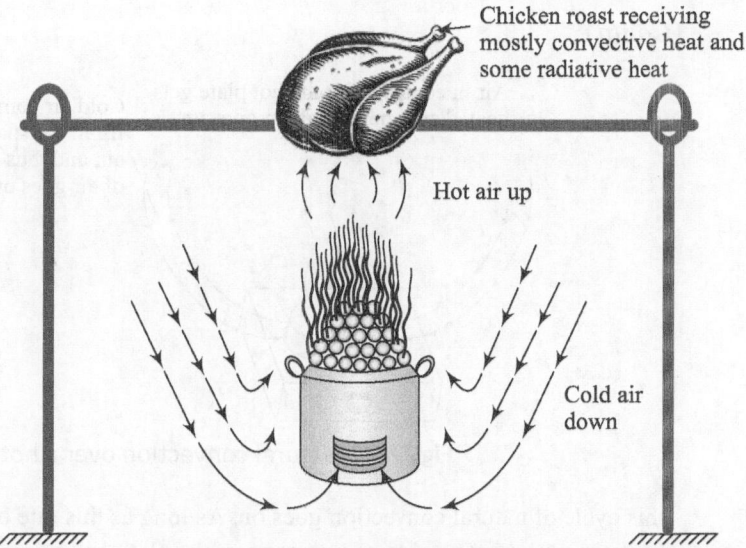

Chicken roast receiving mostly convective heat and some radiative heat

Hot air up

Cold air down

Fig. 5.4   Natural convection due to upward movement of hot air (lighter in weight) and downward movement of cold air (heavier)

Although the heat transfer by convection is quite complex to estimate, fortunately, we have some relationship derived from experiments which can be written as

$$Q = A \times h \times (T_{wall} - T_{stream}) \qquad (5.6)$$

where A is the surface area of the substance to be heated/cooled by convection, T_{wall} is the surface temperature of the substance, T_{stream} is the temperature of the flow far away, and h is called the convective heat-transfer coefficient. The heat-transfer coefficient depends upon fluid properties, the condition, and orientation of the surface with respect to flow, and the velocity of the flow. There are experimental curves and correlations available from which the value of h can be found out.

A 500 watt electric hot plate of 15 cm diameter is switched on and left to be heated up (Fig. 5.5). An approximate value of natural convection heat transfer coefficient for this condition of horizontal plate with hot side on top may be obtained as 20 W/square m/degree Celsius. Atmospheric air temperature is 30°C. We are to find steady-state temperature of the hot plate when the hot plate is bare and nothing is put on it.

As soon as the hot plate is switched on, the plate gets heated up by the electrical coil inside. As the temperature goes up, heat is transferred from the plate to the adjacent air by conduction, which gets heated up and moves up, cold air descends and finds its place, and

(Contd)

(Contd)

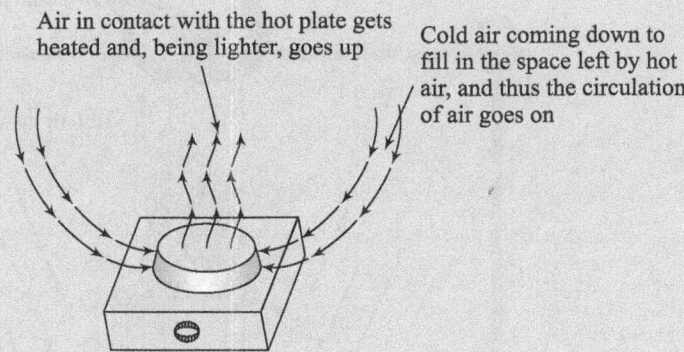

Air in contact with the hot plate gets heated and, being lighter, goes up

Cold air coming down to fill in the space left by hot air, and thus the circulation of air goes on

Fig. 5.5 Natural convection over a hot plate

this cycle of natural convection goes on. As long as this rate of heat loss is less than the heat generation rate inside, temperature of the plate will go on increasing. The higher the temperature of the plate, higher is the heat loss, according to Eqn. 5.6. Eventually, heat loss will be equal to the heat generated and the temperature of the plate will be steady. We are to find this steady-state temperature.

We have, $Q_{\text{convection}} = Ah(T_{\text{plate}} - T_{\text{stream}})$

Under steady state, $Q_{\text{convection}} = Q_{\text{generated}} = 500$ W

Area, $A = \pi d^2/4 = 3.14 \times 15 \times 15/4 = 176.6$ sq cm $= 176.6 \times 10^{-4}$ square m

or $500 = 176.6 \times 10^{-4} \times 20 \times (T_{\text{plate}} - 30)$ or $T_{\text{plate}} = 1445.6°C$. Undoubtedly, we do not expect such a high rise in temperature. Apart from the convective heat loss, there will be radiation loss from the plate and the steady-state temperature will be much less.

Considering the hot plate to be made of plain carbon steel, having a mass of 300 g, a little higher mathematics with heat conduction shows that it will take a little less than 10 minutes of time for the hot plate to reach this temperature.

Radiation

The process by which heat is transmitted from one place to another without the help of any material medium or without heating the intervening medium is called radiation.

Heat from the sun is transmitted to the earth by radiation. Every matter radiates heat and every matter receives heat through radiation from every other matter surrounding it. The net heat exchange will determine whether a body will get heated up or cooled down in this heat transfer process.

Radiant heat is essentially an electromagnetic wave spreading from a body by virtue of its temperature. Radiant heat from a body is proportional to the fourth power of absolute temperature of the body and a property of the surface of the radiating body called emissivity. So, mathematically, we can write

$$Q_{\text{radiation}} = A\varepsilon\sigma T^4 \tag{5.7}$$

where $Q_{\text{radiation}}$ is the radiation heat per second emitted from the surface area, A, of a body at a temperature, T. Note that the temperature must be in absolute unit of kelvin, K. ε is the emissivity of the body and σ is called Stefan–Boltzmann constant.

Equation 5.7 explains why we feel more radiation heat when we stand near a bed of red hot coal than near an LPG stove flame (Fig. 5.6), although the temperature of LPG flame is higher than that of the coal bed. The flame of LPG burner has a very low emissivity than a red hot coal bed. Hence, the heat radiated from a coal bed is much higher and we feel hotter. In the discussion on convection heating/cooling, we remarked that the chicken gets heat from the red hot bed of coke by convection mechanism. While this statement is true, it reflects only a part of the heat transfer from the oven bed to the roast. Most of the heat received by the roast is by direct radiation from the bed. Thus, even if the coal bed could burn without any air (which cannot happen normally) and be in vacuum, the roasting would have been accomplished as desired and more or less at the same pace. Normally, the intervening air/gas is not heated up by radiation and the heat directly reaches the intended surface. As such, this mode of heat transfer is very efficient and very fast. However, the heat emitting surface has to be of high emissivity (for example, red hot coke bed and not gas flame) and a reasonably high temperature.

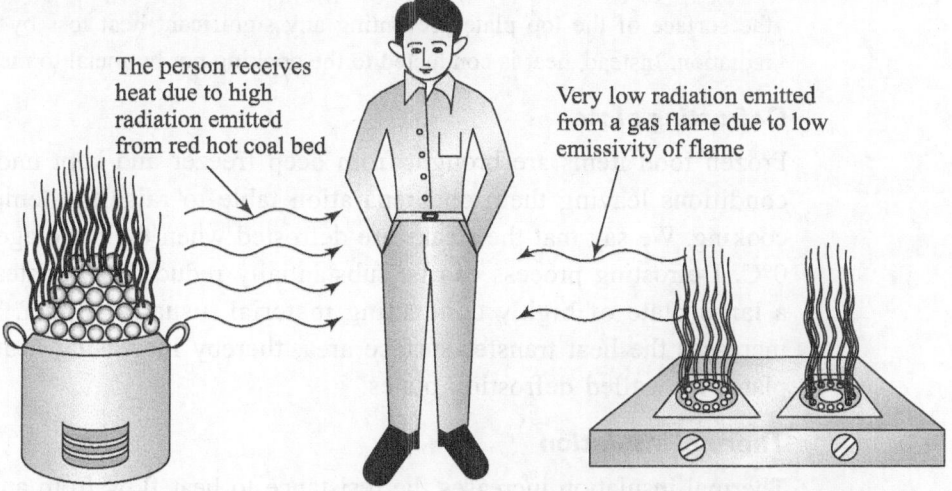

The person receives heat due to high radiation emitted from red hot coal bed

Very low radiation emitted from a gas flame due to low emissivity of flame

Fig. 5.6 High radiation from burning coal and low radiation from LPG flame

Let us solve a problem.

We consider the hot plate problem once again. This time we consider heat loss from the plate due to natural convection as well as radiation loss. We neglect any heat received by the hot plate from other hot neighbouring objects that it can 'see' by radiation from them. We take emissivity (ε) of rusted steel plate as 0.8. Stefan–Boltzmann constant, σ, is 5.67×10^{-8} W/square m.

This problem has to be solved in the absolute unit of temperature.

$T_{air} = 273 + 30 = 303$ K

Total heat loss from the plate = Heat loss due to natural convection + heat loss due to radiation

$$= Ah(T_{plate} - T_{air}) + A\varepsilon\sigma T_{plate}^4$$
$$= 176.6 \times 10^{-4} \times 20 \times (T_{plate} - 303) + 176.6 \times 10^{-4} \times 0.8 \times 5.67 \times 10^{-8}(T_{plate})^4$$

or $\quad 500 = 176.6 \times 10^{-4} \times [20 \times (T_{plate} - 303) + 0.8 \times 5.67 \times 10^{-8} \times (T_{plate})^4]$

This equation is not a linear equation and T_{plate} can be calculated only with special techniques. However, using such techniques, we obtain

$T_{plate} = 527°C$ (approximately).

The problem becomes much easier, if we consider heat loss only by radiation. In this case

$$500 = 176.6 \times 10^{-4} \times 0.8 \times 5.67 \times 10^{-8} \times (T_{plate})^4$$

or $\quad T_{plate} = 888.8$ K $= 615.8°C$

So, we find that the temperature of the hot plate drastically reduces when radiation loss is taken into account compared to when only natural convection loss is considered.

While cooking on a hot plate, the cooking surface almost completely covers and touches the surface of the top plate preventing any significant heat loss by convection and radiation. Instead, heat is conducted to the cooking pot by metal to metal contact.

Defrosting plate

Frozen food items are brought from deep freezer and kept under atmospheric conditions leaving them on preparation table to raise the temperature before cooking. We say that the steaks are defrosted when their average temperature is 0°C. Defrosting process can be substantially reduced if the steaks are kept on a large plate of highly conducting material, usually copper. This, in effect, increases the heat transfer surface area, thereby increasing heat transfer. Such plates are called defrosting plates.

Thermal insulation

Thermal insulation increases the resistance to heat flow from and into a system by wrapping the system or covering the system with additional layer of materials, which are bad conductors of heat. Hot water pipelines in a hotel are wrapped with glass wool and asbestos rope to minimize heat loss and consequent drop in temperature of the water being conveyed. Steam pipelines from boiler to laundry and other places are also insulated. Refrigerated water lines for air-conditioning units are also insulated with Styrofoam or the likes, to reduce heat flow from the outside hot environment. Air-conditioning ducts carrying low temperature air are also insulated with proper insulation. The manifold advantages of thermal insulation may be listed as follows:

- Energy conservation by reducing unwanted heat transfer
- Protection of people from burn, e.g., in bare hot pipes, furnace walls, etc.
- Maintenance of proper temperature of objects as in chilled water line, AC duct, refrigerator, etc.
- Protection of water from freezing in very cold weather condition in water pipe line
- Reduction of noise and vibration
- Fire protection as most insulation materials are fire resistant

Thermal efficiency

Not all the heat supplied to the food being cooked, including the cooking utensil, goes to heat it. A part of the heat is lost to the surrounding, from the body of the food item and the utensil. This loss of heat is due to convection and radiation loss, mainly to the surrounding air.

Thermal efficiency of cooking during the preparation of food is the ratio of the heat utilized by the utensil and the item being cooked to the heat supplied to it by the burner/heating source.

Similarly, the hot flame or the hot plate cannot transfer all the heat it produces from burning of gas or from electricity to the cooking utensil in contact with it. There will be loss of some heat to the surrounding again by convection and radiation. This is the reason why a hot plate whose top heating surface is not fully covered by the utensil will lose more heat to the surrounding than when it is covered by a bigger utensil, and hence the same hot plate will have different efficiencies, depending on the area of the hot plate being covered by the utensil placed on it. Thermal efficiency of the heating unit is, thus, the ratio of the heat received by the utensil and the item being cooked to the heat energy received by the heating unit. So broadly, we can define thermal efficiency of any thermal system as

Thermal efficiency = Heat gainfully utilized/heat supplied

A 1200 W electric hot plate with an efficiency of 80 per cent runs for 10 minutes. Determine the total heat received by the utensil and the food item in it. If the utensil and food item lose 10 per cent of the combined heat they receive from the burner, find out the net heat received by the cooking system.

Electrical energy input converted to heat in 10 minutes is

H (electrical heat) = Power × time = 1200 × 600 = 720 kJ.

$$\text{Efficiency of hot plate} = 0.8 = \frac{\text{Heat given to the utensil with food stuff}}{\text{Heat received from electricity}}$$

Or heat given to the cooking system = 0.8 × 720 kJ = 576 kJ

The cooking system loses 10 per cent of this heat, i.e., 57.6 kJ.

So, it gainfully utilizes (576 − 57.6) kJ = 518.4 kJ of heat.

5.3.2 A Comparison of the Three Modes of Heat Transfer

The foregoing discussions indicate the mechanisms of different kinds of heat transfer and also elaborate to some extent the application of this knowledge to the heating process in heat appliances and related efficiency of such appliances. We now summarize the three different modes of heat transfer in relation to their commonalities and differences.

1. Conduction and convection require some medium to take place while radiation can transmit heat without help of any material medium.

2. Conduction and convection are slow processes. Radiation is a very rapid process. Being an electromagnetic wave, it travels with the velocity of light.

3. Radiant heat travels in a straight line like light but heat due to conduction or convection may travel in a curved path. Heat from the sun travels in a straight line (although phenomena like diffraction and interference indicate that these waves can also bend to a little extent).

4. Radiant heat does not warm up the medium through which it passes (unless the medium contains radiation absorbing emitting material such as carbon dioxide, water vapour, etc.). Heat transfer through conduction and convection warms up the medium through which it travels.

5.4 | FUELS AND COMBUSTION

Fuels and combustion play a crucial role in the operation of hotel industry. Starting from boilers for producing steam and hot water, to the gas ovens in the kitchen, fuels are burnt to produce heat. To ensure safe and energy-efficient operation of these sets of equipment, a sound knowledge of different types of fuels, their relative merits and demerits, and the mechanism of combustion and procedures for handling them safely are of utmost importance. In the following sections, we discuss their different aspects, separately.

5.4.1 A Few Terms Describing the Character and Performance of Fuel

In order to understand the phenomena of combustion in cooking appliances and other thermal devices, such as boiler, calorifier, etc., it is necessary to have some familiarity with the various terms used in connection with fuels and combustion. A few common terms are discussed below.

Fuels

Fuel is a substance that readily combines with oxygen to burn and in the process gives off heat energy, which is used for heating a room, cooking food, making steam in a boiler, etc. Firewood, coal, petrol, diesel, liquified petroleum gas (LPG), natural gas are few fuels used commercially.

Combustion

Combustion is a phenomenon where a substance (elements such as carbon, hydrogen, or compounds, or mixture of compounds/elements such as petrol, kerosene, diesel, LPG, coal gas, gunpowder, etc.) combines with oxygen and reacts vigorously producing heat and light. In most of the fuels carbon and methane contents burn. We write a few representative combustion equations as follows:

- C (carbon) + O_2 (Oxygen) = CO_2 (carbon dioxide) + 32.8 kJ (heat released per gram of carbon
- CH_4 (methane) + 3 O_2 = CO_2 + H_2O (water) + 58 kJ heat released per gram of methane

Here, carbon and methane are the fuels for the combustion reactions. Oxygen needed for the combustion comes from air.

Combustion is often accompanied by a flame.

Flame

Flame is the region in which chemical interactions among gases occur accompanied by evolution of heat and light. The flame occupies only a small portion of the combustible mixture at any time. Flame speed and thickness is largely affected by temperature, pressure, and the type of fuel used.

A flame has the following two zones:

1. Pre-heat zone, where little heat is released
2. Reaction zone, where the bulk of chemical energy is released

Ordinary fuels are hydrocarbons containing hydrogen and carbon. Hydrocarbon flames are characterized by visible radiation. A fuel needs a certain amount of oxygen to burn completely, if all the conditions are ideal. This oxygen requirement can be calculated from chemical reaction balance equations. Accordingly, the amount of air needed to supply this amount of oxygen can also be calculated. This amount of air is called theoretical or stoichiometric air requirement. However, quite often, air is needed to be supplied in larger quantity and the air–fuel mixture then is called lean mixture. Sometimes, air quantity may become less than the theoretical quantity. The air–fuel mixture is then called rich (fuel-rich). When there is excess air (lean mixture), the reaction zone appears blue. This blue radiation results from excited CH radicals present in the high-temperature zone. Blue flame indicates good efficient burning. When the mixture is rich, the flame appears blue-green. This is due to radiation from excited carbon particles. If the mixture is still made richer, soot (carbon particles) will form. Soot is absolutely black in colour and at high temperature, being a black body, will emit high thermal radiation (usually lost to atmosphere in case of open flame). To human eyes, this radiation appears as bright yellow or dull orange in colour, depending on the temperature. Combustion and heating become very inefficient under such conditions. Efficiency of burner and cooking can be achieved by proper control of fuel and air supply.

Calorific value of fuel

It is the heat released by a fuel during its combustion. For a solid fuel, it is expressed in terms of J/kg or kcal/kg. For a liquid fuel, it is expressed in Joule/ litre or kcal/ litre, while for a gas, it is in J/cubic metre or kcal/ cubic metre.

Almost all fuels contain water stored within them, as also hydrogen. During combustion, hydrogen reacts with oxygen to form water. A part of the heat released by the fuel goes to heat these components of water to steam. If we could re-convert this steam to water by condensation, the latent heat could be recovered and used for heating purposes. Accordingly, there are two measurements of calorific value. One, called the net calorific value (NCV), is the useful heat as we finally get by burning the fuel. The other is called gross calorific value (GCV), which is the NCV plus the heat that we could get back by condensing the steam as it comes out with the gaseous product of combustion. As a result, GCV is always higher than NCV. As far as the consumers are concerned, NCV is the parameter for comparing the heating values of different fuels.

Calorific value of coal varies considerably with the composition of coal such as ash content, carbon content, etc. and type of coal. Fuel oils have much more consistent calorific values.

Density

This is defined as the mass of the fuel divided by the corresponding volume occupied. This is a basic property of any matter of given composition. Its unit is clearly kg/cubic metre. For a liquid, it is measured by an instrument called hydrometer. Knowledge of density is important for quantity calculations and assessing ignition quality. Additionally by comparing densities of two gases, we can say that the gas with higher density will sink when released in a gas of lower density. For example, carbon dioxide having a lower density than air will tend to move towards the floor when released in air.

Relative density or specific gravity

This is a measure of how heavy a matter is with respect to water. Relative density of water is, therefore, 1. It is a pure number. Relative density of kerosene is 0.8. Relative density of a solid or liquid would normally indicate whether it will (in case of liquid, if it is immiscible in water) float on water or sink. If it is less than 1 (like kerosene, etc.) it will float.

Viscosity

We observe that glycerine is very sticky and does not flow easily as water. This property of internal resistance to flow is called viscosity. For a liquid, viscosity increases with a decrease in temperature and hence during winter, at a very low temperature, fuel oil will be reluctant to flow out of its storage tank. Viscosity is, thus, a very important property of fuel oil and influences the degree of pre-heat required for handling, storage, and satisfactory atomization (a process by which liquid fuel is disintegrated to fine misty particles needed for combustion in burner nozzles).

Flash point

Any liquid always gives off vapour at any temperature (called evaporation process), which is why we do not see any trace of water kept on a plate after some time. The higher the temperature, higher is the amount of vapour it gives off in a given time. Flash point is the lowest temperature at which a liquid will give off sufficient vapour in air so that it ignites momentarily (without sustained burning) in presence of an external igniter like a flame or spark. Table 9.2 in Chapter 9 will furnish data on flash point of various substances.

Pour point

The pour point of a liquid fuel is the lowest temperature at which it will pour or flow when cooled under prescribed conditions. It is an indication of the lowest temperature at which the fuel oil is still readily capable of being pumped without getting excessively thick when cooled.

Specific heat

As already explained, it is the heat required to raise the temperature of 1 kg of substance through 1°C (in SI system). For fuels it is usual to express as kcal/kg°C. This quantity is more important for liquid fuel as it is relatively high for them and the heat needed to heat it from room temperature to a desired high temperature for it to be effectively burnt. Light oils have low specific heat, while heavier oils have higher specific heats. These values usually vary from 0.22 to 0.28 for fuel oils.

5.4.2 Types of Fuels

We may broadly categorize different types of fuels as shown in Fig. 5.7.

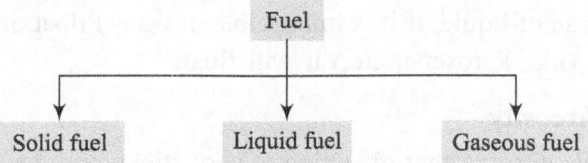

Fig. 5.7 Different types of fuels

Solid fuel

Commonly used solid fuels are firewood, coal, and coke. Nowadays, agro-residues such as rice husk, bagasse, etc. are also used as solid fuels. In sophisticated applications such as missile, rocket, etc., solid magnesium is used as a fuel. The use of locally available agro-residue is on the rise. However, still the solid fuels used in industries in India are mostly coal and coke. Coal can be classified into three major groups, namely

1. Bituminous coal
2. Anthracite
3. Lignite

Anthracite is geologically the oldest grade of coal. It is a hard type of coal consisting mainly of carbon with little volatile content and practically no moisture. Lignite is the youngest from geological perspective. It is a soft coal composed mainly of volatile matter and moisture with low fixed carbon. The common coal used in Indian industries is bituminous and semi-bituminous. Bituminous and lignite coal normally produce huge quantity of smoke. Anthracite has less volatile matter than bituminous coal and is smokeless. It is expensive but has higher calorific value.

Another solid fuel that is used in industries as fuel is coke. Coke is produced by heating coal in a closed oven until its volatile content is driven off. Coke has a higher heating value and emits very little smoke but is costly. Its calorific value is about 16,000 BTU/lb or 10,000 kcal/kg.

Liquid fuel

Liquid fuels used in industries include
1. Kerosene
2. Light diesel oil (LDO)
3. High-speed diesel (HSD)
4. Petrol
5. Furnace oil

These are derived almost entirely from petroleum.

Gaseous fuel

Gaseous fuels commonly used in India are
1. LPG
2. Natural gas
3. Producer gas or coal gas

As cooking gas and gas ovens have become most widely used fuel and heating devices, we discuss them in some detail. Gaseous fuels in common use are LPG, natural gas, producer gas, coke oven gas, or coal gas etc. The calorific values of gaseous fuels is expressed in kcal per normal cubic metre (kcal/Nm^3), i.e., the volume corresponds to normal temperature (20°C) and atmospheric pressure (760 mm of mercury). We discuss below a few gaseous fuels in some detail.

Liquefied Petroleum Gas (LPG)

Liquefied petroleum gas (LPG) or common cooking gas, which is used in our houses and in catering industry, is a liquefied mixture of propane and butane gases, which are obtained as a petroleum by-product.

It is a predominant mixture of propane (C_3H_8) and butane (C_4H_{10}) with a small percentage of other hydrocarbons. It is obtained as by-product of a petroleum refinery. Although it is a gas under normal temperature and pressure conditions, it can be liquefied by the application of moderate pressure and is stored and transported as liquids for ease of handling. It also lakes about 250 times less space than that taken by the same mass in gaseous state. The pressure of LPG gas cylinders is of the order of two to three times normal atmospheric pressure. However, when the stored liquid comes out through the cylinder opening and comes under the reduced pressure in normal atmosphere, it again evaporates to become gas, which people burn in the gas oven burner outlet.

Liquified petroleum gas vapour is denser than air, butane being about two times and propane about one-and-a-half times heavier than air. Consequently, in case of a leak, the gas may flow down toward a low level, and hence may catch fire in presence of a naked fire even at distance from the cylinder from where it has leaked out. To help detection of even a small amount of LPG leaks, LPG is required to be odorized. Traces of odorous organic sulphide known as beta-mercaptan ($C_2H_5 SH$) are mixed with LPG for the purpose and emit the typical smell one gets when gas comes out of domestic cylinders. There should also be adequate ground level ventilation where LPG is stored. For this reason, LPG cylinders should not be stored in cellars or basements, which have little ground level ventilation.

Natural gas

Mother earth is the main source of natural gas. In many locations during the exploration for petrol reserves, natural gas is discovered and explored for use as a very good fuel. Methane (CH_4) is the main constituent of natural gas and accounts for 95 per cent by volume. Other components include ethane (C_2H_6), butane (C_3H_8), propane (C_3H_8), pentane (C_5H_{12}), nitrogen (N_2), and carbon dioxide (CO_2). Very small amounts of sulphur compounds are also present. Natural gas is a fuel with high calorific value and does not require any storage space, as it comes via pipelines from gas supply companies. It readily mixes with air and does not produce any smoke or soot while burning. Calorific value of natural gas ranges from 12,000 kcal/m^3 to 14,000 kcal/m^3.

Coal gas or town gas

This is obtained as a by-product during the production of coke from coal in coke-oven plants. Coke is used for making iron from iron ore. It is also used as a very good (although costly) solid fuel. Coal gas is mainly a mixture of hydrogen, carbon monoxide, methane, together with nitrogen, oxygen, and carbon dioxide, besides some other hydrocarbons, hydrogen sulphide and cyanide. This gas is transported from the coke-oven plant to the premises of the consumer through pipelines. In earlier days, street lamps used this gas to burn. This gas is toxic because of the presence of carbon monoxide. Hydrogen sulphide, which is one of the constituents, smells like rotten egg, and hence acts as an indicator in case of gas leak, if any. Calorific value of coal gas is about 4,900 kcal/m^3.

A comparison of different categories of fuels (solid, liquid, and gas) fuels are given in Table 5.5. Table 5.6 furnishes a comparison of gross calorific values and other properties of a few of the fuels commonly used in industry. Calorific values (or heat values) are expressed in kcal/kg for solid fuel, kcal/litre for liquid fuels and kcal/cubic metre for gaseous fuel fuels.

Table 5.5 Comparison of different categories of fuels

Solid fuel	Liquid fuel	Gaseous fuel
Calorific value to weight ratio is least	Calorific value to weight ratio is higher than solid fuel	Calorific value to weight ratio is highest
Easily available and cheap	More costly than solid fuel but cheap in countries of origin	Except natural gas, other gaseous fuels are costly
Ash and smoke are always produced during combustion. Disposal of ash becomes a problem. Smoke causes pollution and greenhouse effect.	No ash problem, burning is clean but high carbon and aromatic liquid fuel (for example, furnace oil, kerosene, etc.) may produce smoke and soot	Clean fuel; produces neither smoke nor ash
Least risk of fire hazard	Higher risk of fire hazard	Highest risk of fire hazard
Combustion is slow and its control is not easy	Combustion is quick but its control is easy	Combustion is rapid but its control is very easy
Requires a lot of costly space for storage	Requires less space than solid fuel	There may be a central space for storage and supplied conveniently through pipe line. In case of town gas /natural gas supply, there may be no need of storage at all; gas pipeline enters the premises from street supply line
Storage environment is unclean	Storage environment is unclean	Storage environment is clean

Table 5.6 Comparison of grass calorific values and other properties of different types of fuels

Types of fuels	Heat output	Cleanness	Smoke emission	Calorific value	Cost
Bituminous	Fairly good	Not clean	Considerable smoke emission	5,000–8,000 kcal/kg	Cheap
Anthracite	As above	Not clean	Smoke	9,000 kcal/kg	Costly
Lignite	Relatively poor	Not clean	High smoke emission	4,500	Cheap
Coke	Good	Less clean	Much less smoke	10,000 kcal/kg	Costly
Kerosene	Good	Fairly clean	A little smoke	11,100 kcal/kg	Costly and restricted availability
Diesel oil	Good	Fairly clean	A little smoke	10,800 kcal/kg	Costly
LDO	Good	Fairly clean	A little smoke	10,700	Costly
Furnace oil	Good	Less clean	A little smoke	10,500	Costly

(Contd)

Table 5.6 (*Contd*)

Types of fuels	Heat output	Cleanness	Smoke emission	Calorific value	Cost
Coal gas	Very good	Very good	Smokeless	4,900 kcal/m^3	Costly
Natural gas	Excellent	Clean	Smokeless	9,500 kcal/m^3	Less costly
Electricity	Very good	Cleanest	No smoke at all		Costliest
LPG	Very good	Very clean	Smokeless	27,800 kcl/m^3	Very costly

5.4.3 Fuels used in Hotel Industry

Solid fuel

Coal: It is used in boiler to produce steam.

Coal/coke/wood: It is charcoal used in *tandoor* to make chapattis and kebabs.

Liquid fuel

Kerosene: It is not very widely used and may supplement main fuel in case of deficiency.

High-speed diesel: It is used in oil-fired boilers and water heaters.

Gaseous fuel

LPG: It is used in most of the cooking ovens in hotel industry.

Natural gas: It is used in many places across the globe where cheap natural gas is available.

Coal gas: If a city is supplied with coal gas, it is a cheap fuel to use.

5.5 | STOVES AND BURNERS—TYPES AND PRECAUTIONS FOR USE

Stoves and burners are the appliances that are two of the arterial equipment in the kitchen of a hotel or a catering establishment. The correct choice of the type of ovens and burners are critical for energy efficiency, safety, and functional performance of a kitchen. Burners are also used in boilers and many other heating equipment in the hotel industry. This section introduces this set of equipment and also discusses the precautions and care required to operate them.

5.5.1 Oil Burners

Oil stoves are generally not used for cooking in the hotel industry. However, oil burners are, of course, used in oil-fired furnaces of boiler and water heaters in a hotel. The purpose of an oil burner is to make fine particles of fuel oil, which are burnt at the mouth of the burner. This process of breaking bulk of any liquid to very small droplets is known as atomization. It is different from vaporization in that in the former the liquid remains liquid but in the latter, liquid state is changed to gaseous state. This atomization facilitates very rapid evaporation of liquid fuel, which then burns in association with oxygen. If a high velocity jet of air is passed over a sheet of liquid, the liquid sheet is broken into small droplets causing atomization. This is accomplished in a burner nozzle. Normally, atomization is carried out by primary air and complete combustion is ensured by additional air called secondary air. Figure 5.8 shows a simplified oil burner-head.

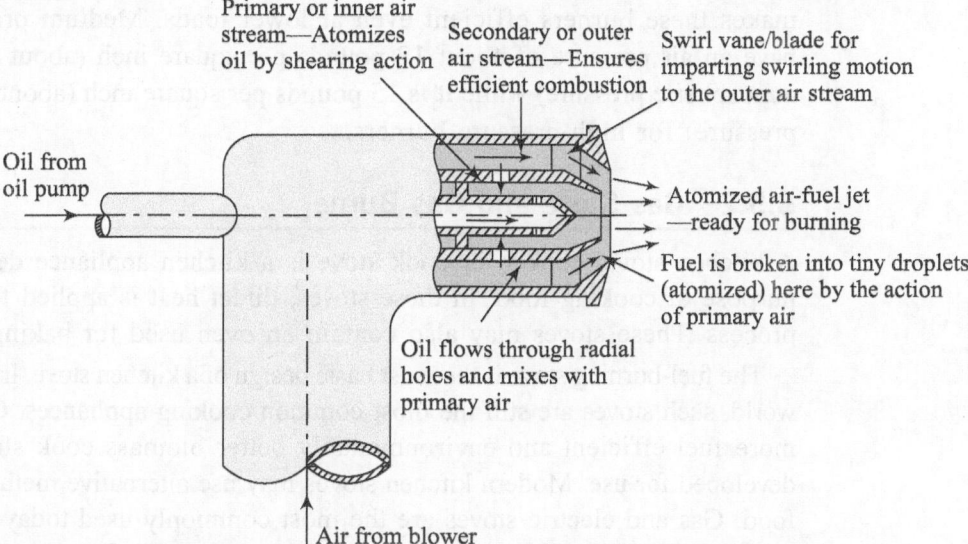

Fig. 5.8 A simplified view of a low-pressure oil burner-head

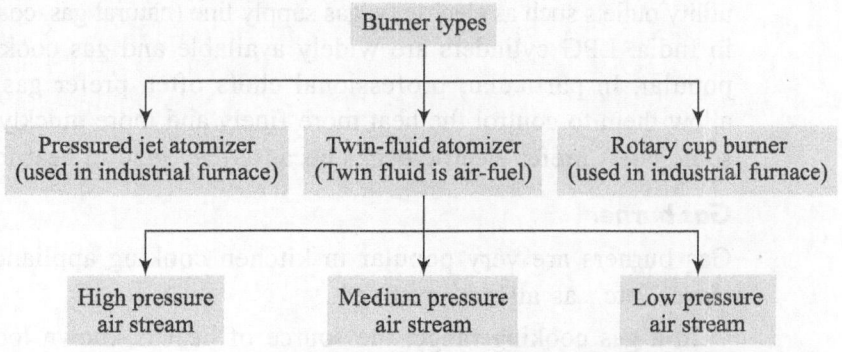

Fig. 5.9 Types of burners

Burners have been categorized into various types as shown in Fig. 5.9. We shall discuss only twin-fluid atomizer type burner here.

Low air pressure burner Air at low pressure is provided by a blower. Air also acts as an atomizing agent. Oil is injected through a nozzle at the open end of the burner where it gets atomized. Air and oil flow are regulated by valves. At a low oil flow rate, efficiency of these burners becomes lower. Air pressure in these burners is about 8 pounds per square inch (about half atmospheric pressure).

Medium- and high-pressure burners In these burners, compressor supplies the pressurized air flow. When load changes, the quantity of atomizing air does not change; only the secondary air entering the system changes. This arrangement makes these burners efficient even at lower loads. Medium pressure burners have an air pressure of about 12 pounds per square inch (about 80 per cent of atmospheric pressure) while it is 15 pounds per square inch (about 1 atmospheric pressure) for high-pressure burners.

5.5.2 Gas Stove and Gas Burner

A kitchen stove, cooker, or cook stove is a kitchen appliance designed for the purpose of cooking food. In these stoves, direct heat is applied for the cooking process. These stoves may also contain an oven used for baking purposes.

The fuel-burning stove is the most basic design of a kitchen stove. In the developing world, such stoves are still the most common cooking appliances. Of course, new, more fuel efficient and environmentally better biomass cook stoves are being developed for use. Modern kitchen stoves may use alternative methods for heating food. Gas and electric stoves are the most commonly used today in India. Both are highly developed in engineering and control and safe to use, and the choice between the two is largely a matter of personal preference and already existing utility outlets such as electricity, gas supply line (natural gas, coal gas etc.,). However, in India, LPG cylinders are widely available and gas cooking ranges are very popular. In particular, professional chefs often prefer gas cooktops, as these allow them to control the heat more finely and more quickly. On the other hand, some chefs prefer electric ovens because they tend to heat food more uniformly.

Gas burner

Gas burners are very popular in kitchen cooking appliances such as kitchen stoves, etc., as already outlined.

In a gas cooking range, the source of heat is known loosely as a burner to which the flow of gas is directed by turning a knob or a valve handle. In between, air is sucked in and mixed with the fuel and then burnt at the end of the burner. Depending on the working gas pressure, gas burners can be classified as

1. High-pressure gas burners
2. Low-pressure gas burners

Low-pressure burners use gas at a low pressure of less than 0.15 kgf/cm^2 or 2 pound per square inch (psi) (about 1/7 of one atmospheric pressure). They are usually of the multijet type, in which gas from a manifold/cylinder is supplied to a number of small single jets or circular rows of small jets centred in or discharging around the inner circumference of circular openings in a block of heat-resisting material.

In a high-pressure gas mixer, the energy of the gas jet draws air into the mixing chamber and delivers correctly proportioned mixture to the burner (Fig. 5.10). When the regulating valve is opened, gas flows through a small nozzle into a venturi tube (a tube with a constriction at some section). At the narrow section, the gas velocity becomes high and consequently pressure drops, and hence sucks in air from outside through openings at the narrow section at the end of the pipe. The gas-air mixture is then piped to the burner. Air flow can be controlled by turning a revolving shutter, which can be locked at any desired position. We now discuss a common gas oven burner we use in domestic and small volume cooking applications

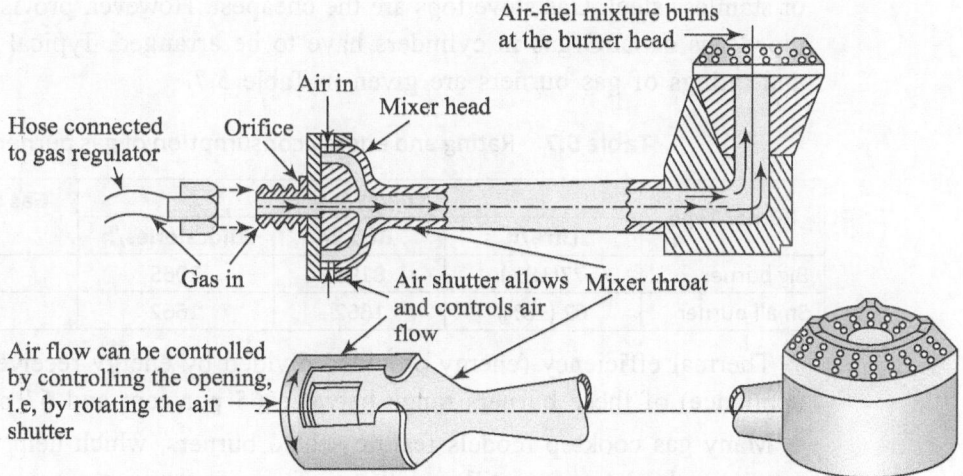

Fig. 5.10 Major components of an LPG burner system

The burner has a mixer head, mixer throat, and burner head, cast in one piece to prevent gas leakage, as shown in Fig. 5.10. The mixer head carries an air shutter and a hole for the gas orifice, which is the opening through which the gas flows and by means of which the flow is controlled. There are openings in the form of small holes in the burner though which gas-air mixture comes out and is burnt into gas flames. These openings are called ports, which are distributed horizontally in series of rows from the inner portion of the burner top to the outer rim. The inner rows, having four or five ports, heat the central portion of the utensil bottom and frequently form the simmer section of the burner.

Different types of gas burners Burners are of five types, namely

1. Giant: For rapid heating and for large utensils; Heat rate: not less than 12,000 BTU per hour
2. Regular or standard:For general cooking in average size pans; Heat rate: Not less than 9000 BTU per hour
3. Simmering: For domestic cooking in medium size pans; Heat rate: Not less than 1200 BTU/hour
4. Pilot: For small families with small size pans; Heat rate: Not more than 300 BTU/hour
5. Mini pilot: For very small use with very small size pan; Heat rate: Not more than 125 BTU/hour

Nowadays, big-size ranges are available with a heat output as high as 40,20,000 BTU/h while some heavy duty fryers have heat output of 1,20,000 BTU/h.

How quickly the food gets heated or cooked depends upon the power of the burner. Cooking surfaces of gas burner are available in enamel, glass, aluminium, or stainless steel. Gas stove tops are the cheapest. However, provisions for either piped gas or LPG gas in cylinders have to be arranged. Typical specifications and ratings of gas burners are given in Table 5.7.

Table 5.7 Rating and energy consumption of gas burners

Burner Type	Energy Input			Gas Consumption (g/h)
	Litre/h	BTU/h	Kilocalories/h	
Big burner	77(±8%)	8195	2065	189
Small burner	62 (±8%)	1662	1662	153

Thermal efficiency (energy provided divided by energy received by standard appliance) of these burners range between 65 per cent and 67 per cent.

Many gas cooktop models feature sealed burners, which help to make clean up easier by keeping spills on the surface. The burners are surrounded by snugly-fitted rings, which prevent spills from flowing underneath and becoming trapped under the burner.

5.5.3   Safety Precautions when using Gas Equipment

Some safety precautions when using gas equipment are given below.

1. The equipment should be cleaned regularly, especially since gas contains sulphur, which corrodes metals.
2. Gas cylinders should be placed in upright vertical position and never in horizontal position.

3. The gas cylinder should not be titled to an inclined position while being used, in order to completely utilize the gas.

4. The cylinders must not be hammered.

5. Cylinder and ovens should be placed in well-ventilated spaces.

6. Inflammable material should not be kept very near the gas bank.

7. The flexible metal hose connecting the gas valve and oven should be checked, at regular intervals, for crack or damage.

8. Empty gas cylinder should be immediately removed and kept in a proper place away from fire or other source of heat.

9. After the work is over, the gas regulating valve and the burner switch are kept in 'OFF' position.

10. Every day burner heads should be removed from the ovens and pots cleaned with steel wire brushes.

11. Since there are possibilities of fire hazards and explosions due to leakage of gas, it is important to open doors and windows before lighting burners.

12. While putting off the oven, first the regulating value of the gas line is closed and then the gas stove knob is put to the 'OFF' position.

5.5.4 A Few Steps for Efficient Operation of LPG/Fuel Gas Stove and Burners

A few steps for efficient operation of LPG/fuel gas stove and burners are given below.

1. Correct size of burner/nozzle should be ensured.

2. If there is no pilot flame, a flame has to be ensured from the igniting torch.

3. Attention should be given to fading or pulsation of flame.

4. Flame size has to be adjusted to suit the requirements.

5. Proper shut down procedure is to be followed.

Burners should be dismantled and cleaned periodically, preferably before start or end of each shift. Spare burners should always be kept ready.

5.6 | ELECTRIC OVENS AND ELECTRICITY AS FUEL

Electricity has many advantages over chemical fuels and, as a result, there are many heating devices that use electricity in catering industry. Very common examples of such devices are electric oven, hot plate, rice cooker, electric toaster, etc. for cooking, geyser for heating water, and portable room heater among others. Microwave ovens also use electricity for its functioning although it heats and cooks food by different physical principles compared to ordinary heating by electricity. Figure 5.11(a) shows typical electrical heaters in the form of hot coil top and hot plate top. The components of a typical electric hot plate are shown in Fig. 5.11(b). Ceramic cooktop electric ranges are very widely used in cooking activity, particularly in countertop heaters, inside restaurants of hotels. They radiate heat very little, and hence do not tend to heat up the atmosphere. Figure 5.12 (a) and (b) show two types of typical electric ovens.

Fig. 5.11 (a) Typical coil top and hot plate electric heaters

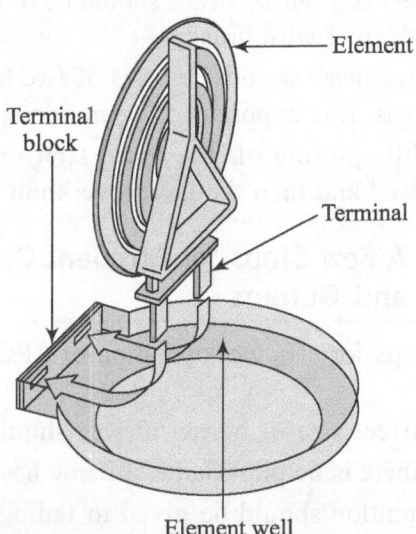

Fig.5.11 (b) Heating coil and contact terminals for an electric hot plate

(*Source:* 2006 Publications International Ltd., www.howstuffworks.com)

Fig. 5.12 (a) A typical electric oven   **Fig. 5.12** (b) A typical cooktop electric oven

5.6.1 Microwave Heating

Microwave oven, dryers, and heaters are some common sources of microwave radiation. The source emits electromagnetic radiation with frequency (or wavelength) corresponding to microwave region. When this radiation falls on an object, the molecules of the object start vibrating vigorously and the effect becomes thermal in nature, resulting in a rise in temperature. A fairly low-intensity microwave radiation has heating value of 20–60 mW/cm^2 for 1 s. The stipulated frequency of microwave radiation is 10^7 Hz (cycles/second) to 10^{11} Hz with a maximum energy density of 1 mWh/cm^2 during any 0.1h period. The frequency/wavelength of radiation is such that it is reflected from metal surface but transmitted through cookware surfaces such as glass, ceramic, or plastic. Hence, metal cookware should not be used for cooking/heating in microwave ovens. Radiations is absorbed by food items, especially water, sugar, and their fat content, and hence the food items get heated up in the process. Microwave ovens must not leak radiation of more than 1 mW/cm^2 at the time of buying and no more than 5 mW/cm^2 at any time during the lifetime of the oven. Like any other heat sources and consequent heat transfer process, here also, the skin of the food is first heated up by the microwave radiation and rest of the bulk is heated up through heat conduction inside. However, compared to conventional convection ovens, there is no intermediate air to heat up and the heating here is very fast, almost instantaneous and at the same time more energy efficient.

Advantages of electricity as kitchen fuel

Electricity has many advantages as a fuel and these are as follows:
- It is the cleanest of all fuels.
- There is no need of any storage space.
- Fine control of heating is possible.
- Continuous supply is almost ensured.
- Wastage of energy can be made almost nil by a little planning.
- Microwave heating is very fast.
- Best for countertop heating.

However, electricity is always costlier than chemical fossil fuels such as LPG, natural gas, etc. and a cost-benefit analysis should be made in this regard. Maintenance of electrical devices needs skilled expertise.

5.7 | GAS BANK

All hospitality units have large kitchen for daily service of food items. Instead of having individual gas supply to the ovens, a gas bank is often installed in a centralized arrangement usually outside the kitchen. In this gas bank arrangement, individual gas cylinders are connected to a common pipeline called manifold or header (Fig. 5.13a). One or more gas supply lines are drawn from this manifold into the kitchen and these pipelines again serve as common supply line to a group of ovens. Short pipelines connect each individual oven to this common supply line. One or more gas banks may be designed to connect numbers of cylinder to one common delivery line. Depending upon the process, 1. duty or primary and 2. stand by or reserve manifolds are installed. There are arrangements to automatically switch gas supply from the primary cylinder bank to the reserve cylinder bank, whenever there is such a requirement (Fig. 5.13b).

Fig. 5.13   (a) A typical gas bank for a hotel kitchen

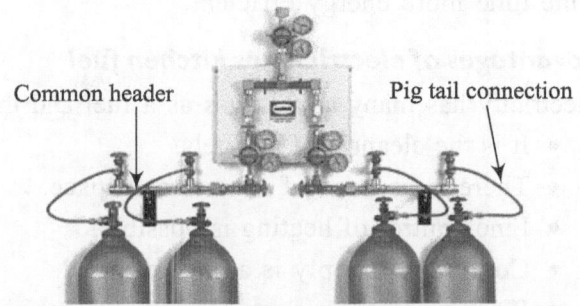

Fig. 5.13   (b) Gas manifold header with automatic pressure differential switch over

(*Source:* www.hithermboilers.com/multi_gas_cylinder-installation.html)

Regulators must be manually reset to change the primary and reserve bank. Proper safety installations are critical while using gas banks. Cylinders can be connected or taken out even when the systems are in use. Non-return valves (or one-way valve, which allows gas flow only in one direction, e.g., a cycle-tube valve, which allows air flow from air pump into the tube but does not allow air to flow from tube inside to outside) are provided for each individual cylinder so as to ensure that gas from a cylinder travels to kitchen supply line instead of going to other cylinders having lower gas pressure, thus trying to balance the pressure in coupled cylinders. Additionally, it also ensures safety in case of pigtail failure. Flame arrestor is installed to work as a safety device. Cylinder and manifolds are connected through high-pressure flexible pigtails and click-on adaptors. The operating pressure of the supply gas can be set by primary and secondary pressure regulators provided for the purpose. Pressure gauges are fitted on each cylinder as well as manifold to indicate cylinder and line pressures.

A fully automatic gas bank is essential for medical requirements and is increasingly used in hotel industries too. In this gas bank, the installed system automatically switches to the reserve bank when the primary bank is depleted. When the depleted cylinders are replaced with full cylinders, the system will automatically reset itself in preparation for the next bank change. The primary side is the bank in use and the remaining side is the bank on reserve. A light emitting diode (LED) indicator would indicate a depleted bank by turning red. An alarm buzzer will also sound to indicate an empty bank or banks. The indicator turns green as soon as the system is reset by replacing the depleted cylinder. And the buzzer also stops. If the manifold is connected to the central alarm system of the premises, this will also indicate that the bank was depleted. This is very important for oxygen cylinder banks in a health care unit. The only manual activity in a medical gas manifold should be changing of the depleted cylinders.

Changeover is performed by solenoid valves contained in the control cabinet. In the event of an electrical power failure, solenoid valves will automatically open to provide an uninterrupted gas flow and this is called 'fail safe' arrangement, in engineering parlance.

Under normal operating conditions, the gas will leave the high-pressure cylinders through the pigtails into the header bars. The pigtails have one-way valves (check valves) to allow the replacement of depleted cylinders without gas pressure back flow into the remaining depleted cylinders on that bank. In the event a safety relief device on an individual cylinder should activate or a pigtail should leak excessively, the local check valve will also prevent loss of gas from the rest of the cylinders on that bank.

The gas flows through the header bar to the control unit. A manually operated shutoff valve is installed downstream of each header bar. These valves are located outside the control cabinet, one on the right and one on the left side. The valves are normally left fully open to allow unrestricted flow of gas from the two cylinder banks. In case of an emergency, they can be closed to isolate one or both banks.

The gas flows through the manually operated shutoff valve into the primary regulator. This regulator reduces the high cylinder pressure to an intermediate pressure. The intermediate pressure gas flows through the solenoid valve to the line regulator for its final (line) pressure reduction for use in the health care facility. Two line pressure regulators are installed in parallel and each is capable of maintaining a constant dynamic delivery pressure at the maximum designed flow rate of the system.

The solenoid valve is the key to the 'automatic' mechanism in the manifold. A solenoid valve is an electrically operated valve. This component ensures that the flow of gas is not interrupted and the pressure does not fluctuate during normal operation. When the operating bank pressure fails to a predetermined level, which is controlled by the preset pressure switches (high and low), the switches will activate the electronic control board, causing the solenoids to switch to the fresh (reserve) cylinder bank.

The manifold control cabinet has three means of giving continuous information on the system status. They are pressure gauge, LED indicator, and alarm buzzer. Pressure gauges indicate the gas bank pressures and the delivery pressure. In one design, six numbers of indicator LEDs are installed to indicate the service status of the cylinders. Two green LEDs indicate which cylinder is in service, two yellow indicate reserve ready, and two red indicate that a bank is depleted. A loud audible buzzer gives an alarm when either or both banks are depleted.

The LED indicators are controlled by sensing the bank pressure. Replacing the depleted cylinders on the empty bank resets the system, changing the indicator from red to yellow. At the same time, the yellow LED will change to green to indicate in service status from the reserve ready status.

5.8 | SOLAR COOKER

A solar cooker is like a hot box, in which food can be cooked without any cooking gas or kerosene, electricity, coal, or wood. This cooker works with solar energy. In a solar cooker one can boil, bake, and roast food. Bigger size solar cookers are nowadays available for cooking food for up to 15 persons. However, solar cooker is still not an alternative cooking process for hotel and restaurants for reasons of timing, size, and speed of cooking.

There are many different types of solar cookers. All solar cookers use the sun's heat and light to cook food. The basic principles of solar cookers are as follows:

- Concentrating sunlight by a reflecting mirror, making the heat more intense
- Converting the collected radiation into heat available for cooking efficiently, by using blackened surfaces, both at the interior of the cooker and the exterior of the cooking vessel

- Trapping heat by isolating the air inside the cooker from the air outside the cooker, thus reducing heat loss to the surrounding

Most solar cookers use two or all three of these strategies in combination to get temperatures high enough for cooking. Figure 5.14 shows a typical solar cooker.

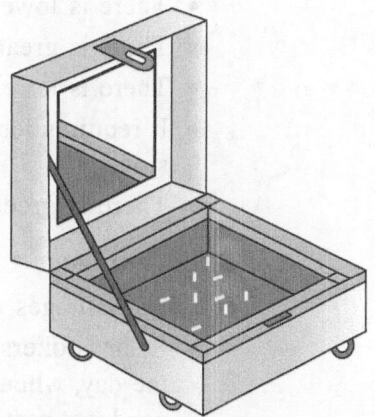

Food is prepared as it would be for an oven or stovetop. It is well known that food cooks faster when it is in smaller pieces and so, smaller pieces of food are used in solar cooking, e.g., potatoes are usually cut into pieces rather than being cooked whole. Bread is usually baked as individual rolls instead of large loaves. When a food item, such as rice, has to be cooked in water, the minimum necessary amount of water is used. This is done to save the energy needed to heat extra water.

Fig. 5.14 A typical solar cooker

(*Source:* Gujarat Energy Development Agency)

Solar ovens can be used to prepare anything that can be made in a conventional oven or stove—from baked bread to steamed vegetables to roasted meat. One can cook a large number of items, such as pulses, rice, *kheer, khichdi,* vegetables, cereals, etc., which are needed for our daily meals, in a solar cooker. Some special dishes such as, idli, *dhokla, lapsi, pulao,* and soup, can also be cooked. One can even bake bread, biscuits, cake, etc. Non-vegetarian items, such as fish, chicken, meat curry, kebabs, etc., which are boiled or roasted can also be prepared in a solar cooker. The type of food, time of the day, intensity of solar radiation, the quantity of food that needs to be cooked, wind factor, etc. decide the rate at which solar cooker and the food item receive and absorb heat, which, in turn, determines the time taken for cooking. The following data will give a rough idea as to the time taken to cook some of the dishes in a small solar cooker for two people:

- Rice—About two hours
- Vegetables—Two to three hours
- Black gram and *rajma*—About four hours
- Cake—About one hour
- Melt butter—Fifteen minutes
- Bake cookies—Two hours

All the figures above are only indicative in nature and depending on the local conditions and the type of solar cooker being used, the time estimates could be half as long, or twice as long.

Advantages

The advantages of using a solar cooker are as follows:

- There is lower cost compared to firewood or cooking gas
- There is greater safety for the cook
- There is lower likelihood of starting a fire that could destroy a house
- It requires less hands-on time for cooking compared to tending a fire or stove
- There is green and clean cooking.

Disadvantages

The disadvantages of using a solar cooker are as follows:

- Solar cookers provide hot food during or shortly after the hottest part of the day, when it is already late for main meal. However, a thick pan that conducts heat slowly (such as cast iron) will lose heat at a slower rate, and that, combined with the insulation of the oven or an insulated basket, can be used to keep food warm well into the evening.
- Solar cookers take longer to cook food compared to an oven
- Solar cookers cannot be used to make chappati or any fried item
- Solar cookers do not work during a cloudy or rainy day
- It is not yet possible to employ solar cooker as a replacement for conventional ovens in a commercial kitchen as in a hotel, etc. because of very low heating rate, lack of control of the cooker and uncertainty regarding climatic conditions, among others.

5.9 | CONCLUSION

The subject of generation of heat, heat transfer, and utilization of heat in heat appliances is in the domain of specialized technical expertise. A variety of heat appliances are used in the hospitality industry for kitchen and guest-care requirements. Determining the exact requirement of such appliances and properly specifying them are very important for efficient functioning and need knowledge of heat and heat transfer. The engineer sits with the chef for kitchen requirement and decides on the type and specification of such equipment. As such, students of hospitality management would be benefitted if they go through the basics of the science and technology of heat appliances, as discussed. It is also observed that the operators of the heat devices have to possess enough knowledge to properly handle the devices, particularly with respect to safety and energy efficiency. The basic heat appliances as used in a kitchen are discussed and serve to indicate the potential advantages regarding the correct choice of the devices for particular situations.

KEY TERMS

Burner	It is a device in a stove that atomizes the fuel and makes a mixture of fuel and air ready for burning.
Conduction	The process in which heat is transmitted from a hotter part of a body to a colder part without any visible movement of the material of the body is called conduction.
Convection	The process in which heat is transmitted from a hotter part of a body to a colder part by the actual movement of the material of the body is called convection.
Flame	Flame is the region in which chemical interactions among gases occur accompanied by evolution of heat and light.
Fuel	Fuel is a substance that readily combines with oxygen to burn, and in the process gives off heat energy.
Gas bank	It is an arrangement of gas cylinders connected to a common manifold from where supply lines to appliances are branched out.
Heat conductor	Those substances which can conduct heat easily are called heat conductors.
Heat	Heat may be regarded as a form of energy, whose absorption either makes a body hot or changes its state (solid to liquid and liquid to gas) and whose rejection either makes a body cold or changes its state (gas to liquid and liquid to solid).
Latent heat	Heat which causes change of state of a body without any change of temperature is called latent heat.
LPG	It is liquefied petroleum gas comprising mainly butane and propane gases.
Microwave	It is an electromagnetic radiation falling within a specific range of wavelength.
Radiation	The process by which heat is transmitted from one place to another without the help of any material medium or without heating the intervening medium is called radiation.
Sensible heat	Heat which causes a change of temperature in a body is called sensible heat. It can be measured by a thermometer.
Solar cooker	It is a cooker that uses sunlight as the source of energy for cooking food.
Specific heat capacity	Specific heat capacity is the measure of the heat energy required to increase/decrease the temperature of a unit mass of a substance by unit temperature difference.
Temperature	Temperature is defined as the degree of hotness or coldness of matter body.
Thermal efficiency	It is the ratio of heat utilized to the heat received.
Thermal insulation	It is the process of adding additional layers of materials on pipelines, furnace and boilers walls, etc. to reduce heat loss.

REVIEW QUESTIONS

5.1 Define heat and temperature, giving their units of measurement.

5.2 Discuss the different units of heat.

5.3 Define specific heat capacity, mentioning its units.

5.4 Distinguish between sensible heat and latent heat.

5.5 Explain convection and radiation heat transfer.

5.6 Explain why convection is not possible in solids.

5.7 Make a comparison of different modes of heat transfer.

5.8 Discuss thermal insulation, stating its general advantages.

5.9 Explain thermal diffusivity, indicating its importance in roasting.

5.10 Which mode(s) of heat transfer would you expect to dominate the process of roasting?

5.11 Define thermal efficiency, giving examples.

5.12 State the advantages and disadvantages of different types of fuels.

5.13 Which are the different types of fuel used in catering industry?

5.14 Write a short note on LPG.

5.15 Explain functioning of a gas burner and list different types of gas burners.

5.16 Draw an LPG gas stove and explain its working.

5.17 Discuss the precautions and points to check while dealing with LPG gas stove.

5.18 Discuss flame and explain how the appearance of the flame in a burner can indicate the condition of burning.

5.19 State the advantages of electricity as a fuel.

5.20 Discuss the working of a microwave oven.

5.21 Discuss, in brief, the arrangement of a gas bank.

5.22 Discuss the advantages and disadvantages of using a solar cooker.

REVIEW PROBLEMS

5.1 A piece of turkey breast of mass 200 g is at room temperature of 30°C. It is then put in an oven and heated to a temperature of 60°C. Calculate the heat to be supplied by the oven (ultimately, the electrical energy if the oven is an electrical oven). The specific heat of the turkey breast may be taken as 3590 J/kg-K.

[21540 J or 5146 cal]

5.2 If the heating, as in Problem 5.1, is to take place in 25 seconds, determine the power rating of the oven. Consider that all the heat energy produced in the oven by electricity is available for heating the turkey breast [861 W]

[*Hint:* The rate of heating is J/s and is basically work done (in form of heat supply) per second which is the power requirement.]

5.3 Chicken soup is to be prepared with 5 kg water, 1 kg shredded chicken, and 500 g of corn, besides other minor constituents. All these components are mixed and the mixture is heated from 30°C to 100°C so that the mixture boils. It is kept boiling for 5 min for good cooking and the water content of the mixture reduces by 500 g during evaporation. Determine the heat input rate to the soup to achieve this. Hence, determine the time required for heating (sensible heating) the mixture from 30°C to 100°C and the total time of heating for the soup preparation. Assuming 80 per cent thermal efficiency of the cooking range, determine the heat rating of the burner to be used for the purpose. Take latent heat of evaporation of water to be 540 cal/g. Specific heat values for water, chicken, and sweet corn may be taken as 1000 cal/kg°C, 850 cal/kg°C, and 423 cal/kg°C.[3240 kcal/h, 6 min 52 sec, 11 min 52 sec, 4050 kcal/h or 4.71 kW or 16068 BTU/h]

[*Hint:* Amount of water evaporated during boiling is 500 g. Latent heat of evaporation to be taken from the burner is, $m \times L = 500 \times 540 = 270,000$ cal.]

So, heat rate is 270,000/5 = 54,000 cal/min. Ideally, it is sufficient that the burner supplies heat at this rate. But since thermal efficiency of the heating and cooking utensils is not 100 per cent, the burner heat rate must be more than this. The sensible heating also utilizes heat at this rate of 54000cal/min. Calculate sensible heat requirements separately for 5 kg water, 1 kg chicken, and 500 g of corn using their individual specific heat values and a temperature rise of 70°C. This amounts to 370755 cal. Then determine the time required to supply this energy using the heat rate calculated earlier. Efficiency of heating is 80 per cent and, hence, burner heat input rate is 54,000/0.8 cal/min etc.

6 Chapter

Hardness of Water and its Removal

Learning Objectives

Water, essential for human existance, assumes special importance in the hotel industry. Many equipment and processes in a hotel use water for their functioning. These include boilers, heaters, refrigeration condensers, cooking processes, laundry, dish washing, waste disposal, etc. It is in this context that water needs to be categorized vis-à-vis its physical effects on the equipment that use it.

This chapter deals with the classification of water according to hardness and methods to remove hardness. After reading this chapter, the students will be able to:

- understand the causes and chemistry of hardness
- know about various methods of removal of hardness
- understand the working and components of a typical water softening unit

6.1 | INTRODUCTION

The main sources of water in India are lakes, rivers, wells, and deep tube wells. Water is seldom available in nature as a chemically pure compound of H_2O. Almost always it will have some salts dissolved in it. The presence of some of the salts have a pronounced effect on the suitability of water for use with soap for cleaning, etc., as also the use of it as industrial water in boilers and similar other utilities, requiring hot water to pass through them. Water is classified as hard and soft depending upon its behaviour towards soap solution. However, apart from these salts, water also contains impurities that include

1. Floating impurities—Dirt, undissolved particles, etc.
2. Minerals
3. Bacteria

The various impurities can be removed and the water made potable by the following methods:

1. Hardness can be removed by water softening plants
2. Floating impurities can be removed by filtration process
3. Water can be made near free from minerals by demineralization process
4. Bacteria are removed by chlorination and ultraviolet (UV) rays, as in Aquaguard water purifier

By water treatment, we mean that water is made free from hardness and in specific situation, free from all minerals.

6.2 | HARDNESS OF WATER—DIFFERENT TYPES, CAUSES, AND EFFECTS

Before we technically define hardness of water and the phenomenon of hardness, let us introduce some related terms commonly used and understandable in simple terms.

Soft water Water that readily forms lather with soap is called soft water. Rain water, pond water, and distilled water are examples of soft water. Soft water is slightly acidic.

Hard water Water that does not easily form lather with soap is called hard water. Sea water and groundwater from tube wells are sources of hard water. Hard water is alkaline in nature. Hard water is undesirable because it may lead to higher soap consumption, scaling of boilers, causing corrosion and incrustation of pipes, making some food tasteless, etc.

6.2.1 Comparison of the Characteristics of Hard Water and Soft Water

It is true that hard water has many disadvantages and is not used for industrial purposes but it has some merits too. Similarly, soft water, notwithstanding its various advantages, has its share of demerits as well. The advantages and disadvantages of hard and soft water are listed below.

Advantages of hard water
- Hard water provides good taste for drinking.
- It provides vital minerals necessary for the body.

Disadvantages of hard water
- Use of hard water causes wastage of soap and detergents. On an average, hard water will destroy about 3 lb of soap per 10 gallons of water. This, in addition, shortens the life of clothing and textile materials in laundry.

- Scale is precipitated out of such water, which forms thick white crust (called scaling) inside pipes that carry them and in cylinders, boilers, etc., which use the water. This crust or scale (a) gradually reduces the internal diameter and size of pipes, cylinders, boilers, etc. (b) lowers the efficiency of the heating system by retarding the transmission of heat from furnace or heating coil to the water inside pipe in the boiler and other water heaters, and (c) interferes with the functioning of flow control valves in the pipeline.
- Hard water is not suitable for paper, textile, sugar, and pharmaceutical industries.
- Some people complain of dry skin after bathing in hard water.

Advantages of soft water

- Soft water consumes less soap for cleaning clothes.
- It is good for cooking, imparting good taste to cooked food.
- Soft water does not produce crust or deposit any scale in the pipe or other equipment that use water.

Disadvantages of soft water

- Soft water is usually less palatable to drink due to lack of minerals.
- Soft water does not provide vital minerals such as calcium and magnesium necessary for the body.

6.2.2 Hardness

Here we discuss the basis of measurement of hardness of water.

Measurement of hardness

The degree of hardness is generally defined as calcium carbonate, equivalent of calcium and magnesium ions, present in water and is expressed in mg/L. In simple terms, hardness is a measure of how much calcium (and to a much lesser extent, magnesium) is in the water. Water hardness industry measures it in grains per litre, where 1 grain/litre = 64.72 mg/L.

Total hardness is a test for overall water quality. There is no health concern associated with it. Rather statistical analyses indicate a general reduction in cardiovascular disease with increasing water hardness. Water less than 150 mg/L is considered soft water, while water with more than 200 mg/L of hardness is considered hard water.

We now define a term called pH value of water to correlate hardness with pH so that a very cheap and rapid means of indicating water hardness is possible.

pH value of water

pH value of water indicates the number of free or active hydroxyl (OH^+) ions or, more precisely, hydronium (H_3O^+) ions in a solution, which, thus, in turn, indicates the degree of alkalinity. pH value usually ranges from 0 to 14. A value of 14 indicates fully alkaline solution, while pH 7 indicates a neutral solution of pure water.

The pH value of water is calculated as the logarithm of the reciprocal of hydrogen ion concentration (power of hydrogen and hence the term pH) present in water. It is, thus, an indicator of the acidity or the alkalinity of water.

If pH of water is above 7, it will be alkaline and if it is less than 7, it will be acidic. The maximum acidity will be at zero value of pH and the maximum alkalinity will be at a value of pH equal to 14.

The pH scale is logarithmic. That means each change of one in pH value means 10 times more acidic or alkaline. Therefore, a substance with a pH of 2 is 1000 times more acidic than one with a pH of 5.

Although pH value and hardness do not indicate the same property of water, the presence of carbonate ions in water (sometimes called 'carbonate hardness') pushes it towards alkalinity, and hence hard water is likely to have a high pH. A pH greater than 8.5 could indicate that the water is hard. Since measurement of pH is relatively simple, this could act as a reliable method to test the hardness of water.

The pH values of some common substances are given in Table 6.1.

Table 6.1   Typical pH values of a few common substances

Substance	Typical pH value
Battery acid	0.3
Stomach acid (gastric juices)	1.4
Lemon juice	2.1
Vinegar	3.0
Wine	3.5
Tomatoes	4.2
Pure rain water	5.5
Pure water	7.0 (neutral)
Blood or tears	7.4
Baking soda solution	8.5
Household bleach	12.5

(*Source* : http://www.ilpi.com)

6.2.3 Causes of Hardness

The hardness of water is mainly due to the presence of dissolved bicarbonate (HCO_3^-), chloride (Cl^-), and sulphate (SO_4^{2-}), salts of calcium (Ca), magnesium (Mg), and iron (Fe). Sodium (Na) salts do not cause any hardness. All these salts that cause hardness come from the rocks. These salts get dissolved in the groundwater by way of rainwater percolating through the rock.

Hardness and action of soap with water

Ordinary soaps are essentially sodium and potassium salts of organic fatty acids such as stearic acid, palmitic acid, and oleic acid. When soap is added to hard water, calcium and magnesium salts dissolved in water reacts with the salts in the soap and form an insoluble scum. The reaction occurs as follows:

Na/K stearate/palmitate + Ca/Mg salt $\rightarrow$ Ca/Mg stearate/palmitate + Na salt

(Soap salt)   (Hard water salt)   (Insoluble scum)

Calcium and magnesium salts of chlorides, bicarbonates, and sulphates in hard water react with soap, forming their organic salts, which are insoluble in water. These insoluble salts come out of the soap solution as thick scum and rise to the surface of water. No soap in the mixture (mixture of hard water and soap) will form lather till all the salts present in the hard water are removed as scum. Soap is, thus, consumed without producing lather needed for cleaning clothes, resulting in a great wastage of costly soap. Synthetic detergents largely overcome this problem as they form soluble salts of Ca and Mg and do not form a scum or soap curd.

6.2.4 Types of Hardness

Depending upon the specific nature of the dissolved salts, hardness may be further classified as

1. Temporary or carbonate hardness
2. Permanent or noncarbonate hardness

Temporary (or carbonate) hardness

It is caused by the presence of dissolved bi-carbonates of calcium [Ca $(HCO_3)_2$].

Permanent hardness

It is caused by the presence of chloride and sulphate salts of calcium and magnesium. These salts have the chemical formulae of

- $CaCl_2$ (calcium chloride)
- $CaSO_4$ (calcium sulphate)
- $MgCl_2$ (magnesium chloride)
- $MgSO_4$ (magnesium sulphate)

6.3 | REMOVAL OF HARDNESS AND WATER SOFTENING

The reduction or removal of hardness from water is known as water softening. It is not essential to soften water in order to make water safe for public use. The advantages of softening lies chiefly in the reduction of soap consumption, lowered cost in maintaining plumbing fixture and boiler tubes, improved taste of food preparation, and improved heat transfer in equipment such as boiler and water heater, and saving energy.

The permissible hardness for public supplies normally ranges between 75 to 115 mg/L (14.25 mg/L is equivalent to one degree of hardness). Figure 6.1 shows the different methods of removing different types of hardness in water.

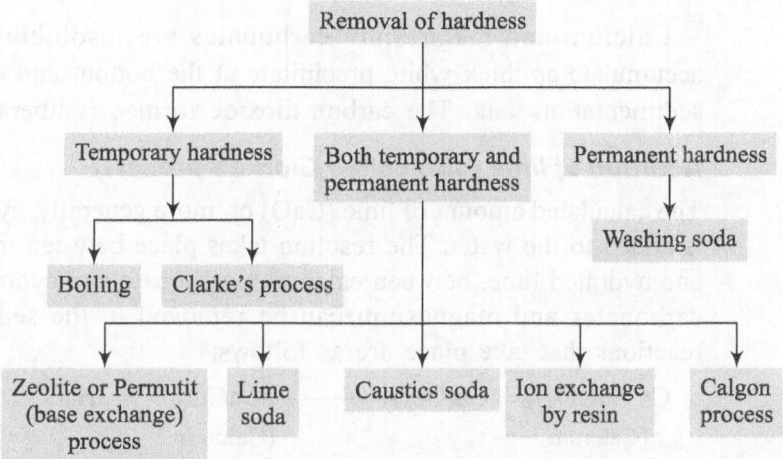

Fig. 6.1 Different methods for removal of hardness

6.3.1 Method of Removing Temporary Hardness or Carbonate Hardness

The temporary hardness of water can be removed either by boiling or by adding lime to the water. This is called temporary hardness because the hardness is removed simply by boiling it. We now discuss both the methods.

Boiling

Rain water when falling absorbs carbon dioxide (CO_2) in air. Calcium carbonate inside the earth, being only slightly soluble, will usually exist in water as calcium bicarbonate because it easily dissolves in the water containing CO_2. When such water is boiled, the CO_2 gas will get out, while insoluble $CaCO_3$ comes out of water as a thick white precipitate, which can be sedimented out in the settling tank. The down arrow sign ($\downarrow$) by the side of a chemical product denotes that it is insoluble in the solution and would form a precipitate and can be removed from the solution (water in this case) by filtering or decantation process. The up arrow sign ($\uparrow$) by the side of a chemical product denotes that it is a gas and goes out of the system as a product of the chemical reaction.

$$Ca(HCO_3)_2 + Heat \longrightarrow CaCO_3 \downarrow + CO_2 \uparrow + H_2O$$
(Calcium   (Calcium
bicarbonate)   carbonate,
                insoluble)

$$Mg(HCO_3)_2 + Heat \longrightarrow MgCO_3 \downarrow + CO_2 \uparrow + H_2O$$
(Magnesium
bicarbonate,
insoluble)

Calcium and magnesium carbonates are insoluble in water, and hence accumulate as thick white precipitate at the bottom and can be removed in the sedimentation tank. The carbon dioxide formed is liberated as gas.

Addition of lime (also called Clarke's process)

The calculated amount of lime (CaO) or, more generally, hydrated lime [$Ca(OH)_2$] is added to the water. The reaction takes place between magnesium bicarbonate and hydrated lime, between calcium bicarbonate and hydrated lime. The calcium carbonates and magnesium can be removed in the sedimentation tank. The reactions that take place are as follows:

$$Ca(HCO_3)_2 + Ca(OH)_2 \longrightarrow 2CaCO_3 \downarrow + 2H_2O$$
(Calcium   (Calcium
bicarbonate)   carbonate,
                insoluble)

$$Mg(HCO_3)_2 + 2Ca(OH)_2 \longrightarrow Mg(OH)_2 \downarrow + 2CaCO_3 \downarrow + 2H_2O$$

6.3.2 Methods of Removing Permanent Hardness

Permanent hardness can be removed by certain special methods, generally called water softening methods. The basic principle of softening is the same as the removal of temporary hardness, i.e., converting soluble Ca and Mg salts in hard water into corresponding insoluble salts and removing the resultant precipitate from water by filtration or decantation. The six methods, which are commonly used for softening permanent hardness, are as follows:

1. Washing soda process
2. Base exchange process called zeolite process
3. Lime soda process
4. Caustic soda process

5. Ion-exchange process or demineralization (DM) process
6. Calgon process

Washing soda process

The chemical formula of washing soda is Na_2CO_3. The reactions that take place on adding washing soda solution to hard water are as follows:

$$Na_2CO_3 + CaSO_4 \longrightarrow CaCO_3 \downarrow + Na_2SO_4$$

$$Na_2CO_3 + MgCl_2 \longrightarrow MgCO_3 \downarrow + 2NaCl$$

Zeolite (or permutit) or base exchange or cation exchange process

Hardness can be very effectively and economically removed by using a chemical called zeolite. Zeolite (a green salt generally termed as green sand) is a naturally available salt ($Na_2O. Al_2O_3. 4SiO_2. 2H_2O$), which reacts with salts in hard water and forms insoluble salts of Ca and Mg. This is also manufactured synthetically and the white coloured salt is sold under the name of 'Permutit' which is, chemically, hydrated sodium aluminium orthosilicate, having the formula $Na_2[(AlO_2)_2 (SiO_2)_2]$, xH_2O.

Permutit is manufactured from naturally available feldspar, kaoline, clay, and soda. In chemical equations, we use the short form of the chemical as sodium zeolite (Na_2Z) where Z represents $[(AlO_2)_2 (SiO_2)_2]$. Hence, the process of removing hardness by using this chemical is known as zeolite process as well as 'permutit process'. Any salt has a positively charged part (or cations) and negatively charged part (or anions). Usually, the metallic part of the salt is cation (also called base). For example, in calcium carbonate, calcium is the base or cation and carbonate is the anion. These zeolities have the excellent property of exchanging their cations (sodium) with those (calcium and magnesium) in hard water, which enables them to perform the softening operation. The sodium ions of the zeolite get replaced by the calcium and magnesium ions present in hard water, forming insoluble calcium zeolite and magnesium zeolite.

Softening reaction

$$\underset{\substack{\text{(Sodium zeolite/} \\ \text{or active zeolite)}}}{Na_2Z} + \underset{\substack{\text{(Hard water} \\ \text{salt)}}}{\text{Ca or Mg salt}} \longrightarrow Na_2 \text{ salt} + \underset{\substack{\text{(Exchanged} \\ \text{or used zeolite)}}}{\text{Ca or Mg zeolite}}$$

After a few days of operation, all the active sodium zeolite should change to used calcium and magnesium zeolite. However, calcium and magnesium zeolite can be converted back into active sodium zeolite by reacting it with 10 per cent solution of sodium chloride (common table salt solution). This process is known as regeneration. The exchange reaction that takes place during regeneration can be represented as follows:

Regeneration reaction

$$CaZ/MgZ \quad + \quad 2NaCl \quad \longrightarrow \quad Na_2Z \quad + \quad CaCl_2/MgCl_2$$

(Used zeolite)   (Sodium chloride   (Regenerated   (Dissolved salt
solution)   zeolite)   removed by
flushing water)

Working of a zeolite softening plant   A zeolite softener (or a cation exchange unit) resembles a sand filter in which the filtering medium is a zeolite rather than sand, as shown in Fig. 6.2. The hard water enters through the top as shown in the figure and is evenly distributed on the entire zeolite bed. As the hard water passes through the bed, reactions as given above take place and the softened water is collected through the strainers at the base. The zeolite bed is increasingly depleted in its active sodium zeolite content with increasing proportion of Ca and Mg zeolites. When a significant portion of the sodium in the zeolite has been replaced by calcium and magnesium, softening process is stopped by stopping flow of hard water and regeneration operation is started first by washing the bed with good soft water and then passing 10 per cent solution of brine through the bed. The excess brine solution retained in the zeolite, as well as the soluble Ca and Mg chloride salts removed from the activated zeolite bed, after the treatment, are removed by again washing them with good soft water. The regenerated zeolite is now ready and can be used afresh for softening.

The rate of filtration through a zeolite softener is about 300 litres per square metre per minute.

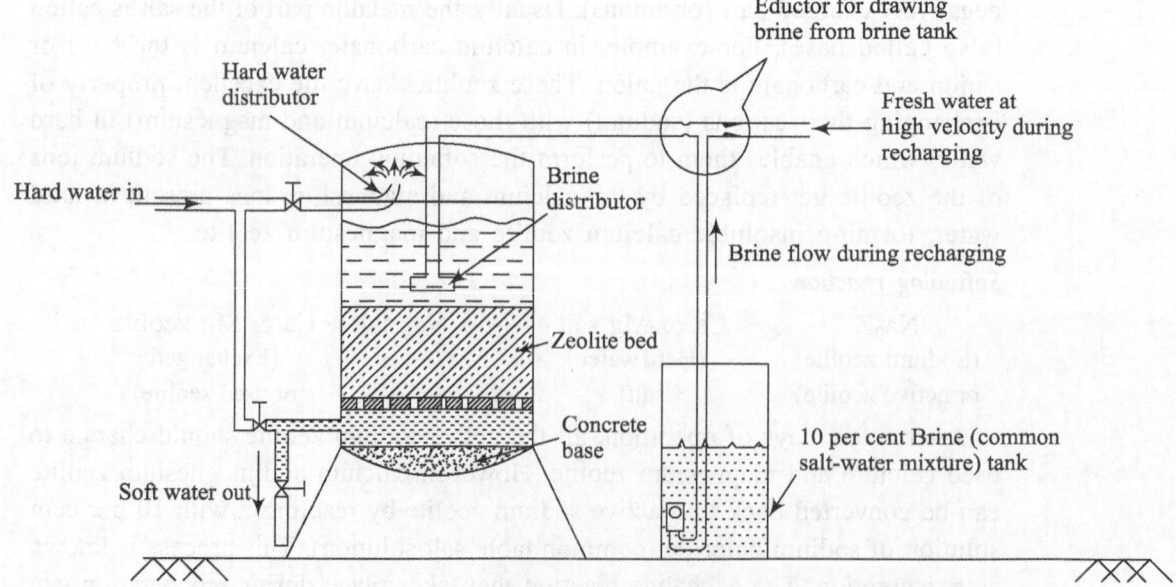

Fig. 6.2   Schematic diagram of case-exchange zeolite water softener

Advantages of zeolite process The advantages of zeolite process are as follows:

- Zero hardness can be obtained and have specific uses in textile industries boilers, etc.
- The plants are compact, automatic, and easy to operate.
- The running, maintenance, and operation (RMO) cost is quite less.
- It also removes iron and manganese from water.
- There is no problem in treating water of varying quality.

Lime soda process

In this process, in addition to washing soda, lime is added to remove temporary hardness. Hydrated lime reacts with bicarbonates of Ca and Mg, as already outlined.

Caustic soda process

Caustic soda can be used to remove both carbonate (temporary) and non-carbonate (permanent) hardness. Calcium and magnesium hydrogen carbonates react with caustic soda to produce insoluble calcium carbonate and magnesium hydroxide.

$$2NaOH + MgSO_4 \longrightarrow Mg(OH)_2 \downarrow + Na_2SO_4$$

$$2NaOH + CaCl_2 \longrightarrow Ca(OH)_2 \downarrow + 2NaCl$$

$$4NaOH + Mg(HCO_3)_2 \longrightarrow Mg(OH)_2 \downarrow + 2Na_2CO_3 + 2H_2O$$

$$2NaOH + Ca(HCO_3)_2 \longrightarrow Ca\,CO_3 \downarrow + 2Na_2CO_3 + H_2O$$

From the above reactions, we find that this process not only removes permanent hardness but also temporary hardness. This process is very efficient for low alkalinity water.

Ion exchange by resin or demineralization process for removing hardness

There are many pieces of equipment in industry, such as water boiler and other process equipment and scientific apparatus, which would need water free of any minerals. The process by which the minerals are removed is known as demineralization (DM) and the plant is known as DM plant. Since this process helps us in completely removing or reducing the mineral content to any desired extent, it is very suitable for producing water of any desired hardness or even mineral free water. The demineralized water is sometimes called deionized water and is as pure as distilled water.

The process consists of passing the water through cation exchange resins, which produce almost similar effects as are produced in the zeolite method, except that hydrogen (instead of sodium) is exchanged for the basic metallic ions. The cation exchange resins in fact are phenol aldehyde condensation products.

Their chemical formula may be represented by H_2R, which represents the hydrogen ion and R represents the organic part of the substance.

$$Ca\,(HCO_3)_2 \quad + \quad H_2R \longrightarrow CaR \quad + \quad 2H_2O \quad + \quad 2CO_2 \uparrow$$

(Fresh cation   (Exchanged

exchange resin)   or used resin)

$$Na_2CO_3 + H_2R \longrightarrow Na_2R + H_2O \quad + \quad CO_2$$

Thus, we find that these cation exchange resins exchange sodium ions also, which come out of the water forming exchanged resin (Na_2R), making water free from minerals. However, absolutely pure water can be very corrosive if it comes in contact with air, thus absorbing oxygen from it as this water has a strong affinity for oxygen.

Calgon softener

Calgon is the trade name of a complex salt, sodium hexametaphosphate $(NaPO_3)_6$. It is used for softening hard water. Calgon ionizes to give a complex anion, which subsequently combines with Ca and Mg ions in hard water.

$$(NaPO_3)_6 \text{ or } Na_2(Na_4P_6O_{18}) \longrightarrow 2Na^+ + \quad Na_4P_6O_{18}^{2-}$$

(Complex anion)

The addition of Calgon to hard water causes the calcium and magnesium ions of hard water to displace sodium ions from the anion of Calgon according to the following reaction:

$$CaSO_4 + Na_2\,(Na_4P_6O_{18}) \longrightarrow Na_2SO_4 + Ca\,(Na_4P_6O_{18})$$

This results in the removal of calcium and magnesium ions from hard water in the form of a complex compound with Calgon. The water is, thus, softened. Sodium salts are released into water without causing any hardness. It is being dosed into water used for washing machine, dyeing work, etc.

6.4 | WATER AND COOKING

Cold water should be used for drinking, cooking, and especially for making baby formula. Hot water is likely to contain higher levels of lead. However, nowadays lead pipes are seldom used for water plumbing work.

The hardness of water, as determined by the presence of minerals, such as calcium and magnesium, can have a considerable effect on cooking. For example, cooking of many food items, including meat, with hard water can impart a bitter, astringent taste to the food due to the presence of magnesium in hard water. Soft water has a mild, sweet taste, and, more importantly, when used for cooking, helps bring out the flavour and aroma of the ingredients used. However, this is an area of active research and hard water has, of course, important minerals that could be vital for children.

Further, when fruits and vegetables are cooked in water, the amount of calcium ions in the water influences the textural properties of the products. For example, calcium ions may form insoluble salts (calcium pectates) that help maintain firmness in cooked fruits and vegetables. However, an excessive amount of calcium ions in the water may make the fruits and vegetables excessively tough, and dried beans and dried peas will be difficult to rehydrate when cooked.

Abstracts of two relevant papers in this respect are included below.

Cowpeas were cooked in water made hard (or soft) by the separate addition of similar concentrations of certain salts ($CaCl_2$, $MgCl_2$, or $NaHCO_3$). The beans were also cooked in hard tap water and in double distilled water before and after soaking in water. Hard water caused a significant decrease in softness, led to reduced water absorption, and also decreased solids loss in the cooked product, but it increased the cooking time and discolouration of the beans. Hard water also gave rise to a significant ($P<0.05$) increase in mineral content, but it had less effect on the proximate composition of the cooked products.

(*Source*: Stella G. Uzogara et al. 2007, 'Effect of water hardness on cooking characteristics of cowpea (*Vigna unguiculata* L. Walp) seeds' *International Journal of Food Science & Technology*, 27(1), 49-55, Wiley Interscience).

The effects of calcium sulphate (at levels typical of 'hard' and 'soft' water supplies) in the cooking water on losses of pectic substances from potato, carrot and swede(yellow turnip) after boiling and pressure cooking were also investigated. Water hardness did not influence losses during pressure cooking but samples boiled in hard water had decreased losses of pectic substances compared to those boiled in soft water. The increase in retention of pectic substances due to hard water was about 8 % for potatoes (cv. Pentland Dell), about 33 % for sliced carrots (cv. Berlicum, approved maintenance Perfecta) and about 23% for diced swedes (cv. Acme). The interaction of water hardness with cooking method was significant for potato ($P<0.05$) and swede ($P<0.001$) and approaching significance ($P=0.055$) for carrot.

(*Source*: Donald E. Johnston et al. 2006, 'Losses of pectic substances during cooking and the effect of water hardness' *International Journal of Food Science & Technology*, 34(7), 733-736, Wiley Interscience).

The chemical identity of the various pectic substances is not definitely settled. The pectic substances play an important role in plant life. The primary function of the pectic substances is the cementing together of the individual cells that compose the plant. The pectic substances are found in the leaves, bark, roots, tubers, stalks, and fruits of plants. In fruits, the pectin is usually found in the pulp and not in the juice, though there are some exceptions. For example, currant juice often contains pectin. The skins and cores of fruits, such as apples, contain large proportions of pectin. Some of the root stocks, such as sugar beets, carrots, rutabagas, and turnips, contain appreciable amounts of pectin.

6.5 | CONCLUSION

It may be concluded that though hard water does not pose any health problems, it is definitely very costly to use for laundering, as consumption of soap and detergent is very high. Cooking with soft water in many cases provide a better taste to some food items. Boilers and other heating equipment necessarily require soft water for efficient functioning. There are many standard methods for softening of water and the choice of a particular plant depends upon the ground conditions and requirements prevailing.

KEY TERMS

Demineralization	It is process of removing the mineral bases from water to make it fit for use in boilers, etc.
Hard water	Water that does not easily form lather with soap is called hard water.
Permanent hardness	It is the hardness caused by dissolved chloride and sulphate salts of calcium and magnesium and can be removed by adding chemicals.
Permutit	It is a synthetic chemical similar in action to natural zeolite and is made from feldspar, kaoline, clay, and soda. It is used extensively in removing hardness of water.
pH value	pH value of water indicates the number of free or active hydroxyl (OH+) or more precisely hydronium (H3O+) ions in a solution.
Soft water	Water that readily forms lather with soap is called soft water.
Temporary hardness	It is the hardness that is easily removed by boiling and is caused by bicarbonates of calcium and magnesium.
Zeolite	It is a naturally available chemical used for removing hardness.

REVIEW QUESTIONS

6.1 Explain what is meant by hardness of water.

6.2 Explain soft water and hard water and state their relative advantages and disadvantages.

6.3 Classify different types of hardness and state the methods of removing them.

6.4 Explain 'zeolite' process of removal of hardness.

6.5 Draw a neat sketch of a zeolite softening plant and label the components.

6.6 State the necessities for a demineralization plant in a hotel and describe in brief its working principle.

6.7 Discuss the effect of hardness of water on food quality.

7
Chapter

Water Distribution System

Learning Objectives

In this chapter, the students are introduced to both cold and hot water distribution system used in the hospitality industry. Waste disposal and sewerage system are also briefly discussed. The primary objectives of this chapter are to:

- introduce cold- and hot-water distribution systems used in the hospitality industry
- discuss the solar water heating
- give a brief description of a few common pipe fittings and pipe materials

After reading the chapter, the readers should become familiar with general systems and working of water distribution and sanitary system.

7.1 | INTRODUCTION

Water is a fundamental necessity—for basic day to day life and agricultural processes as well as for various industrial processes. It is essential in any establishment—domestic, industrial, commercial, or hospitality—and has diverse uses. Water services is one of the most important services in the hospitality industry. In hospitality industry, water is used in a number of ways starting from supplying bacteria-free clear drinking water, hot and cold water in bathrooms and toilets in boarders' room to water in kitchen for cooking purposes. Water is needed in laundry, boiler (if any), for staff requirement, for cleaning floors and articles, for fire sprinkler system etc.—water is required in almost all spheres of hotel activities. There is, thus, a cold-water and a hot-water distribution system comprising pumps, piping, valves, fittings, and other related units such as boiler, hot-water calorifier, etc. In addition, water lines are also used to collect and transport wastewater, storm water, and sanitary waste out of the premises into the municipal main sewerage or to the waste treatment plant inside the premises for recycling. The topics of wastewater collection and sewerage system are taken up in Chapter 8. The selection of proper plumbing equipment and materials of construction is crucial for continuous service of water and maintenance of the highest level of hygiene in the premises. We discuss all these aspects, in brief, in subsequent sections.

7.2 | COLD WATER—SUPPLY, STORAGE, AND DISTRIBUTION

Usually, in cities and towns, water is supplied by civic bodies, where public water lines are available, near the premises. It is required to have a storage reservoir usually on top of the building. The civic body charges the establishments for water consumption either lump-sum or through recoding of consumption by water meters. But many hotel establishments, particularly in isolated areas, find it economical and convenient to have their own borewell pumps within the premises that provide raw water, which is further appropriately treated for consumption. But indiscriminate lifting of groundwater poses a serious threat to the reserve of ground water, which is fast depleting, as a result, in many places. This is a matter of serious concern and permission must be obtained from government agencies for digging such wells. A hotel establishment needs a lot of water for various essential functions. An average figure will be 200 litres (about 50 gallons) per person per day.

Cold water is used for diverse functions in a hotel, which include the following:

- Drinking
- Cooking in the kitchen
- Kitchen, restaurant, and lavatory wash sinks and washbasins
- Lavatory flush
- Laundry
- Estate and floor cleaning purposes
- Fire-sprinkler system
- Cooling the diesel generating set, refrigeration plant, etc.
- Hot-water and boiler make-up water
- Gardening, etc.
- In swimming pool and other water bodies, if any.

In India central-room heating is not required and hotels normally do not use such systems. As such, hot water is used for laundry purpose and kitchen, throughout the year and for personal use of the guests, particularly during winter season in hotels in the plains and throughout the year for hill station hotels.

Water quality requirements are different for different uses in the hotel industry. For example, water lines for drinking, kitchen, and wash basin/sink must be supplied with bacteria-free soft water, while laundry and other functions can work with plain soft water. In modern establishments, however, laundry line is connected to the same line as the drinking water line. However, lavatory wash basin and water closets (WCs) may be provided with only soft water. In many modern installations, wastewater is collected, treated, and recycled. The basic source of water supply can be deep tube well sunk within the premises or tapped directly from the corporation/municipality, as already stated. There may be a combination of both as well. Rain-water harvesting has assumed great importance and in many establishments this may supplement the main water supply system. We discuss these systems as follows.

7.2.1 Treated Water Supply from Municipal Corporation Source

We first discuss a system where continuous supply of drinking water is available from the corporation. The most general components of the cold-water distribution system are the following:

1. Rising main
2. Storage cistern
3. Discharge piping
4. Stop valves

Rising main A supply pipe line, called the 'rising main', rises from ground-level to the storage cistern on the rooftop. For high-rise buildings, there can be intermediate booster pumps and multiple storage cisterns for serving a group of floors. At the lowest point inside the building, there should be a 'stop valve' and drain-off tap; the latter enables the pipe to be emptied for repairs.

Storage cistern The water is stored in the storage cistern through the rising main. The storage cistern is situated at the highest point in the building (at different levels in case of high rise building), to ensure a reasonable pressure in the distribution pipes.

Discharge or service piping Water for drinking, kitchen work, and wash basins is directly taken off the rising main by branch pipes at different floors and flows into the service taps. Each floor branch and sub-branches have stop valves at suitable locations so that maintenance work can be carried out at some particular place without affecting supply to any other guestroom or kitchen. From the cistern, water is distributed throughout the building by means of 'downcomer pipes' to cater to the needs of flush water and laundry work. They also serve boiler and hot water make-up. Stop valves should be fitted immediately after the exit from the cistern and also at each branch point to isolate the downcomer pipe from the cistern as well as individual branches from the main downcomer line. The branches, as usual, lead to the service taps and other appliances at each floor.

Stop valves   Stop valves are used to completely open or close the water lines and they are usually screw-down type. Closure of this valve in the draw-off lines from the storage cistern allows maintenance work to be carried out in cold-water lines and fittings downstream.

7.2.2   Storage Cistern

We now discuss the storage cistern in some detail. A storage cistern will have the following major components for functioning:

1. A ball valve at the end of the inlet pipe that opens out when the water level falls and closes as the water level rises to a pre-set value.
2. A silencer pipe is fitted at the end of the inlet pipe. This is a short pipe attached to the inlet to lower the point of discharge a little above the bottom of the cistern. This reduces noise of water while falling in the cistern.
3. An overflow pipe is needed if the ball valve does not operate satisfactorily so that the water comes out through an overflow pipe. When such a condition occurs, the water line maintenance and operation crew would close a stop valve in the rising main just before the cistern inlet, manually, to prevent wastage of water. It should have a hinged flap at the end to prevent insects and dust from entering the cistern.
4. Cisterns must have a close fitting lid (but not air tight) to keep out dust, rodents, birds, tree leaves, etc. from contaminating water. They are opened for regular periodical cleaning of the cisterns.

Whenever the liquid level in the cistern rises to a level set by the user, the float, which rises as the level rises, pushes a rod inside the ball valve, which closes the valve and the water flow stops. The cistern should be well protected from contamination and sabotage. The size of the cistern should be decided based on the consumption pattern and maximum expected number of guests and employees and the capacity of restaurants. Its capacity should be sufficient for one day's consumption, calculated on the basis of 15 litres per head for hotels and 8 litres per meal in restaurants, considering a hot tropical country like India.

Maintenance of the cistern

If the ball valve fails to shut off the flow of water completely, the cause is likely to be one of the following:

- Worn-out washer
- The arm carrying the ball being in need of adjustment by slight bending

- A punctured seat
- A pitted value seating
- A low-pressure valve being used instead of high-pressure valve

Water supply from own source

However, the circuit will be a little different when the establishment has its own deep well-pumping system. All the wastewater including semi-solid waste from privy are collected and treated in the in-house sewage treatment plant (STP), which again generates fresh water and this fresh water is recycled for the same services. Water from the treatment plant is fed to a common header pipe. From these header pipes, a group of rising mains leads up to the top floor. At each floor, a pair of rooms is served by two branch pipes taken from a particular rising main. Suppose there are three floors in a hotel with each floor having twenty guest rooms, there will be ten drinking water rising mains going straight up. Each of the rising mains will have two branch lines for two rooms on a particular floor. Kitchen cooking lines are also served by treated water in this line via branches, taken off the rising main at the kitchen floor level. Water from STP is also pumped through its own common manifold and, as in drinking water line, each floor will be served by branches taken off the rising mains to supply water to flush lavatories. Laundry is also served by this water-supply main. Return line of all basins and WCs and other waste lines join to, say, ten downcomer pipes served by two return lines from each guest room in a floor, in a similar fashion. Water in waste return lines is then collected and treated in STP for onward journey.

Rain-water harvesting system

In recent years, with the cost of civic-body water going up and restrictions on lifting of water from underground sources, rain-water harvesting has become important, and a duty of the citizens, and government is providing incentives for such an activity. A hotel has a large roof and open space, which permits collection of rain water. Rain water thus collected is either recharged to the ground or used in-house. As such rain water contains some physical and chemical impurities, which make it unsuitable for direct use as potable and service water, unless it is treated. This rain water may join the main water supply input to the treatment plant or it may be separately treated and used separately.

Water lines are often served with booster pumps to provide sufficient pressure. In many situations, two parallel water lines run, each served by pumps with two different pressure heads, one usually at a pressure of 6 atm and the other at 3 atm. If a pump generates water flow with about a pressure of 1 atm head of water, the water with normal flow velocity can approximately push up to a height of 8 to

9 m straight up (two-three storey building). Pushing it though different pipe fittings such as valve, coils, bends, etc. needs extra pressure. Further, there will be energy loss in transit to force the water through pipes due to friction. All these make up the pressure requirements for a particular layout and application. Higher pressure lines are required for bath showers, fire sprinkler, and for reaching to higher floor services, in general. Different service taps need different flow rates and the required flow rate can be obtained by suitably selecting the particular water line and the type of valve/tap, etc. at the service point. Sometimes a pressure-reducing valve may have to be fitted in the line before the particular appliance to reduce the flow rate to the desired level. Table 7.1 furnishes typical flow-rate requirements for various plumbing fixtures in a hotel.

Table 7.1 Typical flow rates for fixtures in litres per minute (lpm)

Fixture type	Kitchen sink	Pastry sink	Laundry sink	Dish-washer	Lavatory	Bathtub	Shower
Flow rates	4–6	6–8	8–11	4–11	2	8–16	6–12

7.2.3 Swimming Pool

Swimming pools are normally installed in clubs, hotels, and resorts, some schools and colleges, and also in some private-lodging properties. Different aspects of a swimming pool such as design, construction, and maintenance are subject to local public-health regulations. Although some of the guests may not use the pools in a hotel, such pools enhance the status of the property and in some cases a must for accreditation for a particular star category.

Swimming pool construction

There are many forms of construction of a swimming pool. These are as follows:

- The most durable and also the costliest are the steel-reinforced concrete pools. However, during winter time to avoid cracking due to freezing of water, the pool has to be either emptied or the water has to be heated. Figure 7.1(a) shows such a pool.
- Then there is the semi-concrete type. It has a concrete base and stainless-steel sidewalls.
- Vinyl-line pools, which have dirt-packed bottom, covered with vinyl or plastic liner and stainless steel sidewalls. Figure 7.1(b) shows a vinyl pool.
- The overground container-type swimming pools have a plastic bottom and stainless-steel sidewalls. This is least costly and reasonably durable. Figure 7.1(c) shows one such swimming pool.

Fig. 7.1 (a) A common steel-reinforced concrete-type swimming pool

Fig. 7.1 (b) A swimming pool with vinyl base

Fig. 7.1 (c) An overground container-type swimming pool

Pools can be built in many shapes and sizes. A very good size could, e.g., be 35 ft × 70 ft × 12 ft deep.

There is a variety of options available as to the layout, design, construction, depth, surface area, etc. Ideally, they should depend upon the budget, the approximate number of people using the pool at a time, and location of the pool within the property.

Strict regulations are to be followed and displayed for the users, as regards the use of the pools. It is always advisable to have swimming pool guards having certified life-saving training.

Water supply and maintenance of pools

Filtered water must be used in the pool. This keeps water free from dirt and makes it transparent and sparkling. Both cartridge filters and sand-gravel type filters are used. While the former is costlier, the latter type takes longer time to filter due to lower flow rate of water through such filters. After a specified period of time, usually as recommended by local public health authority, entire water content should be cycled through the filter. After a number of cycles of filtration, the filters accumulate the dirt they separate out from the water and may be clogged. To ensure proper functioning, the filtration equipment must be cleaned periodically according to a maintenance schedule. The process is known as backwashing, whereby fresh water is pumped through the filters in a reverse direction and the dirty wastewater at the outlet is drained into the wastewater pipe joining the premises', drainage system.

Leaves, insects, and debris all collect in the water. Some of them float, while some go to the bottom. There are many ways to remove them. Different types of nets are used to remove floating objects and also large objects near the bottom. Soil and debris collected at the bottom may be manually swept towards the pool drain hole at the bottom. In some arrangements, the pool water is continuously kept in motion to make them mix in the bulk and made to pass trough the filter during the filtration cycle and trapped in the filter. The swimming pools have an opening at the water surface level, leading to the drain at the outside, which canalizes the water to the filter plant. This opening is called weir and a strainer is placed at this opening which arrests the objects floating in water, which is subsequently removed.

Chemicals are added to the pool water to maintain the proper hygienic conditions. Disinfectants are added to make it bacteria-free. The dosing can be done manually or may be automated. To maintain the correct level of acidity/alkalinity appropriate chemicals are added. The pH level is usually kept between 7.6 and 7.2 (both being slightly alkaline; remember, pH 7 is neutral). Algae, which are plants that grow in water, are also very difficult to completely eliminate. However, disinfectants added to water also kill algae. If their growth becomes excessive, they have to be manually removed.

The initial water as well as the make-up water for swimming pool comes from the central water-treatment plant for a hotel.

7.3 | HOT-WATER GENERATION AND DISTRIBUTION

Hot water is necessary in hotel for the boarders for their personal use for bathing and washing during cold season, for laundry purposes, and also for kitchen.

Cold water raised to a temperature of 100°F (38°C) is usually considered hot. Energy conservation and safety are key factors in selecting the temperature range of hot water being used. Most often, the operating temperature range is between 120°F (49°C) and 160°F (71°C), while in most of the cases, temperature does not exceed 140°F (60°C), which is the tolerance level of an average human being. This temperature is also normally sufficient for automatic clothes washer and dishwasher use.

In the hotel industry, hot-water is generated and supplied by the following two systems:

1. Central hot-water generation and distribution system and
2. Localized hot-water generation and distribution system

In hotels of small size, the second method is more popular, while large- and medium-sized hotels prefer the first system due to per capita less cost of generation and supply. As a thumb rule, maximum daily demand of hot water in a hotel will be 115 litres (about 25 gallons) per person and a storage requirement of 45 litres (about 10 gallons) per person.

7.3.1 Central Hot-water Generation and Distribution System

Large hotels would usually employ a central hot-water generation and supply system. There are three different principles by which hot water is generated in central hot water generation and supply system, namely

1. Direct heating system where hot water is produced directly in the boiler
2. Indirect system of hot water
3. Central solar water heating system, employed by many hotels in India and abroad as a supplement to existing systems of hot-water generation and in many cases slowly replacing the existing system

Direct heating system for hot-water supply

In such a system hot water from a boiler circulates through a hot water tank called cylinder. Water pipes inside the boiler may be heated by firing coal or more often by burning high-speed diesel. Circulation continues because hot water being lighter rises up to the cylinder from the boiler and relatively cold water descends down to the boiler. This type of circulation is called natural circulation. Service pipelines are taken from the cylinder to the service taps. So, whenever taps are opened, hot water flows to them. As water is consumed at the taps, it must be replenished and a cold-water cistern is connected to the cylinder for the purpose, from where a cold-water feed line runs into the cylinder via a stop valve. Further, whenever water gets heated up, it increases in volume and if it does not have any space to accommodate this increase in volume (e.g, when all the service taps are closed), the resulting pressure rise would lead to a possible bursting of pipelines. To take care of this fact, a pipeline, called expansion pipe, is taken from the draw-off line at the top of the cylinder opening into the overhead cistern. In this arrangement the same water circulates through boiler, cylinder, and the service tap lines.

Although the arrangement is simple and attractive, it suffers from a number of disadvantages. They are as follows:

- Heating of water liberates oxygen from cold water. Therefore, when cold make-up water from the cistern enters the boiler and is heated, continuous liberation of oxygen occurs and this would cause rusting of boiler tubes, leading to problems of maintenance and heat transfer.

- Unless the cold water is demineralized, there would be severe solid scale formation inside the tubes of the boilers, grossly affecting heat transfer and water-carrying capacities of the lines. On the other hand, demineralization is a costly process and as such there is no such requirement for the rest of the systems like cylinder and water lines to the service taps, but, since, as explained earlier, the same water passes through all the units they also receive demineralized water and, thus, the whole volume of hot water supply is to be demineralized, which is otherwise not needed if an indirect heating system is used.

- If the service taps are situated far away from the hot water cylinder, stagnant water in the line up to the particular tap may become cold, if not insulated properly and when the taps are opened, for some time, cold water will flow till fresh hot water reaches it from the cylinder. This results in wastage of water and delay in delivery of hot water.

- To assist natural circulation, the calorifier is placed a little above the boiler.

To overcome the problem of wastage of water and delay in hot water supply, a secondary circulation may be devised in the cylinder, which ensures hot water always running through the loop by natural circulation. Any service tap when opened would, by this arrangement, provide instantaneous supply of hot water.

As such, direct system of heating is used only for small consumption requirement and is employed in small hotels. They are also less costly than their indirect heating system counterpart.

Indirect heating system

The problems of direct heating system are overcome in the indirect-heating system. In this system, the cylinder is replaced by a heat exchanger called calorifier where hot water/steam from the boiler enters the calorifier through a coil of tube and heats the return line service water and cold make-up water passing through the calorifier. Thus, boiler water line and service water line are separate and do not mix and form two separate circuits of water circulation. These two circulations are as follows:

- Primary circulation that takes place between the boiler and the calorifier
- Secondary circulation that takes place between the calorifier and the service taps

Primary circulation This is the path taken by the heating water/steam generated in the boiler that goes into the calorifier, heats the water for distribution, and after consumption at different taps, becomes relatively cooler by giving away heat and comes back to the boiler for reheating and the cycle continues.

Secondary circulation The secondary circulation takes place between the calorifier and the taps. Water, after being heated by boiler steam/hot water, leaves the top of the calorifier, passes through various shafting, much in the same layout as cold-water supply lines round the building, close to the various fittings, and goes back to the calorifier at a point lower down. Short branches run from this main pipe to the various taps. We compare the cold-water system and observe that the cold-water line vertical shafts rise straight to each floor, where branches are taken to individual pair of rooms. The main rising shaft terminates at the top without going back. Here also the same arrangement could be made, but in that case, the stagnant water in the vertical rising main would become cold if a particular group of taps are not used and when they are opened, one would get the cold stagnant water for some time before hot water from the calorifier would reach it. Return of the rising main to the calorifier ensures that hot water is always flowing in the supply main and it is available whenever a tap connected to the supply branch is opened.

So, in this arrangement, the primary circulation and the secondary circulation are physically separate. As water gets consumed, the water content of the calorifier decreases and it is necessary to replenish this by additional water called make-up water. Thus, the make-up water for the calorifier for replenishing consumed water does not pass through the boiler tube and, thus, ordinary soft water can be used in the make-up cold water storage cistern, while demineralized water is used for make-up for the boiler water, which will be minimal unless the boiler is a centralized boiler catering to the need of generation steam and heating water for central-room heating. Secondary circulation ensures instant hot water supply at service taps. In many cases, particularly for large establishments, such as hotels, etc., the secondary circulation is aided by a pump, which not only enhances heat exchange between boiler hot water/steam and calorifier water inside the calorifier, but also ensures good delivery pressure at the service taps. The pumping arrangement also eliminates the restriction on the placement of the calorifier. Most often, we do not need a separate boiler exclusively for such use and as indicated earlier, the boiler used for producing steam and hot water for central heating system for room heating in hotels and buildings in cold regions can be utilized to share a part of its heating area to heat the coils carrying the hot water for the calorifier.

From the preceding discussion, it is apparent that indirect system of hot water supply offers better flexibility of boiler use and more control on operation of hot water distribution system and easier maintenance of boiler tubes.

A stop valve is fitted in each cold water feed downcome pipe near the exit from the cistern, while a drain cock is fitted at the end of the pipe. Stop valve is used for closing the main supply line for any maintenance work in the pipe lines and other fittings, while the drain cock is opened to drain off water from the cistern for maintenance of the cistern. Stop valves should also be fitted to every branch pipe supplying a tap or group of taps. Such arrangement facilitates maintenance of equipment while not disturbing other parts in the system.

This system is usually employed in medium- and large-sized hotels.

Hot-water generation by solar water heater (SWH)

Today, the hotelier has various options to get hot water by using electricity, firing woods, coals, waste materials, LPG gas, and also by using solar energy. One of the cheapest and very green (environment friendly) ways to get hot water is by using solar energy. Solar water heating has the following advantages:

- Energy is available free of cost.
- There is no need for an operator.
- There is no chance of short circuit, blasting, and accidents, as in case of localized electrical storage water heating system (discussed later).
- Instant and continuous hot water supply is available.
- It is an eco-friendly system.
- Once properly installed and properly maintained, solar system has a much longer life.

Water can be heated up to 95°C during summer and up to 80°C during winter in a solar heater.

Working of solar heaters   A solar heating system typically consists of a collector-absorber of solar energy, a fluid, which in turn absorbs heat from the collector-absorber surface, and a storage tank for storing and re-circulating hot water. While there are many ways of classifying solar heaters, we group them according to the following ways:

- Mechanism of fluid circulation through the collector-absorber: thermosyphonic (or natural) circulation or pumped (or forced) circulation, also termed as passive and active systems, respectively.
- Mechanism of heating water: direct or indirect, also called open loop and closed loop systems, respectively.
- Type of collector-absorber: flat plate solar collector or vacuum tube-type solar collector.

Figure 7.2 shows a solar heater which uses a flat-plate type solar collector with natural circulation and direct heating of water, for the purpose. Water circulates through a series of copper/aluminium/steel tubes kept over a blackened flat surface (hence the term flat plate) in a thin insulated metal box, covered by glazed or unglazed glass surface. The whole unit is kept under the sun on the roof, hung on an external wall facing the sun or placed on the ground depending on the conditions. The blackened surface absorbs solar radiation and gets heated up. Colder water enters the bank of tubes through a manifold, gets heated up, gets lighter, moves up, and collects into a hot water cylinder placed above the tubes and is stored there for direct draw-off (or, direct type) and water at the bottom of the cylinder becomes a little less hot due to losses and being heavier, comes down, and enters the tube again through the header and, thus, the natural circulation continues. Any draw-off from the cylinder is replenished by a fresh supply of cold water from any dedicated overhead tank, as already explained.

Fig. 7.2   Flat-plate collector based solar water heaters

(*Source:* Ministry of non-conventional energy
(MNES, INDIA) website)

Direct or open loop and indirect or closed loop systems   The transfer of heat from the solar collector may be directly to the water to be heated, which is collected in the cylinder above. However, during cold nights, if natural-circulation system prevails, a reverse flow of hot water in the cylinder to the cooler collector plates may occur. To prevent this, the indirect water-heating system is devised, where a thermic fluid (an efficient heat absorbing fluid) is circulated through the collector (primary circuit) while water flows through a heat exchanger cylinder where it gets heat from the thermic fluid and forms what is called the secondary circulation. Since the primary fluid, i.e., the thermic fluid moves in a closed circuit, it is called the closed-loop system.

Natural or passive and pumped or active circulation   Since natural circulation depends upon density difference that in turn depends upon temperature difference, which is not much in a solar heater, commercial establishments such as hotels need to force water through the collectors to get more water flow rate. For this purpose, a pump is installed in the water/fluid circulation line. In many installations, the pump is driven by electric motor that is run by electricity generated from photovoltaic cells that produces electric power from solar radiation, thus making the complete set-up fully solar in nature.

Vacuum tube collector   A flat-plate type solar collector efficiency highly depends upon atmospheric temperature. If the external temperature is low, they do not work efficiently. To overcome this problem, vacuum-tube collectors have been developed. Figure 7.3 shows such a collector. Two concentric glass tubes are used here. The space between the outer glass tube and the inner glass tube is evacuated, which allows solar radiation to reach the absorber material but prevents any heat loss from inside to escape into the outside cold climate. The outer wall of the inner tube is coated with selective absorbing material. In some design, additional copper lining is placed after the inner tube. In the inner core is placed the tube carrying the water/thermic fluid, which finally receives the solar energy by conduction from the inner hot tube or copper tube. Vacuum-tube collectors are much more expensive than flat plate collectors but are easy to maintain and offer good service round the year.

Fig. 7.3   Vacuum tube collector based solar water heater

(*Source:* Ministry of non-conventional energy
(MNES, INDIA) website, mnes.nic.in)

However, electrical back-up should be available during days of continuous rain or cloudy sky, which happen, of course, only for a few days in a year, except in a few areas in the eastern zone of the country during monsoon.

There often arises a question like 'Can a solar heater supply hot water on a large scale?'

The answer is an emphatic 'yes'. In fact, a number of solar water heaters can be connected in series or in a parallel manner to produce and supply hot water production in large-scale for commercial establishments, such as a hotel or office building, and there is practically no limit to the size of the system. However, individual bank of collectors must not have more than 150 tubes in series; otherwise, the water may boil and cause dangerously high pressure rise in the system.

Fact sheet of solar heater   Ministry of non-conventional energy (MNES), Government of India provides the following information to encourage use of solar energy for heating purpose.

- Hot water at 60°C–80°C can be very efficiently generated and used in hotels, hospitals, restaurants, dairies, homes, industry, etc.
- Solar water heaters (SWHs) of 100–300 l capacity are suited for domestic application
- Larger systems can be used in restaurants, canteens, guest houses, hotels, hospitals, etc.
- A 100 l capacity SWH can replace an electric geyser for residential use and saves 1500 units of electricity annually
- The use of 1000 SWHs of 100 l capacity each can contribute to a peak load saving of 1 MW
- A SWH of 100 l capacity can prevent emission of 1.5 tonnes of carbon-dioxide per year
- Approximate life span ranges from 15 to 20 years
- Approximate cost for a 100 l capacity SWH is around Rs 22,000 and for higher capacity systems, the approximate cost is Rs 110–150 per litre capacity
- Payback period: 3–4 years when electricity is replaced; 4–5 years when furnace oil is replaced; 6–7 years when coal is replaced.

7.3.2   Localized Hot-water Generation and Distribution System

These units are installed close to the points of consumption as in a kitchen or a bathroom and one unit caters to the need of only a few utilities located in one room. Two types of localized heat-generation systems are in use. They are as follows:

1. Storage-tank water heater (popularly known as geyser)
2. Tankless water-heating system

We discuss each of them, in brief, as follows.

Storage tank water heater

These units are compact heating units, which are normally wall-mounted at a height of about 7 ft or so. The source of heat could be electricity, gas, and oil. In India, electrically heated tanks are more common (Fig. 7.4). An electric switch/ gas valve knob/oil valve activates the heater. Thermostat control switches off the heating when the desired temperature is reached. Automatic control again starts the heating if temperature of the stored water falls below a set temperature.

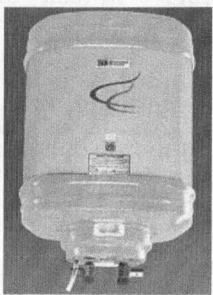

Fig. 7.4   An electric storage tank water heater
(*Source:* Crompton & Greaves,
http://cglonline.com/)

All types of storage tank heaters have the following features for safety and efficient working:

- Thermostat and thermal cut out
- Fusible plug: extra safety device against thermostat failure
- Pressure relief valve: safety against excessive pressure build up
- Vacuum relief valve to prevent collapse of inner cylinder
- Non-return valve to ensure that the heating element is always immersed in water, thus avoiding dry heating of the element.

The capacity of such heaters usually ranges between 5 and 50 litres with higher capacity up to 100 litres in special cases. Power rating ranges from 2 to 4 kW. They are available in both metal and plastic bodies.

Storage tank water heaters have been in use since the early twentieth century. However, they suffer from a few serious drawbacks. These are the following:

Reduced capacity   As water is drawn from the tank, cold water begins to immediately flow into the tank to fill in the place. The cold water instantly starts to take away the heat of the existing water in the tank, effectively reducing its capacity. If we have, for example, a 100-litre tank, we can get a supply of only about 70 litres of hot water at the most.

Limited duration of supply   Duration of supply is limited by the volume capacity of the tank.

Higher temperature setting   People normally set the temperature very high to quickly heat water once the geyser runs out of hot water, a process that is highly energy-inefficient.

Heat loss to the surroundings (stand-by heat loss)   Conventional heaters continuously lose heat to the surrounding by radiation and convection (in spite of thermal insulation) because the tank is always full of hot water, whether or not it is drawn for consumption.

Cumbersome and bulky   The heating tank requires quite a substantial amount of space. It is generally located in closets, utility rooms, etc. It consumes valuable space that could be put to better use.

Problems with hard water scale   If the water being heated is not soft, there will be severe problem of scale and sediment formation in the tank, severely affecting heat transfer, which implies higher energy consumption, apart from maintenance problems.

Tankless water heater (or instantaneous or demand heater)

This type of water heater (Fig.7.5) is a modern development, wherein the traditional hot water tank in the localized heating system is eliminated. It has virtually eliminated all the drawbacks of a conventional type storage-tank water heater. These are compact units that provide hot water when needed. Cold water enters these compact units whenever a hot water tap is turned open. A sensor detects the water flow into it and activates a heating system (either gas burner or electric immersion heating coil as in a geyser) and delivers the hot water at a pre-set temperature. The sensor puts off the heater as soon as the water flow stops. Instantaneous and continuous supply of hot water is achieved with close control on temperature and these heaters are, therefore, also called 'no-patience' water heaters. They have pressure-relief valve and a thermal cut-out. Such a system avoids the stand-by heat loss from a conventional hot-water storage tank (even though the tanks are thermally insulated, there is always heat loss) and thermostatic arrangement provides better control on temperature. In hotels, this is fast replacing the conventional localized hot-water system like geyser. It may even supplement other forms of centralized hot-water generation system, particularly the solar water heater, which has a varying water temperature that can be made more uniform by mixing with tankless hot water whose temperature can be very easily adjusted. Typically, electrically-heated units can supply hot water at a rate of 5–15 litres per minute (lpm), the lower rates being more common. Gas-fired units can supply even at a higher flow rate. Typical power for a 5 lpm system could be of the order of 3–4 kW.

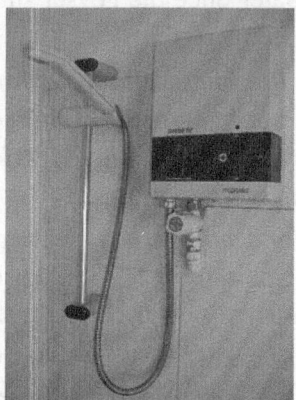

Fig. 7.5   A tankless water heater

(*Source:* en. wikipedia.org)

Advantages of tankless water heater are as follows:

- Hot water is available on demand
- There is no delay in supply of hot water when needed
- Close control of water temperature is possible
- Hot water supply is continuous
- Tankless system can reduce energy bill by up to 30 per cent in relation to a conventional tank storage unit.

7.4 | PIPING MATERIALS FOR CONVEYING WATER

Piping materials for conveying water include (a) lead and lead alloys, (b) cast iron or CI pipes, (c) galvanized mild steel or GI pipes, (d) copper, (e) plastic (for cold water circulation), and (f) asbestos cement.

We briefly discuss about the uses and features of these materials.

Lead and lead alloys   These materials are used chiefly for small pipe work, water-service pipe, cold-distribution pipes, hot-water pipes, waste pipes, and complex-soil pipes or where high durability is a consideration.

Being relatively soft, they are easily worked out and can be laid into complex bends and shapes and, hence, are popular for short and complicated connections.

But they need support at regular intervals when laid horizontally or while carrying hot water because of their tendency to deform and sag. They generally do not corrode and are suitable for hard water. They are expensive and alternative materials and are lighter and easier to handle.

Galvanized mild steel or GI pipes   Galvanized mild-steel pipes or popularly called GI (galvanized iron) pipes are used for transporting chiefly hot water but are also used for cold-water distribution and waste and stack-ventilation pipe work. They are very suitable for pipe work installed above the ground and are free from health hazards. Zinc coating (or galvanization) helps prevent corrosion of such pipes. They are cheaper and mechanically very strong. They are easily welded and small diameter tubes can be bent without the need to heat.

They are prone to corrosion, particularly when water is soft, because there is no protective scale deposition on the internal side of the pipe. Working life is also shorter than lead and copper pipes.

Copper   Copper tubes are used with light gauges (small thickness tubes and pipes) and can be used for all hot and cold water pipes, including heat exchanging coils inside calorifier and other heating systems, waste, and ventilating pipes as also soil pipes.

They are becoming increasingly popular for all purposes because of their smaller diameter compared to equivalent lead or steel pipes. Because of their excellent anti-corrosion property, they are very attractive for all hot water applications involving soft water. They have fairly good strength and the property of ductility enables them to be drawn into thin wall tubes, which becomes lighter and cheap. The smooth surface of copper tubes offers less resistance to flow. Small pipes can be cold-bent.

However, care must be exercised in using dissimilar metals in water pipeline, which, may, in presence of some salts in water start galvanic action where one metal dissolves in water in preference to the other.

Cast iron   Cast iron (CI) pipes are mostly used for waste and ventilating and soil pipes, as also rain water pipes. They are cheap and strong. A large range of fittings are available to build up sanitary installations quickly with reasonable accuracy of fitment without any need for additional fabrication work to be done at site.

However, the material is heavy, cumbersome, and does not present a good sight for viewing. This material is very weak against hammering or similar kind of impact loading.

Plastic or PVC Plastic materials are basically polyethylene, popularly known as polythene, and plastic tubes are now widely used for conveying water and waste. They are very good for cold water services and waste conveying while hot water can be handled only at low pressures.

They are very tough and light, cheap, flexible, and easy to work. They are free from any corrosion attack and can be used underground and overground. Being non-conductors of electricity, it is much safe to work in conditions where they are laid by the side of electric wires. Also, there is no chance of galvanic corrosion in plastic pipes.

However, their principal disadvantage is that it softens even at a temperature of 70°C and melts at 115°C, and hence should not be used at temperatures more than 60°C although they may be used for wastewater lines where hot water may run for very short periods. In horizontal runs, they may need supports at intervals. However, they are not still approved by all water authorities as yet.

Asbestos cement Asbestos cement pipes are similar to cast iron pipes in their areas of applications, for example, soil and wastewater pipes, rain-water drainage, and gutters.

There is no corrosion problem in these pipes and they are cheaper and lighter than CI pipes but are less strong than CI pipes.

Figures. 7.6 (a), (b), (c), (d), (e), and (f) illustrate pipes made of materials as mentioned.

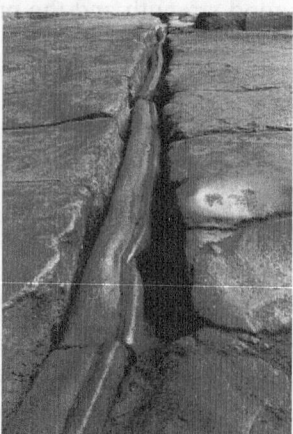 

Fig. 7.6 (a) Lead pipes used to carry water to the Great Roman Bath   **Fig. 7.6** (b) GI pipes stacked on store rack ready for despatch

(*Source:* Shri Aagarsen Steel Industries, Sembudoss Street, Chennai, Tamil Nadu)

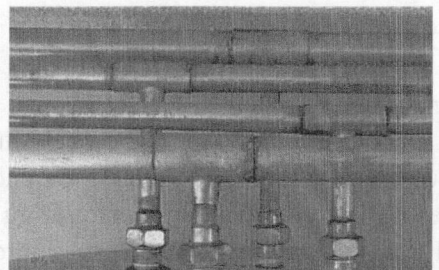

Fig. 7.6 (c) Copper pipelines for conveying water

(*Source:* www.solarexpert.com)

Fig. 7.6 (d) Cast iron (CI) pipelines for conveying water and sewage

Fig. 7.6 (e) Fitting of PVC plastic plumbing is in progress

Fig. 7.6 (f) Asbestos cement pipes for carrying storm and wastewater

7.5 | FITTINGS IN WATER DISTRIBUTION LINE

In this section, we discuss about a few important water line fittings such as valves, taps, cocks, tee, socket, nipple, bend, and flushing cistern. All these fittings are almost always present in any water-distribution system in a hotel. Valves and cocks are kept to a minimum in hot water line.

7.5.1 Valves, Taps, and Cocks

These are fittings in water lines to regulate flow through the lines and are integral part of any water-distribution system. Although usage of the different terms varies, essentially their functions are the same, while the term 'tap' is exclusively used in plumbing work. The function of a valve is the same as a tap, i.e. to open or close or control flow of water through a line. The word tap is used when the valve is small in size and usually fitted just before any service utility such as basin, sink, shower, etc. The flow of water in larger pipelines is controlled by 'valves'.

Valves, taps, etc. are normally of three types: screwdown, plug, and slide. Screw-down valves operate by rotation of a knob or wheel fitted at the top of the valve, which pushes down a screw through a nut. At the end of the screw is fitted a washer, which presses against a metal seat, thus offering resistance to the flow. If the knob is not fully turned water pressure pushes through the contact area and restricted water flows. If the knob/wheel is fully turned out, the washer lifts off the seat and full water flow occurs. Screw-down valves/taps normally work on the principle that if a screw is rotated inside a nut and if the nut is prevented from rotation and any other movement, the screw will slide up and down, thus opening or closing the tap for flow through it. Fibre and leather washers are used for cold-water taps and rubber compositions for hot-water taps. In case of plug valve, a solid cone plug is slotted and normally the slot opening will be in line with pipe allowing full flow. Quarter turn of the plug will rotate the slot opening by 90° and water in the pipe will face solid portion of the plug and water flow will be completely stopped. In a sliding type valve, a plate moves up and down across the pipe to completely open, close, or control the flow depending upon the position of the plate. Sliding gate valves offer very little resistance to flow in fully open condition and hence save pumping energy in the process.

Taps

A tap is what we see at the draw-off end of water service line. They are normally screw-down type valves. Depending upon the directions of water entering the tap and coming out of the tap, they may be categorized as follows:

Bib tap   It has a horizontal inlet and free outlet in the form of a bent tube called bib (Fig.7.7a). The bib prevents dust from entering into the free end and contaminating water when it comes out.

Pillar tap   It has a vertical inlet and horizontal outlet through a bib (Fig.7.7b). They are designed for use in lavatory basins and baths.

Globe tap   It has a horizontal inlet and vertical outlet (Fig.7.7c). Previously, these were used in baths and now largely replaced by pillar taps.

Fig. 7.7   (a) Bib Tap   **Fig. 7.7**   (b) Pillar tap   **Fig. 7.7**   (c) Globe tap

Taps are normally made fully open, however, they can also be used to control flow at the outlet.

Spring or self-closing tap

An alternative type is the spring or self-closing tap, which remains open only while pressed down or up. Left to itself, it will always close by the spring action. It greatly helps in reducing wastage and contributes finally to energy economy.

For the convenience of guests, the hot and cold taps can be distinguished by a red spot and a blue or green spot, respectively.

Cocks

Normally cocks are plug-type valves and are quickly closed by a quarter turn of the knob as in a gas cock. This is used in a cold-water system at the entry point of the civic water line into the premises and is called stopcock (Fig. 7.8). All stop cocks/valves are put in line with the water line to provide either free passage of water through it or completely shuts it off. It is not used for controlling the amount of flow through the line. A stopcock is also sometimes called stop tap.

Fig. 7.8 Stopcock

Valves

The term 'valve' is normally used to indicate controlling and most often for stopping or fully opening supply to a line and fitting. While stopcocks are plug type, valves may be screw-down type or sliding-plate type (for example, gate valve). They are put in line with the pipe line. Figure 7.9 shows a gate valve water-supply line. Such valves are fitted throughout the water distribution line to facilitate closure of a particular line for maintenance work. Normally, before any tap a stop valve/stopcock will be fitted.

Fig. 7.9 A gate valve

Check valve

A very special and important type of valve is a check valve or non-return valve. This type of valve allows free flow of water/gas in one direction while blocking the flow if there is a possibility of reverse flow condition.

Water hammer

We all notice that that when we press a self-closing type tap for drinking water and the release it, the tap suddenly stops the flow of water and in many cases, particularly when the pipeline is a metal pipe and of larger size, there is sudden jerking and vibration of the entire pipeline accompanied by sharp rapping sound. This phenomenon of jerk, vibration, and noise in a pipeline due to sudden closure of a valve is known as water hammer. Any quick closure of tap or defective ball valves can cause water hammer. The vibration may weaken the joints of the pipe work and leakage of water may start through them. In bigger pumping installations with large flow, the resulting pressure rise may burst out the pipeline.

7.5.2   Tees, Sockets, Nipples, and Bends

These are pipe fittings to facilitate layout of pipelines for branching (tee), joining of two pipes (sockets, nipples) and change of direction of pipelines (bends).

7.5.3   Flushing Cisterns

Flushing cisterns are normally of two types. They are

1. Plunger and siphon type
2. Bell type

Both the types of cisterns are supplied with cold water through ball tap as in cold-water cistern and fitted with an overflow discharge so as to cause noise to attract maintenance people to attend them in case of overflow.

Plunger-type flushing cistern   The content of the cistern are flushed out mainly by the syphon action of the flush. The flush pipe is in the form of an inverted 'U' tube (see Fig. 7.10). A pull on the lever causes the plunger to come down and simultaneously by impulse action, the hinged valve fitted on top of the plunger opens out, and throws a volume of water over the crown of the siphon to start siphon action down the flush pipe. This onrushing water create a partial vacuum behind it in the shorter leg of the flush pipe and water in the cistern is sucked down to continue the flow till the water level in the cistern falls to the level of the inlet to the shorter leg of the flush tube. The hinged valve is a non-return valve, which allows water to flow upward but not downward through it.

Bell-type flushing cistern   In this type of a cistern, the pull of the lever lifts the bell and hence causes a reduction in pressure in the space A as shown in Fig. 7.11. Atmospheric pressure in the region B above the cistern water surface forces water into the space A, thus starting the flow through the down-comer flush pipe till the cistern content is emptied to the level of the inlet to the flush pipe.

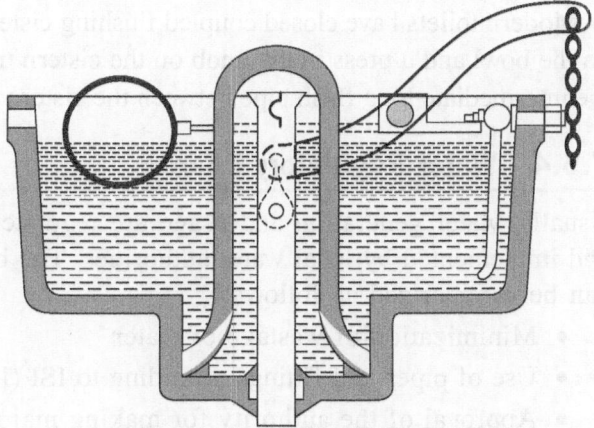

Fig. 7.10   Bell-type flushing cistern

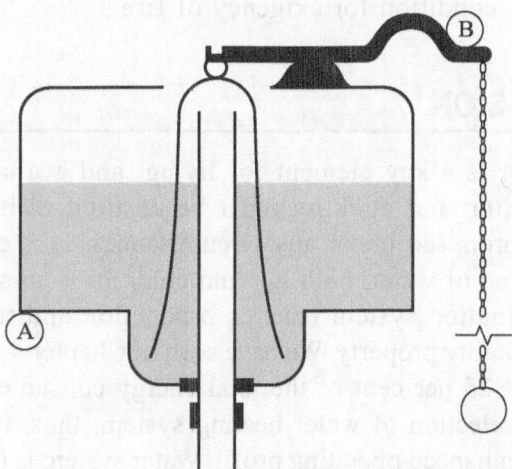

Fig. 7.11   Plunger-type flushing cistern

The capacity of the cistern is limited by local water regulatory boards and is usually 12.5 litres for WCs and 5 litres for automatic flushing cistern for urine pans. High-level and low-level cisterns are installed approximately 1.5 m (5 ft) and 0.3 m (1 ft) above the water-closet pan.

Normal individual flushing cistern takes some time to fill. To overcome this problem, flushing trough is devised, which is a long cistern serving several pans simultaneously and allows more frequent flushing. In many installations, flushing cistern is eliminated wherein a central flushing water line caters to a series of urinals and water closets. Whenever flushing is needed, a knob is pressed, which opens a valve in the branch line connecting the common header flushing line to the particular urinal or WC) pan and flushes the appliance and closes after a measured amount of water (say, 12 litres) flows through it.

Modern toilets have closed coupled flushing cisterns which are directly mounted on the bowl and a press in the knob on the cistern triggers flushing. In this system, the intermediate long flush pipe between the cistern and the bowl/pan is eliminated.

7.5.4 Water Regulatory Board

Usually, water generation and usage are regulated by norms and rules framed and implemented by local water authorities. The basic principles to be followed can be summarized as follows:

- Minimization of wastage of water
- Use of pipes and fittings according to ISI (Indian Standard) specifications
- Approval of the authority for making major change in the system
- Provision of adequate water supply and maintaining it in good working condition for exigency of fire

7.6 | CONCLUSION

Water is a key element for living; and availability of safe, potable water for drinking and cooking must be ensured in hotels, where hygiene cannot be compromised under any circumstances. It is evident that hotels require a huge volume of water, both hot and cold, for a number of essential functions. Water-distribution system ensures production and transport of water throughout the hospitality property. We have seen in Chapter 4 that water heating alone consumes about 25 per cent of the total energy consumed in a big hotel. Efficient design and selection of water heating system, thus, is critical for energy conservation and enhanced operating profit. Water system is thus a very specialized engineering system and should be designed and maintained by professional engineers.

KEY TERMS

Bib tap/cock	It is a very common and popular type of valve at the end of a pipeline over the sink/basin, etc. having a bib to prevent entry of contaminants and ease of use.
Boiler	It is a closed thermal device, which produces hot water/steam using the heat of combustion of fuels such as coal, high speed diesel, furnace oil, natural gas, etc.
Calorifier	It is a heat exchanger in which hot service water is produced by indirect heating by hot water/steam from boiler.
Check valve	A special type of valve which allows flow through it in one direction only is called a check value.

Cistern	Containers for holding liquid, usually water and subsequent supply for various use are called cisterns.
Cock	Valve that fully closes/opens by quarter of a turn of the handle/knob is called a cock.
Downcomer	It is a long vertical pipe that is usually coming down from the storage cistern high up, from which branch lines serve different floors.
Expansion pipes	Pipes extending from the boiler and clarifier outlet line, going up and draining in the make-up cistern to accommodate for the expansion of water during heating are called expansion pipes.
Flushing cistern	A cistern which is placed over a urine pan and water closets for storing and discharging water with sufficient force to clean the bowls is called a flusing cistern.
Primary circulation	The circulation of hot water through boiler and calorifier heating coil is called primary circulation.
Rising main	It is a long vertical pipe, which conveys water to the top of a building usually storing water in a cistern and also, in many cases, from which branch lines serves different floors.
Secondary circulation	The circulation of hot service water through calorifier, service lines, and return line to the calorifier is called secondary circulation.
Solar water heater	Water heater that uses radiation energy of the sun to heat water is called solar water heater.
Stop valve	It is a valve used for completely stopping the flow of water in the line beyond the valve; used for isolating a line for repair/maintenance work.
Tankless water heater	Localized water heater that provides instant and continuous supply of hot water without the need of storage tank is called a tankless water heater.
Tap	Tap is a variant of valve, which is usually fitted at the end of pipeline for consumption and is usually either fully open or fully closed.
Valve	It is a pipe fitting used to control flow of liquid or gas.

REVIEW QUESTIONS

7.1 List the common uses of water in the hotel industry.

7.2 List the most general components of cold-water distribution system.

7.3 List the essential elements of a cold-water cistern.

7.4 Show a typical water-balance flow chart for a hotel of moderate size.

7.5 Draw a line sketch of a typical cold-water distribution system supplied by civic body water mains.

7.6 Discuss, in brief, about the maintenance procedure of a swimming pool.

7.7 With reference to indirect heating system for service hot water, with the help of a neat sketch, explain primary circulation and secondary circulation.

7.8 List a few different types of solar heating system.

7.9 Describe a vacuum-tube collector type solar heating system, stating its relative merits and demerits.

7.10 Describe, in brief, different localized heating systems.

7.11 List the advantages of a tankless water-heating system over its storage tank counterpart.

7.12 List the different piping materials mentioning their particular areas of use.

7.13 With the help of a neat sketch, describe a bib tap/cock.

7.14 Describe, in brief, the phenomenon of water hammer.

7.15 With the help of a neat sketch explain the working of bell-type flushing cistern.

8 Chapter
Sanitation, Waste Disposal, and Pollution

Learning Objectives

In the hotel industry, sanitation and disposal of waste play an important role. This chapter covers the various aspects of these two areas. Pollution caused due to various functions in a hotel is also discussed, along with regulatory norms.

After reading this chapter, the students will be able to:
- identify various sanitary systems and fittings used in hospitality establishments
- understand treatment and disposal of waste in hotels
- know about various kinds of pollution
- understand the sources of pollution and relevant regulations for pollution control

8.1 | INTRODUCTION

Sanitation, in broader terms, means providing a hygienic living environment. It is usually referred to as providing potable water, providing sanitary appliances, collecting and safe disposal of wastewater, soil, and solid waste, etc. This is a vital area in hotel service, as the slightest neglect in design and maintenance of sanitation is capable of playing havoc with public health which may lead to the loss of reputation of the hotelier. So, utmost care is to be exercised in this aspect. Sanitary appliances and fittings of the highest quality should be used for the purpose. We discuss these aspects of sanitation in the sections to follow.

Many epidemics in the past had been the result of contamination of water with sewage. Environment and ecology have been seriously affected by rampant disposal of sewage and solid waste. As such, we also discuss about various methods of sewage treatment and disposal as well as solid-waste disposal. Pollution has become a cause for global concern and hotel activities also contribute to this phenomenon to some extent. There have been certain rules and regulations as imposed by both central and state pollution control boards, which are to be followed by hotel establishments as well. We discuss, as well in the later part of the chapter.

8.2 | SANITATION AND SANITARY SYSTEM

Sanitation engineering is a very important branch of science under public-health engineering and specialist engineers must be employed for design and installation of this system.

As discussed earlier, sanitation can be viewed as comprising the following functions:

- Providing potable water
- Installation of properly functioning, good quality sanitary fittings
- Installation and maintaining safe waste-collection system
- Making provision for draining of the collected waste to either municipal disposal system or in-house disposal and treatment plant for possible recycling of recovered water

8.2.1 Sewage and Waste Collection and Drainage System

Waste material is produced in any property, including hotels, through various activities and they may be broadly categorized as (a) liquid waste and (b) solid waste.

Liquid waste is commonly known as sewage. Although the bulk nature of sewage is liquid, it also contains some solids produced by humans called sullage and would primarily consist of wash-basin water, faeces, urine, laundry waste, and other material that flow into the drainage system of the holding. Sewage is a major source of pollution and epidemics and assumes highest importance in urban areas. It is estimated that contamination of drinking water due to faeces is by far the biggest cause of death globally. Sewers are the conveying pipes for the wastewater and sewerage is the total system of collection, transportation, and safe disposal of wastewater. Well-organized sewerage system prevents pollution of the environment due to sewage by managing the collection, treatment, and recycling or safe disposal of sewage in the environment.

In big cities, the sewage from a holding comes out of the premises drainage and joins directly with corporation main sewers on the street. The city main-sewer load is then treated in treatment plant and disposed in some water body such as rivers. In places where a municipal sewerage system has not been provided, owner of the holding drains the sewage into either (a) septic tanks, where it is treated and clean water output from them joining the city drainage system or (b) cess-pits, from where it is collected in corporation vehicles and taken for treatment or disposal. The second method is acceptable for a very small household and is to be permitted by local civic-body laws. In case of a large holding, such as a hotel, septic tanks are replaced by a large sewage-treatment plant (STP), which produces fresh water of various degrees for recycling into drinking, cleaning and toilet use, and irrigation and gardening purposes. Treatment of sewage is discussed in some detail in a later section of the chapter. In India, sewage collection and treatment is typically governed by regulations and standards made by local municipal bodies, state and union governments.

8.2.2 Sanitary System and Sewage Draining in Hotels

Before transporting the sewage for treatment, there must be a system that collects and conveys all the waste from various points in a holding. The function of a sanitary system is to carry the effluent (wastewater and solid matters) from the kitchen, restaurant and lavatory wash basins, urinals, and water closets (WCs) to the sewer. In many hotels, the outlets of sewage join with in-house sewerage pipes and are directly put into the municipal sewerage for disposal, if this facility is available. In other cases, these wastes, both solid and liquid, will pass through an STP, which produces fresh water for reuse. The latter arrangement, although capital-intensive, saves water, and attracts incentives from the government.

The system of piping through which the effluent flows has two main vertical pipelines through which soil and wastewater come down to the collection trap or gully trap at the bottom. These pipes are called stacks. Branch pipes lead the soil and waste from the sanitary appliances to these stacks. One of the stacks is called a waste stack, which collects wastewater effluent. Branch pipes from bathtub, lavatory and other basins, kitchen wastewater, laundry wastewater, etc. carry the wastewater from the respective appliances to this stack, which is finally discharged into the gully trap and subsequently to the sewerage drain. Similarly, branch pipes from WCs carry the effluent from WC, bidets, etc. and are connected to the other stack called soil stack, which is directly connected to the sewerage drain.

Sewerage drain is a near horizontal pipe laid underground inside the premises, leading from the foot of a stack to the public sewer system in the main road.

Water seal and anti-syphonage pipe

Each of the appliances used in a piping system is fitted with a U-shaped piping at the outlet. This shape ensures that there is always water in the U-shaped portion. The presence of this water in the fitting prevents passage of foul air from the stacks into the room via the appliances. This arrangement is therefore, called U-seal. However, this water seal may break by continuous flow of water under certain conditions of pressure during the movement of soil/water from other branches connected to the same stack. When soil water from a WC above another passes through the branch point of the lower WC, the pressure inside the branch pipe of the lower branch pipe may fall below atmospheric pressure and the atmospheric pressure on the water in the lower WC pushes water through the water seal, thus completely draining it. This phenomenon of draining of water from the U-tube is called syphonic drainage. At the beginning of such action, the pressure at the top of the U-seal becomes less than atmospheric. This pressure may be brought back to atmospheric pressure and thus prevent loss of water seal if the top portion of the U-seal is connected to a free air space. Anti-syphonage or ventilation pipes are provided at the top of each water seal (except for the appliance at the topmost floor), which are joined to the vent-pipe portion of the respective stacks at a height above location of the highest appliances connected to the concerned stack.

Anti-syphonage pipes provide passage for a free current of air throughout the system, thus, helping (a) prevention of foul air in the piping and (b) loss of water.

Different types of stack arrangements

There are a few variants of the stack arrangement. In a two-pipe sanitary stack system soil effluent and wastewater effluent have separate stacks.

One-pipe system has been also developed where both soil and wastewater are carried by a single pipe and discharged directly to the drain without the need of gully trap (discussed later). However, in this arrangement, very secure anti-syphonage pipes are required for all the appliances as these are only traps between the drain and the rooms. This arrangement is much cheaper than the two-pipe system.

In a single-pipe system, even the need of the anti-syphonage pipes is eliminated by grouping the appliances closely around the stack. Such systems are mostly used nowadays in the hotel industry.

Recycling of sanitary waste

In arrangements where the waste effluent is recycled, the drain line of the sanitary waste, instead of joining to the public sewage, goes to the waste-effluent treatment plant, which has a solid-waste treatment plant also. The treated water then again goes to all the appliances via cold-water distribution system, as explained earlier.

8.2.3   A Few Common Sanitary Fittings

A few common sanitary fittings are discussed below.

Traps

As already explained, traps are depressions or bends in a sanitary fitting, which always remain full of water, and thereby do not allow foul air to enter a room from outside sewer pipe or drain via waste-pipe connection through the sanitary fitting inside the room. We have already discussed about traps or water seals in different sanitary appliances and there are other places that require such traps such as in the drainage system of a building. We briefly discuss them as follow:

Gully traps   These are openings into the drain incorporating a water seal and are usually at ground level. Gully traps are used after different branches of local drain pipes, for example, waste stack, soil stack, rain-water drain, etc. joining below the main drain or sewer. A wide variety of shapes are available for different purposes as discussed. The two common features in all of them are (a) the stacks drain at a level higher than the trap water level and (b) the space between the downpipe and the seal is ventilated for atmospheric air through grating or otherwise.

Plain gully traps   These are suitable for rain water; the down pipe drains the rain water collected and discharges directly over the gratings

Back inlet gully traps   These are suitable for waste pipe entries through a separate inlet at the back and discharges underneath the gratings. This avoids splashing of dirty water on the surface if plain gutter were used

Inspection gully traps   These are used instead of inspection chamber for single branch drains. They have rodding eye for inserting rod to clean the trap, if required.

Grease and sludge traps   These are used for trapping grease and sludge. This is required on the waste outlets from kitchens, where there is a heavy discharge of greasy water and cooking sludge. In a grease trap, cold water in the trap solidifies the grease and is collected on the screen tray, which prevents the solidified grease from escaping into the drain. The tray is to be taken out and cleaned periodically for smooth functioning

Potato peelers produce heavy sludge in the kitchen and a sludge trap is used for the purpose. A sludge trap is very much similar to a grease trap with a little modification in the tray screen.

Inspection chambers

Inspection chambers are walled openings built round the drain within which the effluent from all branch drains flows in. Their depth may vary between 1 and 3 ft. They are fitted with a removable cast iron cover.

The objective of installing inspection chambers is to give access to the drains for clearing blockages and they should be placed at every point where the drain changes direction or where it is joined by branch drains.

Water closets and urinals

We discuss here, in brief, some basic types of water closets and urinals, as commonly found in public urinals, including hotel urinals.

Water closets   Starting from a crude humble beginning, modern water closets (WC) have evolved to practically self-cleansing design. Water closets in modern days may be broadly classified into the following two types:

1. Wash-down type   2. Syphonic type

In the wash-down pattern, water from a flushing cistern put at some height comes down and forces out through the opening round the rim of the WC and scours all the exposed surface within the pan and forces the contents through the trap, which is an integral part of the fitting. Obviously, greater the volume of water and greater the height of the flushing cistern, better is the cleaning. The water seal should not be less than 2 inches or 50 mm. Wash-down closets are available with 'P' or 'S' trap outlets, S-type being mostly used in ground floor applications, while P-type is used in higher floors.

Syphonic closets are developments of wash-down type and have resulted in lower-height cisterns and a much quieter operation without noise. Syphonic closets cleanse pan very effectively, and hence are very suitable in hotels and public places where these are used frequently. However, they are more costly than the wash-down type.

Urinals   Urinals are the most difficult fittings to maintain clean. The two forms are slab urinals (Fig. 8.1) and stall urinals (Fig. 8.2). They should be continuously flushed automatically by branches of a common inlet manifold. Urinal bowls (Fig. 8.3) can be used in certain circumstances particularly in places where careful use is envisaged. These bowls make considerable savings in cost.

Fig. 8.1 View of a slab urinal
(*Source:* www.twyford
bathrooms.com)

Fig. 8.2 View of a stall urinal
(*Source:* www.zeek.net)

Fig. 8.3 Urinal bowl in a public urinal
(*Source:* www.pce.ie)

Fig. 8.4 A modern bidet by the side of WC
(*Source:* en.wikipedia.org)

Bidet These are very special lavatory fittings and have a water spray coming upward from a spout placed at the centre of the bottom and a hollow rim through which water flows into the bowl (Fig. 8.4). They are provided with both hot and cold water supply. Usually, these are used for ablution purposes and people with problems of rectum, such as piles or fistula, use them for washing. Since they are used for washing excretory organs, they should be placed near WC. The outlet of bidets is connected to the soil waste pipe. Many hotels in the west are now incorporating these fittings in the lavatory.

Wash basins—Lavatory basin and kitchen sinks

Wash basins are integral part of lavatories and kitchen. Kitchen basins for washing and cleaning of food items and utensils are of special design where they are deeper with flat bottom for keeping materials for washing. Lavatory wash basins are shallower with sloping bottom. Each of the two types are described in some detail as follows:

Lavatory wash basins   Commonly used materials for lavatory basins are vitreous china and glazed fireclay. Vitreous china is considered most desirable as it is comparatively thin and non-porous. Even when the glaze is lost, it remains hygienic due to its non-porosity. In schools and public places, where rough handling is expected, cast iron basins enameled with porcelain may be used in place of the brittle vitreous china or glazed fireclay. There are many shapes and layout of basins. While some are fitted on a pedestal, others are fixed on wall with the help of brackets. Some basins are so shaped that they can be accommodated in the corner of two walls. In some arrangements, basins are installed in a group of four or six or more. In general, basin height should be between 750 mm (about 30 inches) and 825 mm (about 33 inches) at the front. Wash basins are usually provided with two square holes into which hot- and cold-water taps (usually pillar taps) are fitted. The drain hole in the basin at the base is usually provided with a plug, which holds water in the basin, if necessary. Both plug-and-chain type and pop-up type plugs are in use for the purpose. To maintain the basin hygiene, a spray tap may be installed that delivers a given restricted amount of water supply into the basin throughout the period of use. An overflow slot is provided in the basin, which drains overflow water in the basin into main basin waste outlet drain through a passage inside the thickness of the basin at the back side.

Sinks   Sinks are essential items in kitchen-plumbing fittings and used for keeping things inside them and washing them thorough. Small jugs and buckets are also quite often filled with water by keeping them under the tap of the sink. Pan-washing sinks are deeper than normal sinks. They are made in a very wide range to suit different kinds of washing and other secondary needs. As such, a sink almost always a combination sink and drainer. Sink fashion has changed considerably from old-fashioned shallow-type to the modern deep basins made of the much desired stainless steel. Porcelain or china clay sinks are also used but they are a little too hard for some crockery and glassware. Sinks are fitted with plug-and-waste and have a secret overflow arrangement. Normally, sinks are fitted with bib cocks fitted usually at a height of about 13 inches from the bottom of the sink to accommodate a jug or bucket filling without difficulty at the same time ensuring discharge of free water into the sink and not splashing outside. Double sinks are used for washing and rinsing purposes with arrangement of hot water in the rinsing sink for sterilization purpose.

Bath

Two types of baths are normally used. They are the conventional long bath and shower bath. Most baths are made of (i) vitreous-enameled cast iron and (ii) acrylic sheet or glass fibre reinforced plastic (FRP) materials. Normal length of long baths is about 1650 mm (5 ft 6 inches). Due to space limitations, shorter baths are also used. People now prefer low-height baths because they are more convenient to enter. They are fitted with panels on any exposed sides. Although globe taps are now very common, pillar taps are becoming popular, which can be fitted at a level higher than the overflow rim of the bath and thus reducing a chance of water pollution in the service pipes due to back syphonage.

Shower bath has a much smaller receptacle and is not a bathtub and is primarily meant for collecting the water and draining it off. It has a shower rose at the top and in many cases fitted with a swiveling arrangement so that one has the option of not making one's head wet. In modern shower chambers, instead of swiveling overhead shower rose, hand-held shower roses are very convenient to selectively control shower on different parts of the body. Shower chambers have waterproof curtain of polythene or glass to prevent splashing of water outside the bottom receptacle or water collection tray at the foot so that the bathroom floor is not wet.

Both hot and cold water-supply lines are provided in baths. Although a little expensive, thermostatically-controlled automatic mixing device for desired temperature of water at the bath is an option in many baths.

Materials for drains

Drain pipes are generally made of glazed stoneware or cast iron, where strength and resistance to leakage are required. Soil stacks, waste stacks, and their branch pipes can be made of lead, cast iron, or copper.

8.3 | TREATMENT AND DISPOSAL OF SEWAGE

The huge volume of sewage collected from households, institutions, industry, including the hospitality industry poses a major challenge in its safe disposal. Usually, the wastewater is disposed of in water bodies such as river, sea, lake, or as landfill. However, most frequently and justly so, sewage need to be properly treated before its disposal.

8.3.1 Disposal of Sewage

Raw sewage goes stale after a few hours and it become a major threat to public health. There are many ways by which the sewage is disposed of. In many areas, the sewage is not treated and released to a river or dumped in soil, while in others they are treated to varying degrees of purification.

Methods of disposal of sewage

We broadly classify various methods of disposal of sewage without or with very little treatment, as follows:

Dilution Raw sewage or partly treated sewage is thrown into natural water bodies such as sea, river, lake, marshy land, etc. Self-purification is the mechanism in this process, which is helped by the following factors:

- Dilution of contaminants by dispersion in flowing water
- Sedimentation of particles to the water bed
- Oxidation of organic matter by dissolved oxygen in water
- Sunlight, which kills harmful bacteria
- Microbial organisms consuming the solid organic matter in sewage

However, this natural treatment process is susceptible to malfunctioning because of varying sewage character, river condition, habitation pattern downstream, etc. and as such sewage should be given a preliminary treatment before disposal to the water body. The degree of treatment would be guided by the local municipal sanitary rules and regulations and the conditions of the river, etc.

Purification Purification process comprises a change in the chemical and biochemical character of the sewage in a treatment plant and is effected through one or a combination of the following processes:

Land fill and irrigation Raw or partly treated sewage is dumped on land. A part evaporates, while a part percolates. Purification is accomplished by natural oxidation. This is a very good method of sewage disposal where land is available. Bright sunshine helps in microbial activities in the sewage. Decomposed sewage becomes an excellent manure for crops. In many cases, sewage is passed through a septic tank, which helps in reducing load on the land. However, there is always danger of harmful bacteria percolating through ground subsoil, reaching the subsoil groundwater, and contaminating the water reserve, which are used by many people by digging wells. Overuse of land lead to clogging of soil pores, thereby preventing entry of oxygen and thus degeneration process of sewage organic matter is hampered. This is called sewage sickness of soil.

Chemical treatment Effected by addition of chemicals such as lime, alum, etc.

Septic tank Used to provide satisfactory disposal of sewage in a plain sedimentation tank where biochemical reactions (bacterial decomposition) also take place. During this digestion process, sewage is purified and the effluent is taken to soak pits for disposal to city sewers. A small solid portion remains and as such, septic tanks are to be cleaned after a period of 5–10 years.

Bio-aeration The sewage is first passed through coarse screens to get rid of large solids. The sewage is then treated in the main aeration tank with compressed air. The disintegrated sewage is passed through the grease-separation chamber.

If the sewage is treated to a good degree of purification, the clear water is usually reused and the solid residue, called sludge, is processed to produce bio-solids or dumped as landfill.

8.3.2 Sewage Treatment

Sewage treatment is the process of removing the contaminants from sewage to convert it to a composition of clear liquid and solid, which are fit for discharge to the environment or for reuse.

8.3.3 Sewage Treatment Plant

We have noted that it is always preferable to have some treatment of sewage before its disposal to the water body or landfill. In the most sophisticated treatment, clear potable water can be obtained in one hand while leaving only 5 per cent to 10 per cent of solids after treatment. This solid part, called sludge, is further processed to produce what is called biosolid, which have many uses. In cities where sewage-treatment plants operate, wastewater and storm water are carried through city sewerage system to the treatment plant. While hoteliers can discharge their sewage to the public sewerage, along with town sewage, for treatment in the city sewage-treatment plant and final disposal, many of them have installed the most modern sewage-treatment plant (STP) in-house, where they get fresh water for reuse and may use the sludge as manure for the garden activities. The fundamental principle of purifying sewage is to completely break down the original organic matter in it by the action of microorganisms (microbes such as bacteria). These microorganisms digest (eat away) the original organic matter leaving a clear effluent and solids, called biosolids, in the complete treatment.

The stages in the treatment of sewage are as follows:

Primary treatment This removes the suspended and floating objects by means of strainer, screens, grit chamber, sedimentation tanks, septic tanks, etc. The typical materials that are removed in this stage include large objects such as stick, rugs, rocks, etc., sand and gravel, and fats, oils, and grease (FOG), typical of a hotel kitchen waste. Many civic bodies discharge sewage for final disposal after this stage.

Secondary treatment This treatment is designed to degrade the biological and organic content of the sewage by means of microbial action. The bacteria resident in the sewage need oxygen to act and hence this process is called aerobic decomposition. These bacteria consume all the organic materials. The various processes in this stage include filtration, land treatment, activated sludge process, septic-tank process, fluidized bed reactor, surface-aerated basins, etc. The common part of most biological treatment process is the use of oxygen (aeration) and microbial action. Biological-aerated filters are new technologies for the purpose and are called bio-filters. It serves the dual purpose of filtering suspended particles as well as supporting highly active biomass. This has become very popular in sewage treatment.

Tertiary process with or without disinfection The final treatment is performed at this stage before making eventual disposal of the treated wastewater. Tertiary treatment comprises many processes and includes filtration, lagooning, removal of nutrients such as nitrogen and phosphorus, which encourage algae formation and disinfection. All or some of them can be incorporated in a particular treatment system depending on the level of purification needed. Disinfection can be done with chlorine, ozone, and ultraviolet (UV) treatment. In many cases, disinfection is done as the last activity (also called effluent polishing). The purpose of disinfection is to destroy residual bacteria/microorganism in the water after secondary treatment, thus rendering it very safe for final disposal. The solid part remaining after treatment is called sludge and it normally forms only 5 per cent to10 per cent of the final product. For the UV to be effective, water should not be cloudy as suspended particles can screen the organisms from exposure to the ray. Biofilters have become a very popular method for sewage treatment.

Degrees of treatment

The clear water coming out of the treatment plant can be used for different purposes such as disposal to water body, agricultural reuse, premises non-drinking water reuse (e.g., sink, lavatory, cleaning, etc.), as also drinking water reuse depending on the degree of treatment and purification being employed. The treatment given to sewage may be a simple treatment, where only those matters which can be settled out are removed through screening and plain sedimentation. The treated waste can be disposed of to the water body or landfill. The effluent is not very good. If, in addition to the above, fine-suspended organic matter are also removed by chemical sedimentation, rapid filtration, or by septic tank, etc., the treatment is called partial treatment. Complete treatment is the one where dissolved organic matters are also removed by filtration, activated sludge, etc. In the complete treatment, both the suspended and dissolved organic matters are removed mostly by microbial (or bacteriological) action. In addition, disinfection process makes the water fit for reuse. Complete treatment is very costly and is seldom seen in Indian public sewage treatment scenario.

Sludge treatment and biosolids

At various stages of wastewater treatment, large quantities of solid matter or sludge are collected. The sludge may be further processed for safe disposal or production of biosolids for further use. Sludge is subjected to what is called digestion process, which reduces the volume of sludge. It is then burnt or thrown into a flowing water body. Digestion is a process by which the amount of organic matter and disease-causing microbes are reduced. The different methods of digestion are anaerobic digestion, aerobic digestion, and composting. In both anerobic and aerobic digestion, microorganisms grow and reproduce in the sewage transforming the original toxic organic materials and disease causing bacteria and both the process require oxygen for the microbial growth and reproduction. Anaerobic digestion is the bacterial process occurring in absence of air and the microorganisms get the oxygen from the sewage itself or from external inorganic oxides. One major feature of anaerobic digestion is the production of biogas (the important component being methane), which can be used as fuel in boilers for generation of electricity and/ or heating purposes. Aerobic digestion occurs in presence of air, the bacteria devouring the organic matter in the sludge. Under aerobic conditions, bacteria rapidly consume organic matter and convert it into carbon dioxide Composting is also an aerobic digestion where the sludge is mixed with sawdust, straw, etc. The large objects trapped during screening can also be treated similarly. For a liquid sludge, water is removed mechanically and also by adding polymers.

Recent technologies are successfully processing the remaining solid part of sludge to form what is called biosolids that form good fertilizers and manures. Biosolids comprise chiefly dead as well as a small quantity of active microorganism, and some solids, such as sand, that can come down the treatment process. These biosolids typically contain 2 per cent nitrogen and 0.9 per cent phosphorus, the elements responsible for its property as a fertilizer.

8.3.4 Measure of the Toxicity of Wastewater and Sewage

There are many parameters for assessing condition of wastewater and sewage. The widely used among them are biochemical oxygen demand (BOD) and the chemical oxygen demand (COD). These parameters are often used to determine whether a given effluent discharged into a water body significant adverse effect upon the lives of fish, aquatic plants, and other aquatic lives. A measure of these parameters on raw sewage is an indicator of the treatment load of the sewage.

Sewage contains inorganic chemicals, organic matter, and bacteria. The decomposition of the organic matter into stable form is accomplished by oxidation, assisted by bacteria and other microbes of air/oxygen, and is called aerobic decomposition. Other oxidizable chemicals directly use oxygen for reduction. So, sewage matter is decomposed both by bacteriological and chemical oxidation. All this means that if the sewage is discharged into water as such, natural process of decomposition would consume amount of oxygen from within water, thus seriously affecting aquatic life and ecology. Therefore, a measure of these oxygen requirements in the waste being discharged would indicate the degree of toxicity of the waste and its effect on aquatic life.

Biochemical oxygen demand (BOD) and chemical oxygen demand (COD)

Sewage and other wastewater contain organic matter and bacteria in them. Additionally, all natural streams contain bacteria and nutrients. The organic matter decomposes (or degrades) by oxidation. So when wastewater is discharged into water stream, this oxidation takes place assisted by bacteria and consumes oxygen of the natural stream, thus depleting its oxygen content affecting aquatic life very adversely. We have a reaction for this oxidation qualitatively, represented as follows:

Oxidizable organic matter + Bacteria + Oxygen $\rightarrow$ Carbon dioxide + Water + Nitrate and Sulphate salts

Biochemical oxygen   BOD is the oxygen demand in this biochemical reaction during degradation of organic sewage matter. It is measured in milligrams of oxygen required per litre of wastewater. A measure of this indicates its load either on the natural stream or the oxygen requirement for proper treatment. Although a little less accurate, it serves as an important measure of the degree of toxicity of sewage and is often a measurement carried out on sewage at various stages of treatment for proper control of treatment process, among other parameters.

Chemical oxygen demand (COD)   In addition to biodegradable organic matter, sewage also contains other chemicals, which are oxidizable and hence consume oxygen, if available. The general qualitative form of the reactions for other chemicals can be represented as follows:

Sulphide chemicals + Oxygen $\rightarrow$ Sulphate salts

Nitrite chemicals + Oxygen $\rightarrow$ Nitrate salts

The total oxygen demand of biodegradable matter and other chemicals together is known as chemical oxygen demand (COD) and is another indicator of the quality of the sewage (or wastewater).

Thus, the BOD test measures the oxygen demand of biodegradable pollutants, whereas the COD test measures the oxygen demand of biogradable pollutants plus the oxygen demand of non-biodegradable oxidizable pollutants.

8.4 | SOLID WASTE AND ITS DISPOSAL

We all are familiar with a very common sight in a crowded city where roadside dustbins are overflowing and the stench rising from it filling the air all around. The dustbins are dumped with what is called solid waste. Each household property generates garbage or waste day in and day out. Items that we no longer need or do not have any further use fall in the category of waste and we tend to dump them away. These items fall under the category of solid waste and pose serious threat to the environment globally. In today's polluted world, learning the correct methods of handling the waste generated has become essential.

There are different types of solid waste depending on their source. The broad ways of classification of solid waste is according to whether the waste is as follows:

- Organic (for example, tree branches, grass, leaves, fruits and vegetable peelings, fish and meat waste, etc.) or inorganic (for example, waste paper, plastics, cotton waste, etc.)
- Biodegradable and non-biodegradable
- Contains toxic chemical or not, for example, some wastes such as computer monitor, fluorescent tubes, and mercury vapour lamps contain toxic chemicals

Each of the different types of wastes needs different treatment for disposal. Segregation is thus an important method of handling municipal solid waste. Figure 8.5 illustrates the various stages of solid waste management, including segregation at source.

As the cities are growing in size and in problems, such as the generation of plastic and e-waste (such as TV and computer monitors, LCD screens, etc.), various municipal waste treatment and disposal methods are now being used to try and resolve these problems.

Garbage generated in households can be recycled and reused to prevent creation of waste at source and reducing amount of waste thrown into the community dustbins.

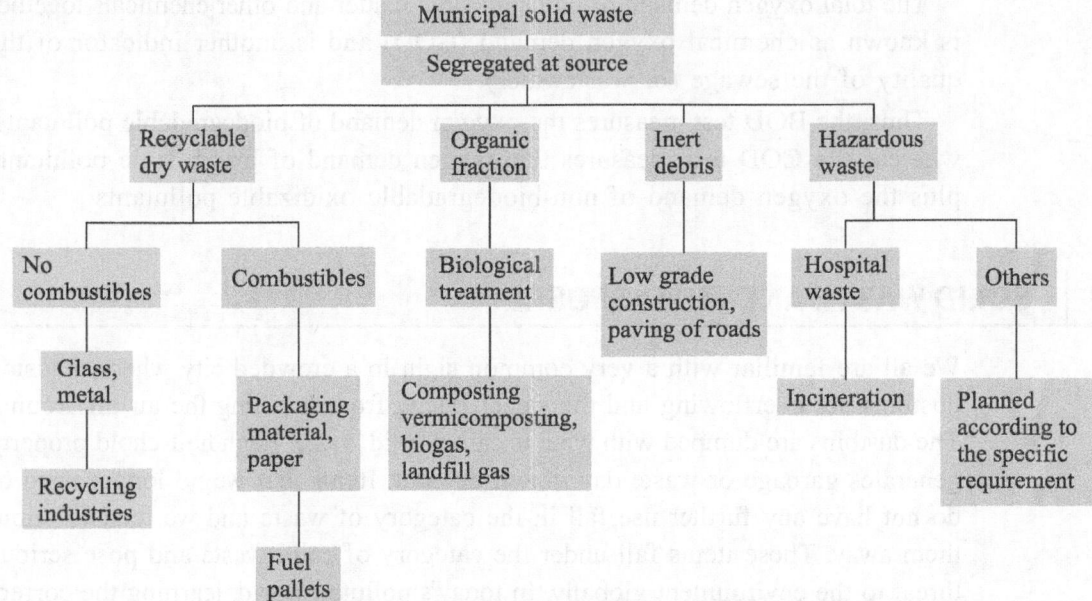

Fig. 8.5 Segregation of municipal solid waste

(*Source*: Central Pollution Control Board report
on management of municipal solid waste)

Solid waste management is a part of public health and sanitation, and according to the Indian Constitution, falls within the purview of the State list. The activity being of a local nature is usually entrusted to the urban local bodies. The urban local body undertakes the task of solid-waste service delivery, with its own staff, equipment, and logistics.

It is estimated that about 1,00,000 MT of municipal solid waste is generated daily in the country. Per capita waste generation in major cities ranges from 0.2 to 0.6 kg. With the growing urbanization, problems of collection and disposal of municipal solid waste are becoming acute and call for immediate and concerted action.

Statistics show that there is hardly any attention being paid to scientific and safe disposal of waste. In several municipalities, the landfill sites have been exhausted and the concerned local bodies do not have resources to acquire new land. Due to lack of disposal sites, even the collection efficiency gets affected.

8.4.1 Solid Waste Management in Hotels and Restaurants

Huge volume of solid waste is generated everyday in a hotel. Many of them, particularly organic wastes, come from kitchen and restaurant residues. Waste papers and other consumables in engineering departments also form a substantial amount of solid waste. They pose a huge environmental and sanitation problem if not collected and disposed properly.

Collection of hotel and restaurant waste

Hoteliers may make their own arrangements for collection of waste individually by big hotels or through their own association for a cluster of hotels, particularly in tourist places. Local bodies may extend help in primary collection of such waste for door step collection of such waste on full-cost recovery basis. Charges for the collection of hotel waste may depend upon the quantity of waste to be picked up from the hotels and restaurants and frequency of collection required.

Solid waste treatment in hotels

In places where the municipal authority disposes of the town solid waste, hotel authorities need only to collect the waste (with possible segregation of different types) store them in bins and deliver them to the municipal collection system. If local authorities are not taking care of this aspect, hotel authorities themselves have to make their own arrangements for such disposal. They may choose one or a combination of following methods for waste disposal.

1. Incineration
2. Pulverization
3. Mechanical compost plant
4. Trenching
5. Controlled tipping
6. Disposal into sea
7. Filling of low-lying area or landfill

We discuss each of methods for waste disposal in some detail.

Incineration   In this method, waste is burnt in the incinerator plant. A recent development is rotary furnace, which can burn practically any kind of waste.

Pulverization   In this method, waste is simply pulverized into powder form without any chemical change. The powder thus formed may be used as manure or discharged through the sewage line.

Mechanical compost plant   A compost plant converts the garbage into manure, which is rich in nitrogen. This is the most hygienic method of waste disposal. However, only organic wastes are candidates for such treatment.

Trenching   In this method, waste is dumped in trench and buried under soil. The garbage is converted to compost.

Controlled tipping   This method is employed where land is available for redevelopment. Waste is tipped from dumper into hollow spaces in the ground about 4 to 7 ft deep and then buried under ground.

Disposal into sea This method is relevant and available only to hotels near a sea. This is quite cheap but in times the non-soluble garbage may come back to the shore and cause problems.

Filling of low-lying areas Waste is dumped into low-lying areas. As the waste lies uncovered, it may produce bad odours and be a potential source of bacteria for a short period before they decompose naturally. During the rainy season, the problem may further aggravate. These areas, therefore, should be far from human habitation. Low-lying areas act as natural draining area for city rain water and thus may cause water logging problem in adjacent areas.

8.5 | REGULATIONS AND NORMS FOR SOLID WASTE MANAGEMENT

There has been a systematic attempt to formalize the various good practices to be observed by citizens, civic bodies and industrial and commercial establishments. Some of them are also in the process of being given legal bindings also. Ministry of Urban Development, Government of India has published *Solid Waste Management Manual* (available in the open domain) for the purpose.

A few selected excerpts from this manual are furnished below.

8.10. GUIDELINES FOR SORTING OF MATERIAL RECOVERY*

8.10.1 General

(a) Sorting of the waste at the source must be accorded the highest priority by the urban local bodies.

(b) The existing system of the *kabadiwala*, which efficiently recovers the dry recyclables and bulky waste (white goods) from the source, must be facilitated.

(c) The role of rag pickers in collecting and recovering recyclables (not taken by the *kabadiwala*) must be recognized and strengthened at the community level by using their services at the household/source level with the help of NGOs/private sector participation.

(d) Municipalities must have separate waste collection, transportation, and disposal streams for:

(i) Biodegradable waste

(ii) Mixed waste (co-mingled waste)

(iii) Construction and demolition waste

(iv) Hazardous waste

(e) Biodegradable waste should be used for biological processing at a central facility/decentralized facility.

(f) Construction and demolition waste should be processed for reuse or stored in landfill cells capable of being mined for reuse.

(g) Hazardous waste should be transferred to hazardous waste landfill or processed appropriately.

(h) Horticulture waste from parks and gardens should be composted at the site or at a decentralized facility to be operated by the municipality.

*Taken from Chapter 8, section 8.10 under the heading 'Sorting and Material Recovery' of *Solid Waste Management Manual*

(Contd)

(*Contd*)

(i) Mixed waste (co-mingled waste) should be sorted into the various streams listed in Section 8.8 either at a transfer station or at a centralized sorting facility. If this is not feasible, mixed waste can be sent to a processing facility which has a well designed pre-sorting/post-sorting facility where the mixed waste can be sorted into separate streams. Mixed waste not found suitable for processing should be landfilled

24.3. PROPOSED LEGAL PROVISIONS**

24.3.1 Prohibition Against Littering the Streets, Deposition of Solid Waste on the Streets, Open Defecation, etc.

No person shall litter public streets or public places or deposit or cause or permit to be deposited or thrown upon or along any public street, public place, land belonging to the local body, State or Central Government or any unoccupied land or on the bank of a water-body any solid waste except in the receptacles specified in 2, 6, and 8 above or resort to open defecation.

24.3.2 Duty of Occupiers of Premises to Store Solid Waste at Source of
Generation

It shall be incumbent on the occupiers of all premises to keep two receptacles, one for the storage of food/organic/bio-degradable waste and another for recyclable and other types of solid wastes generated at the said premises. The domestic hazardous waste, as may be notified by the local body, shall also be kept separately in a suitable container as and when such waste is generated.

24.3.3 Duty of Occupier not to Mix Recyclable/Non-Bio-degradable Waste and Domestic Hazardous Waste with Food Waste etc.

It shall be incumbent on the occupier of any premises to ensure that the recyclable waste as well as domestic hazardous waste generated at the said premises does not get mixed up with the food/bio-degradable waste and stored separately.

24.3.4 Duty of Societies/Associations/Management of Commercial Complexes to Clean their Premises and to Provide Community Bins

It shall be incumbent on the management of co-operative societies, Associations of residents, multistoried buildings, commercial complexes, Institutional buildings, markets and the like to arrange for daily cleaning of their internal streets, common spaces, etc., and provide community bin/bins of appropriate size as may be prescribed by urban local body, for the temporary storage of food/biodegradable waste duly kept segregated by the members of the society/association for facilitating primary collection of food/biodegradable waste from one point by the municipal authorities. A separate community bin may similarly be provided for the storage of recyclable waste where door to door collection of recyclable waste is not practised.

24.3.5 Community Bins to be Kept in Good Condition

Community bins as stated in 24.3.4 above shall at all times be kept in good condition, regularly maintained and shall be provided in such number and at such places as may be considered adequate and appropriate to contain the waste produced by the citizens supposed to be served by the community bins.

24.3.6 Duty of Occupiers to Deposit Solid Waste in Community Bins

It shall be incumbent on occupiers of all premises for whom community bins have been provided as per 24.3.4 above that all segregated domestic waste, trade waste, institutional waste from their respective premises to be deposited in the appropriate community bins.

24.3.7 Duty of Local Body to Provide and Maintain 'Waste Storage Depots'

It shall be incumbent on all Municipal Corporations and Municipalities in the State to:

(i) Provide and hygienically maintain adequate Waste Storage Depots in the city and place large mobile receptacles at such places for the

---

**Taken from Chapter 24, section 24.3 under the heading 'Legal Aspects' of *Solid Waste Management Manual*

(*Contd*)

(Contd)

temporary storage of waste collected from households, shops and establishments as well as from streets and public spaces until the waste is transported to processing and disposal sites.

(ii) Make adequate provision for closed containers in various parts of the city for the deposition by citizens of domestic hazardous/toxic waste material adhering to the provisions of hazardous waste rules of Government of India.

24.3.8 Duty of Occupier of Households/Shops/ Establishment to Hand Over the Recyclable Material/Non-Bio-degradable Waste to the Waste Collectors/ Waste Purchasers/Recyclers

It shall be incumbent on households/shops/ establishments to hand over their segregated recyclable waste/non-bio-degradable waste to the collectors, waste purchaser or recyclers as may be convenient or as may be notified by the local body from time to time. Such waste shall not be disposed off on the streets or in municipal bins or open spaces along with the organic/food/bio-degradable waste.

24.3.9 Duty of Occupier of Households, Shops and Establishments to Deposit Domestic Hazardous/Toxic Waste in Special Bins Provided by the Local Body

It shall be incumbent on households, shops and establishments to deposit domestic hazardous waste/ toxic material in containers provided by the urban local body as per 24.3.7 (ii) above.

24.3.10 Duty of Local Bodies to Collect Waste from Community Bins and to Deposit it at Waste Storage Depots for Onward Transport

It shall be incumbent for local bodies to remove all solid waste deposited in community bins on a daily basis and transfer it to the Waste Storage depots/containers identified in the city and arrange for its expeditious transport to processing or disposal sites.

24.3.11 Duty of Local Bodies to Clean All Public Streets, Open Public Spaces and Slum Areas

It shall be incumbent on local bodies to arrange for cleaning of all public streets having habitation on both or either side, and all slums on all days of the year including Sundays and public holidays.

24.3.12 Duty of Local Body to Transport the Waste Stored at the Waste Storage Depots Regularly

It shall be incumbent for the local bodies to arrange for the transportation of waste stored at waste storage depots before the waste storage containers start overflowing and daily from places where closed containers are not placed.

24.3.13 Duty of Local Body to Arrange for Processing of Food/Biodegradable Waste through Appropriate Technology and Disposal of Rejects

It shall be incumbent for the local bodies to arrange for the processing of food/organic/bio-degradable wastes produced in the city and dispose of the rejects and non-biodegradable waste in an environmentally acceptable manner.

24.3.14 Prohibition Against Deposition of Building Rubbish

No person shall deposit or cause or permit to be deposited any building rubbish in or along any street, public place or open land except at a place designated for the purpose or in conformity with conditions laid down by the municipal corporation/ municipality.

24.3.15 Prohibition on Disposal of Carcasses, etc.

No person shall deposit or otherwise dispose of the carcass or parts of any dead animal at a place not provided or appointed for this purpose.

8.6 | POLLUTION AND THE HOTEL INDUSTRY

Pollution of various forms is a cause of great concern for existence of life in the earth. One of the most disconcerting effects of pollution is global warming. Other direct effects include increasing health problem, mental stress and strain, increase in the number of endangered species, ecological imbalance, among others. Hotels are properties where very high intensity of human and machine activities occur day in and day out. This is bound to produce all sorts of pollution and is subject to very stringent pollution control measures. Also, in tourist destination areas, there will be surface transport carrying them to and from the hotels, thereby causing great pollution from automobile emission, while it is also true that they finally contribute to the developing tourism-based economy. However, hoteliers, in general, are not major offenders in this regard. The various natural systems under the threat of pollution may be listed as follows:

1. Water pollution   2. Air pollution
3. Soil pollution

There is another category of pollution, i.e., noise pollution, which is treated separately in a later part of the section.

Water pollution   Water pollution has assumed a grave situation in developing countries such as India. Rampant discharge of industrial and civic-body effluents into water bodies have played havoc with hygiene and ecology. Wastewater disposal without proper treatment has severely affected marine life and living of downstream people using the water bodies for economic as well as day-to-day use of water. Discharge of hot water also produces pollution in changing the aquatic environment of water bodies.

Air pollution   Components responsible for air pollution are fine suspended solid particles, carbon dioxide, chloro fluoro carbon (CFC) materials (for example, refrigerator gas, materials used in spray dispensers, etc.), oxides of nitrogen and sulphur, etc. All these cause serious problems for human and plant life directly or indirectly. Emission of carbon dioxide causes greenhouse effect and is supposed to contribute to global warming. Fine suspended particles pose a serious threat to human respiratory system and cleanliness of the environment. Oxides of sulphu nitrogen, and carbon dioxide in air combining with rainwater produce acids and cause, what we call, acid rain. It has an adverse effect on aquatic life, trees, and plantation and also damages buildings and materials. Boiler emissions burning coal is the major source of oxides of sulphur. Any combustion device, such as boiler, engine, furnace produce, carbon dioxide. Different oxides of nitrogen are grouped together under the acronym of NO_x. They create harmful particulate matter, ground level ozone (SMOG), and acid rain. Two major sources of NO_x are boilers and engines. So, we observe that chimney emissions in industries including hotel industries, automobile, and stationary engine emissions are the major sources of air pollution.

The discharge of CFC into air is of very serious consequence. It is thought to break the protective layer of ozone gas in the atmosphere, which shields us from the highly harmful ultraviolet radiation from the sun. Hotels make extensive use of refrigerants for freezers and air-conditioning systems which use these CFC materials and are potential contributors to ozone-depleting gases. Ministry of Environment and Forest (MoEF), Government of India (GOI) has identified a list of such materials whose has either been banned or restricted. There is a GOI (Ministry of Environment and Forest) notification titled 'Ozone Depleting Substances of Environment and Forest, dated 19 July 2001. Schedule I of the document furnishes a list of 95 such substances. Table 8.1 furnishes only a partial list of such substances.

Table 8.1 List of ozone depleting substances restricted by MoEF, GoI

S.No.	Name of ozone depleting substance	Chemical composition of ozone depleting substance	Ozone depleting potential
1.	CFC-11	Trichlorofluoromethane ($CFCl_3$)	1.0
2.	CFC-12	Dichlorodifluoromethane (CF_2Cl_2)	1.0
3.	CFC-113	Trichlorotrifluoroethane ($C_2F_3Cl_3$)	0.8
4.	Halon-1211	Bromochlorodifluoromethane (CF_2BrCl)	3.0
5.	Halon-1301	Bromotrifluoromethane (CF_3Br)	10.0
6.	Halon-2402	Dibromotetrafluoroethane ($C_2F_4Br_2$)	6.0
7.	CFC-213	Pentachlorotrifluoropropane ($C_3F_3C_5$)	1.0
8.	HCFC-132	Dichlorodifluoroethane ($C_2H_2F_2Cl_2$)	0.05
9.	HCFC-151	Chlorofluoroethane (C_2H_4FCl)	0.005
10.	HCFC-244	Chlorotetrafluoropropane ($C_3H_3F_4Cl$)	0.14
11.	HCFC-251	Trichlorofluoropropane ($C_3H_4FCl_3$)	0.01
12.	HCFC-252	Dichlorodifluoropropane ($C_3H_4F_2Cl_2$)	0.04
13.	HCFC-261	Dichlorofluoropropane ($C_3H_5FCl_2$)	0.02
14.	HCFC-271	Chlorofluoropropane (C_3H_6FCl)	0.03
15.		Tribromodifluoroethane ($C_2HF_2Br_3$)	1.8
16.	HBFC-124B1	Bromotetrafluoroethane (C_2HF_4Br)	1.2
17.		Pentabromodifluoropropane ($C_3HF_2Br_5$)	1.9
18.		Tetrabromofluoropropane ($C_3HF_3Br_4$)	1.8
19.		Tribromotetrafluoropropane ($C_3HF_4Br_3$)	2.2
20.		Tribromotrifluoropropane ($C_3H_2F_3Br_3$)	5.6
21.		Dibromotetrafluoropropane ($C_3H_2F_4Br_2$)	7.5
22.		Bromopentafluoropropane ($C_3H_2F_5Br$)	1.4
23.		Tetrabromofluoropropane ($C_3H_3FBr_4$)	1.9
24.		Tribromodifluoropropane ($C_3H_3F_2Br_3$)	3.1
25.		Bromotetrafluoropropane ($C_3H_3F_4Br$)	4.4
26.	Methyl bromide	(CH_3Br)	0.6

Soil pollution Solid waste and sludge from sewage-treatment plant are dumped on to the soil. This is putting pressure on land as free land for the purpose is becoming scarce and costly. Further, disposal of carcass creates health problem for people living nearby. Of course, disposal of organic solid waste and municipal sludge makes the soil fertile and many crops are grown on it. Excessive and frequent dumping on a piece of land can lead to sewage sickness, whereby the degeneration of sewage is halted, causing much nuisance and production of foul smelling gas. Also, such land is said to be septic.

8.6.1 Sources of Pollution

The various sources of pollution are
1. Domestic, industrial, and hospitality industry effluent (liquid waste) to water bodies or sewage pollution
2. Domestic, industrial, and hospitality-industry gas emission (furnace, ovens, engine) released to the atmosphere or gas pollution
3. Solid waste and sewage pollution to soil
4. Solid waste and sludge pollution to soil
5. Thermal pollution
6. Noise pollution

We discuss each of these with reference to the hotel industry.

Sewage pollution

The risk and the fall out of sewage contamination with potable water inside the premises is very high. Various methods of collecting and draining sewage have already been dealt with. It is always advisable that before draining the sewage to the municipal main drain (if the system exists), the hotelier should have some initial treatment to reduce the load on the municipal treatment plant and contribute to the societal needs. This should be a must in small towns where the property holders often carry the effluent directly to a water body such as sea, river, or pond.

In many cases, people think that their waste disposal ceases with transporting them to the river or other water bodies without any sort of treatment. This makes the flowing water totally unsuitable for use to the downstream people. Aquatic and marine life can also be severely affected. Hotels produce a lot of wastewater and many of them have now installed their own STP, which contributes greatly in reducing pollution as well gaining economy by way of reuse of clear water. This pollution can be minimized at the source itself by restricting chemicals that mix with water, restricting/eliminating garbage, and oil or unknown chemicals in the waste system (for example, detergent water in the sewage severely affects performance of septic tank), etc. Hotels produce a lot of grease and fat in kitchen and detergent water in laundry, which go to the wastewater. The presence of detergent makes septic tank treatment less effective. Hotels, however, have grease traps in the wastewater line, as already discussed.

Pollution due to gas emission

Hotels liberate gases and contaminated air from various utilities, such as kitchen gas and firewood ovens, fume from materials being cooked, boiler and diesel generating set exhausts, and release of refrigerant CFC, if there is any leakage, as also air from many ventilation ducts throughout the property. Kitchen fume normally will have very little suspended particulate matter because powdered coal, which is the main source of such particles, is not used anywhere in the kitchen. Suction hoods with grease filter are used to suck kitchen exhaust out, which go to the common manifold before being either discharged to atmosphere or, in many hotel kitchens, recycled into the kitchen after being cooled and mixed with fresh air. To keep adjoining area free from smoke and odour, kitchen air pressure is kept slightly below atmospheric pressure so that outside air can get in but not the reverse.

If the boiler is fired with pulverized (powdered) coal, the exhaust gas will have a lot of dust particles and devices such as cyclone separator or electrostatic precipitator (ESP) must be used to separate out dust particles from the chimney gas in order to bring down the level of suspended particulate matter (SPM) below certain size and concentration as stipulated by the pollution regulatory boards.

Diesel engines also produce pollutants such as carbon dioxide, carbon monoxide, and oxides of nitrogen. Better engine combustion reduces carbon dioxide, while catalytic converter installed in the exhaust gas line reduces the oxides of nitrogen to a very low level before being discharged to atmosphere. The release point must be quite away from the guest rooms and at a height of 10 to 20 ft. As regards the release of aerosol compounds, such as CFC from refrigerant, proper maintenance of refrigerant lines would ensure that there is no leakage in the pipeline.

Solid waste and sludge pollution to soil

We have already discussed about solid waste and the problem of its disposal. Excessive dumping of untreated or semi-treated sewage and sludge may lead to sickness of soil where the land may become very clumsy and emit very foul smell, as discussed earlier.

Thermal pollution

The primary source of thermal pollution is discharge of hot water into a water body. It is called thermal pollution because due to the mixing of hot water, the temperature of the water body goes up and may completely change the characteristics of the water body. It may change the nature of plant species and aquatic life. Apart from immediate effects, long-term effects may be drastic and the changes could be irreversible. Release of hot water means wastage of energy that is used to heat the water. Had it been recovered, we could achieve more economy as well as thermal pollution. So, the solution to this problem is to increase energy efficiency through improved technologies.

Noise pollution

We now discuss noise pollution, a kind of pollution which does not have a direct bearing on the environment but can severely affect human health, both mental and physical. It also affects the quality of work environment and can considerably reduce the working efficiency in the zone. High level of noise is unacceptable for public life and workplace. Acceptable levels of noise are worked out for specific work areas and also for public life.

Noise pollution can be reduced by (a) taking measures to reduce the level of noise at the source and (b) using sound proofing arrangements to isolate the area belonging to the source of noise from other places. The effect of sound on human ear and sensitivity depends upon the intensity of sound energy and is measured in decibels (db). Rustling sound of leaves in a tree of a pin falling on the ground can produce a sound level of 10 db and is hardly perceptible to the human ear, while a jet engine produces about 140 db and is quite disturbing to human tolerance. The decibel is a logarithmic scale and a change of sound intensity by 10 db means an energy-level change of 10 times. So an increase of noise level from 60 to 70 db means a 10 times increase in the noise amount to the human ear and its effect on the nervous system. Human response to sound depends not only on the energy content but also on the frequency of the sound waves. Generally, sound produced by a vibrating object consists of many frequencies. There are many sound-absorbing materials that absorb sound depending on the frequency.

Noise pollution in hotels   Hotel industry is not a major generator of noise to the external world and public life, though it has to control noise in-house very closely to maintain good working and living conditions. Hotels have quite a few noise generating sources such as engines, pumps, motors, etc. and sound produced in banquet halls. A speaker at a conference hall can feel disgusted by the background noise of an air-conditioning unit or ventilation system. Proper sound insulations are provided in the walls separating engine and pump rooms, and banquet halls from other places in the property. In fact, increased vibrations of components increase noise and elimination of vibration of the noise-generating units could be a very effective method for reducing noise level. In many cases, anti-vibration pads under the machine foot are excellent sound eliminators. Good design of air conditioning duct for structural rigidity and elimination of fan vibration can reduce noise from air conditioning units. Apart from mechanical vibrations, there could be aerodynamic noise in air duct-fan system, which needs special engineering to be eliminated. Countertop mechanical equipment, such as mixers, as well as some serving utensil, which are frequently kept on stainless steel countertop do have rubber footings at the bottom to reduce noise. Control and maintaining required levels of sound in conference rooms and banquet halls inside and insulating these rooms from noise of the outside world and vice versa are highly technical job and professional sound control engineers design the acoustic systems for these facilities.

8.7 | POLLUTION CONTROL BOARDS AND POLLUTION NORMS

Pollution control boards belonging to both state and union governments have made stringent rule as to the pollution norms of different sources of pollution. Recently, in India, stringent actions are being taken against offenders of pollution norms and it is good to see that industries are installing pollution control equipment for the purpose. Various authorities have set for the quality standard of wastewater being carried to the soil or the water body. The effluent is assessed on the basis of its BOD, COD, lead and other heavy metal content, etc. They have also stipulated the maximum allowable suspended particulate matter (SPM) that can remain in the exhaust gas from equipment prior to its release in the atmosphere. Regarding thermal pollution, regulation is yet to be stipulated. There is also no regulatory directive as to the disposal of solid waste by property holders. The maximum sound level reaching the outer world affecting public life is set to about 65 to 70 db as stipulated by different pollution control boards in the country, depending upon the norm of the particular pollution control board in the region of the country. Hospitality establishments such as any other engineering, process, and service industries have to obtain certificates from appropriate pollution control boards for conformance to the pollution norms. These certificates are renewable every year. Inspectors of the board quite often make surprise visits to the site to see if the pollution control equipment and processes are running properly and the effluent quality and other pollution parameters are being maintained. Willful non-compliance of the norms can lead to discontinuation of the operation of the offending unit. Hotel industry has to guard against this eventuality very closely as this may ruin the business. As such, all the stipulations of pollution control must be very meticulously followed and practised in hotel industries.

8.8 | CONCLUSION

We have seen how important sanitation and waste disposal are and how complex they could be for a hotel industry. Improper sanitation can create havoc for public health and general hygiene in a property. There must not be any compromise as to the quality of components and systems used for the purpose. Sewage treatment plant has revolutionized water conservation by recycling process. Hotel management must be responsive for proper solid waste management as hotels produce a lot of such waste. Waste disposal and pollution control measures should be practised not only as a legal requirement but also as a social responsibility.

KEY TERMS

Biochemical oxygen demand (BOD)	The oxygen demand in this biochemical reaction during degradation of organic sewage matter, measured in milligrams of oxygen required per litre of wastewater, indicating sewage toxicity is called biochemical oxygen demand.
Chemical oxygen demand (COD)	It is the total oxygen demand of biodegradable matter and other chemicals together, another indicator of the quality of the sewage (or wastewater) in addition to BOD.
Inspection chambers	These are walled openings built round the drain within which the effluent from all branch drains flows in and facilitates inspection for blockage, if any.
Sanitation	Sanitation is the hygienic living environment within premises
Sewage	Liquid and semi-solid waste is called sewage.
Sewer	Sewer is a sewage transport conduit.
Sewerage	The system of sewers and waste drainage is called sewerage.
Sewage treatment	It is the process of removing the contaminants from sewage to convert it to a composition of clear liquid and solid which are fit for discharge to the environment or for reuse.
Sewage treatment plant (STP)	It is a plant which receives raw sewage and partially of fully purifies the sewage.
Traps	A trap is a depression or a bend in a sanitary fitting that remains always full of water and thereby does not allow foul air to enter a room from outside sewer pipe.

Water seal   This is an arrangement in a pipe where a particular portion of the pipe, usually in the form of U-tube shape, always remain full of water to prevent passage of gas from one end of the U-tube to the other

REVIEW QUESTIONS

8.1 Define sanitation and list its principal functions.

8.2 Discuss, in brief, about sewage.

8.3 With the help of a neat sketch, explain wastewater collection and drainage from rooms in a hotel.

8.4 List a few wastewater collection and drainage components identifying their uses.

8.5 Explain why water seals are necessary in wastewater lines.

8.6 Compare between two-pipe and single-pipe stacks.

8.7 Explain why anti-syphonage pipes are necessary in wastewater stack.

8.8 Discuss about baths and kitchen sinks.

8.9 Define sewage treatment.

8.10 Discuss about purification method of sewage treatment and disposal.

8.11 List the stages of processes in a modern sewage treatment plant.

8.12 Define BOD and COD, citing their importance as measures of toxicity of sewage.

8.13 Write about biosolids and their uses.

8.14 Discuss about the importance of segregation in solid waste management.

8.15 List the various means of solid-waste treatment and disposal that can be adopted in a hotel establishment.

8.16 Discuss about various types of pollution, with special reference to hotel industry.

8.17 Discuss about pollution control boards and pollution norms in India.

9 Chapter

Fire—Prevention and Control

Learning Objectives

This chapter introduces the topic of fire and firefighting in some elementary terms while furnishing information for practising maintenance engineers, as well as students of hotel management and for general reading. It discusses the mechanism of fire, types of fire, and fire extinguisher and preliminary means of controlling fire. The primary objectives of this chapter are to:

- orient the reader towards the basics of fire, fire hazards, and fire control
- familiarize the reader with different types of fire and fire extinguishers
- impress upon the reader as to the regulatory requirements and need for adherence to them
- present a few case studies of fire in the hotel industry

9.1 | INTRODUCTION

We are all familiar with the term fire. History of civilization is said to have begun the moment man learnt to ignite fire. However, uncontrolled fire is devastating and could play havoc. Many an architectural edifice, in the past, has been totally destroyed by fire. Besides, fire has taken a huge toll on human life. For any industry including hospitality industry, fire means not only damage to human life and property but also huge revenue loss. So, every effort should be made to make the construction and operation of the establishment fireproof as far as practicable, and sufficient firefighting arrangements should be provided. Fire regulations make it mandatory for every establishment to have arrangements for fire prevention and fire extinguishers to control it in case of fire.

9.2 | ELEMENTS OF FIRE AND FIRE PYRAMID

Technically, 'fire' means a rapid oxidation-reduction reaction resulting in the production of heat and, generally, of visible light in the form of a luminous flame.

We now introduce a few terms in connection with fire, which will help us in understanding the mechanism of fire and means to control and prevent it.

9.2.1 Elements of Fire

The basic elements and common terminologies associated with the phenomenon of fire are as follows:

Oxidation-reduction reaction Whenever a material burns, the combustible part of the material reacts with oxygen. The combustible part is said to be oxidized and oxygen is said to be reduced. In simple terms, such a reaction is called an oxidation-reduction reaction. The combustible part is called a reducing agent with oxygen as the oxidizer.

Fuel Fuel is a substance that acts as a reducing agent. A reducing agent gives up electrons to an oxidizer, which is capable of accepting electrons, resulting in a compound consisting of the fuel and the oxidizer. When a fuel reacts with an oxidizer, we say the fuel is burning. This reaction releases heat and visible light. When this happens very fast with rapid and large amount of heat and visible light, we say that fire has occurred. So in common parlance, fuel may be defined as a substance which when burnt gives off heat and/or light for useful purpose.

A fuel may be an element such as carbon, hydrogen, magnesium, or a compound such as carbon monoxide and hydrocarbons such as methane, propane, butane, etc. It may be a complex organic compound such as firewood, coal, jute, cotton, rubber, natural gas, petroleum, etc. Methane, propane, and butane are used as cooking gases. Commercial liquefied petroleum gas (LPG) is a mixture of propane and butane.

Oxidizer Oxidizer is one essential component for burning. Scientifically speaking, an oxidizer is a substance that has an affinity for electrons and acquires them from a reducing agent (tendency to lose electrons) like a fuel as described. Besides oxygen, fluorine, chlorine, nitric acid, sulphuric acid, etc. are strong oxidizers.

Explosive limit The explosive limit of a gas or a vapour in air is the limiting concentration of the gas or vapour in air that is needed for ignition and explosion. There are two explosive limits for any gas or vapour, namely the lower explosive limit (LEL) and the upper explosive limit (UEL). At concentrations below the LEL, there is not enough fuel to continue the reaction for an explosion while at concentrations above the UEL, there is not enough oxygen to begin a reaction.

The explosive limits of some gases and vapours are given in Table 9.1. Concentrations are in per cent by volume of air.

Table 9.1   Explosive limits of some flammable gases

Substance	LEL (%)	UEL (%)	Substance	LEL (%)	UEL (%)
Acetylene	2.5	82	Ethanol	3	19
Methane	4.4	17	Diesel	0.6	7.5
Propane	2.1	9.5	Petrol	1.4	7.6
Butane	1.8	8.4	Kerosene	0.6	4.9
Benzene	1.2	7.8	Hydrogen	4	75

(David R. Lide, Editor-in-Chief, *CRC Handbook of Chemistry and Physics,* 72nd edition; CRC Press, Boca Raton, FL, 1991).

Flash point   Flash point is the lowest temperature at which a liquid fuel will give up just enough vapour to form an ignitable mixture with air, which will burn momentarily but it will not sustain burning.

A fuel does not burn as a liquid until it gives off enough vapour by way of evaporation (the evaporation rate depends on temperature and pressure) to approach its LEL in air. If an external igniter (like an ignited matchstick or some naked flame) is introduced in the environment, the vapour will burn momentarily and die out for want of sufficient release of vapour to sustain the burning. Thus, at flash-point temperature, there will be a flash burning but there is no adequate fuel supply to continue the burning process. The flash point indicates how easy a chemical may burn.

Flammable liquids have a flash point below 37.8°C. Combustible liquids have a flash point of 37.8°C or more.

Fuels and their flash points at atmospheric pressure are given in Table 9.2.

Table 9.2   Some common fuels and their flash points

Fuel	Flash point (°C)	Fuel	Flash point (°C)
Benzene	−11	Diesel fuel (1-D)	38–55
Ethyl alcohol	13	Fuels oils	38–121
Petrol/gasoline	−42.80	Gear oil	190–304
Isobutane	−82.7778	Isopentane	Less than −51
Isooctane	−12	Kerosene	38–72
Methyl alcohol	11	Motor oil	216–252

(Contd)

Table 9.2 (*Contd*)

Fuel	Flash point (°C)	Fuel	Flash point (°C)
n-Butane	−60	*n*-Pentane	Less than −40
n-Hexane	−22	*n*-Heptane	−4
n-Octane	13	Naphthalene	79
Propane	−105	Styrene	32
Toluene	5		

Materials with higher flash points are less flammable or hazardous than those with lower flash points. Table 9.3 gives an indication of the hazard levels associated with a particular flash point.

Table 9.3 Hazard level corresponding to flash point of a substance

Hazard level	Flash point
Very low hazard	Greater than 93°C
Moderate low hazard	66°C to 93°C
High to moderate hazard	38°C to 66°C
Extreme to high hazard	−18°C to 38°C
Extreme hazard	−18°C

An open flame is not always necessary to ignite the gas. A hot surface such as a heating element or warm machine is enough to ignite chemicals with flash points that correspond to high hazard level.

The flammable (explosive) range is the range of a gas or vapour concentration, within which it will burn or explode if an ignition source is introduced. Limiting concentrations are commonly called the lower explosive or flammable limit (LEL/LFL) and the upper explosive or flammable limit (UEL/UFL).

Fire point   We have seen how a flash occurs when a fuel reaches flash point in presence of air or any oxidizer and an igniter. Now, if the fuel is further heated, the duration of such burning increases. The lowest temperature at which burning occurs in the presence of an external igniter and continues for at least 5 seconds is called fire point.

Ignition temperature   If the temperature of air is increased beyond fire point, a temperature would reach when fuel vapour and air will combine and catch fire even without the presence of external igniters.

The minimum temperature at which a substance will continue to burn without application of external heat is called ignition temperature. This is essentially an auto-ignition process. It is also called kindling point. Every ordinary fire occurs when a substance (fuel) in the presence of an oxidizer (for example, air) is heated to this critical temperature.

Table 9.4 furnishes ignition points for commonly used fuels and flammable materials.

Table 9.4 Ignition points of some fuels

Fuel	Ignition Point (°C)	Fuel	Ignition Point (°C)	Fuel	Ignition Point (°C)
Benzene	560	Isopropyl alcohol	399	Rifle powder	288
Bituminous coal	300	Light gas	600	Semi-anthracite coal	400
Butane	420	Light hydrocarbons	650	Styrene	490
Carbon	700	Methane (natural gas)	580	Petrol	280
Carbon monoxide	609	Methyl alcohol	385	Petroleum	400
Charcoal	349	Naphtha	550	Kerosene	210
Coke	700	Propane	480	Gun cotton	221
Ethyl alcohol	365	*n*-Butane	405	Pine wood (dry)	427
Fuel oil	210–260	*n*-Heptane	215	Heavy hydrocarbons	750
Isobutane	462	*n*-Pentane	260		

Fire triangle Fire sets in when heat (or heat of combustion), fuel, and oxygen are present in the right amount. These conditions may be represented by the three sides of a triangle, which is commonly called fire triangle. So, fire triangle may be defined as a triangle, the sides of which indicate the three components ordinarily essential for the occurrence of fire.

Chain reaction The well-known chemical reaction equations for burning of carbon and methane are as follows:

$$C + O_2 \rightarrow CO_2$$
$$CH_4 + 2O_2 \rightarrow 2H_2O + CO_2$$

However, the combustion mechanism is far from what is indicated by the simple reaction equations shown above. Till date, fourteen intermediate reactions have been identified that finally culminate in the familiar chemical reactions as above. These intermediate chemical reactions are known as chain reactions, so named because one reaction triggers another and this process is continued till the final product is obtained. Formally, we may define chain reaction as a self-sustaining chemical or nuclear reaction yielding energy or products that cause further reactions of the same kind.

Some of the reactions are exothermic (releasing heat) while some are endothermic (absorbing heat). The net heat of combustion is the balance of heat among the reactions. If endothermic reactions are predominant, the environment will cool down and fire would stop.

Inhibitors, which cause endothermic reactions, can, therefore, be introduced externally in a fire to suppress it. There are suppressants that cause endothermic reaction to occur at rates faster than exothermic reactions.

9.2.2   The Fire Pyramid and Principle of Extinguishing Fire

From our foregoing discussion, it is apparent that apart from the three elements of fire, we have a fourth one, which is the chain reaction needed to sustain fire. The principle of extinguishing of fire is based on 'fire triangle' and the removal of any of its three sides, each representative of the three essential components. Remodelling the 'fire triangle' into a 'fire pyramid' presents a more realistic concept of extinguishing fire. The pyramid has four planes. These four planes represent the following components necessary for sustained fire.

1. Heat (or heat of combustion)   2. Fuel

3. Oxygen   4. Chemical chain reaction

Removing one or more of the four components would be necessary for extinguishing fire and forming the basis of fire-extinguishing techniques and devices.

To extinguish fire, one or more of the following steps should be taken:

Removing/limiting the supply of oxygen by

- Preventing air from entering the fire-zone air, e.g., blanketing by water mist, non-flammable gas, etc.
- Shutting the lid over a burning fuel tank or putting foam to prevent entry of fresh air
- Diluting air with a non-oxidizing gas such as carbon dioxide, nitrogen, etc.

Stopping supply of fuel   By mechanical means, e.g., diverting the fuel or closing the supply line valve

Removing heat   By way of cooling the burning materials below the ignition point with a suitable cooling agent

Interrupting the chemical chain reaction of the fire   By using dry chemical or Halon extinguishing agents, as discussed later

9.3   CLASSES OF FIRE

Fire is classified. depending upon the type of fuel/combustibles involved and the source of heat. Each of the different types needs separate and specific means of control to extinguish it. Table 9.5 furnishes a summary of such a classification.

Table 9.5   Classes of Fire

Category in the USA	Category in the UK/Europe	Fuel/source of fire
Class A	Class A	Ordinary combustibles
Class B	Class B	Flammable liquids
	Class C	Flammable gases
Class C	Class E	Electrical equipment
Class D	Class D	Combustible metals
Class K	Class F	Cooking oil or fat

We now discuss, in brief, the different classes of fire according to the UK/Europe/Asian nomenclature.

Symbols

Class A   Fire in ordinary combustible materials such as wood, paper, clothing, plastics, trash, etc. Here quenching and cooling effects of water or of solutions containing large percentage of water are of prime importance.

Class B   Fire in flammable liquids, which must be vapourized for combustion. These include gasoline, petroleum, and paints. This category also includes liquids, such as propane and butane, but does not include fires involving cooking oil and grease. Here, exclusion of air and interrupting the chemical chain reaction are most effective in the suppression of fire.

Class C   This refers to flammable gases, Fire that occurs in flammable gases and fuels, e.g., hydrogen, ammonia, acetylene, LPG, petrol, etc.

Class D   Fire that occur in combustible metals such as magnesium, lithium, and sodium. Many car tyre rims are now made with magnesium. Combustion temperature and energy release are high compared to class A, B, or C fire. Special extinguishing agents and techniques are needed for fires of this type.

 *Class E* Fires involving electrical equipment—fires in or near live electrical equipment where the use of non-conductive extinguishing is of first importance. The material that is burning, however, may belong to either class A or class B in nature.

  Class F Class F fires are fires involving cooking oils or fats. High-efficiency deep fat fryers have necessitated the introduction of this class of fire separately. Class F fires differ from conventional liquid fires due to the high temperatures involved. Typical flammable liquids, for example, petrol, have low flash and auto-ignition temperatures and are relatively easy to extinguish. Cooking oil and fat have auto-ignition temperatures exceeding 340°C and the fire thereof is very difficult to extinguish using conventional extinguishers having a class B capability. The use of the incorrect extinguisher can be extremely dangerous in such cases. For example, a water-jet extinguisher directed onto the surface of burning cooking oil will create an explosion, as water is quickly converted into steam, resulting in the expulsion of burning oil possibly spreading the fire and harming the operator. Other pressurized extinguisher types, such as water, foam, powder, or CO_2, must not be used on burning cooking oil, as the pressure jet might carry the burning oil and spread the fire. There are other reasons too for not using conventional fire extinguishers. Conventional foam extinguishers are capable of extinguishing the flame, but the heat involved quickly destroys the foam blanket, exposing the surface of the oil, allowing re-ignition. Carbon dioxide and multi-purpose powder extinguishers are effective in extinguishing the flame, However, surface of the burning liquid is not sealed off from external oxygen environment and, hence, the oil rapidly re-ignites. Recognizing the difficulties and inadequacies of conventional class B extinguishers, a new standard BS7937: 2000 is created to cover the special risks involved.

9.4 | FIRE PROTECTION AND EXTINGUISHERS

Fire protection in a building is a comprehensive system comprising the following:

Active fire protection It is primarily fire detection and fire suppression.

Passive fire protection It primarily compartmentalizes the building by using fire-resistant walls and floors. The whole building is organized into separate smaller zones, called fire compartments, consisting of one or more rooms or floors. This arrangement prevents or slows the spread of fire from the spot of the origin of fire to other zones in the building, thus limiting damage to the building. It also provides more time to occupants for evacuation or to reach a safe area.

Fire prevention This includes various preventive measures such as minimizing fire sources, educating the occupants and operators of the facility, awareness and concern for correct functioning and operation of fire-related systems and emergency procedures.

We discuss active fire protection and fire prevention in the following sections.

9.5 | ACTIVE FIRE PROTECTION

As discussed, active fire protection involves the control of smoke, detection and communication process signaling the occurrence of a fire outbreak and triggers some sort of counteraction towards extinguishing the fire. As such, it comprises fire extinguishers, fire and smoke detectors, water sprinklers, etc. We discuss each of them separately.

9.5.1 Fire Extinguishers

A fire extinguisher is an active fire-protection device. It is normally a manually operated and typically a hand-held device (a cylindrical or conical pressure vessel), which produces fire-resistant agents and can be discharged to extinguish a fire. These are to be used to extinguish or control small fires only, often in emergency situations. It is not designed for use on fire apparently going out-of-control, such as one which has reached the ceiling or puts the incumbents and the user of the extinguishers in danger, for example, there is no escape route or huge smoke is billowing out or there is a risk of explosion, etc. Such large-scale fire requires the expertise of a fire department to control.

Different types of fire extinguishers

Different types of fire extinguishers are designed for use in specific classes of fire. These may be listed as follows.

1. Water type
2. Foam type
3. CO_2 type
4. Dry chemical powder (DCP) type
5. Halogenated hydrocarbon (HALON) system
6. Wet chemical type

We now discuss them in brief as follows.

Water type   In this type, water is forced out of the extinguisher, thereby blanketing and cooling the combustibles. This is suitable for Class A type of fire. There are a few variations of this depending upon the propellant that is used to force water out. They are the following:

- Soda-acid type/wet CO_2 type, where pressurized CO_2 (a product of a reaction taking place within the extinguisher) forces water out
- Gas cartridge type, in which method of expelling water is by pressure in carbon dioxide gas cartridge
- Stored pressure type, in which the method of expelling water is by air or nitrogen pressure, the advantage being the elimination of recurring expenses of refilling of gas cartridges

We discuss in some detail the working of soda-acid type of water-type fire extinguishers. It is a wet-type fire extinguisher, which ejects water with carbon dioxide gas as the propellant on to the fire and extinguishes it by suppression of oxygen and also by cooling the object under fire. The conical cylinder contains sodium bicarbonate ($NaHCO_3$) or baking soda solution and concentrated sulphuric acid (H_2SO_4). Concentrated sulphuric acid is kept in a cartridge or wire gauge to which a plunger is fitted. When fire occurs and the extinguisher is intended to use, the plunger is struck on to the hard ground so that the cartridge breaks and the acid comes in contact with the soda solution. As a result of a chemical reaction ($H_2SO_4 + 2NaHCO_3 \rightarrow Na_2SO_4 + 2H_2O + 2CO_2$), carbon dioxide gas is liberated and becomes pressurized and forces water inside to gush out of the nozzle. This water-CO_2 mixture is applied on to the fire. Carbon dioxide, being heavier than air, sinks and makes a blanket over the fire, thus preventing oxygen from entering the zone of combustion. Water cools the zone as well as masks the oxygen.

It is to be remembered that water-type extinguishers are suitable for class A fire but not for B and E classes of fire.

Foam type   Soda-acid type fire extinguishers cannot be used in oil fire because oil being lighter floats on water. Thus, a modification of soda-acid type is made by adding Saponin to baking soda in the cylinder. Saponins are steroid glycosides that are found in various parts of plants as well as in fruits. Saponins dissolve in water to form a soapy froth. This helps in emulsification (mixing oil, for example, with water forming a cloudy substance) of oil. The wire gauge contains aluminium sulphate, $Al_2(SO_4)_3$.

In case of fire, the plunger is struck against the floor to break the wire gauge, which allows aluminium sulphate to react with soda liberating carbon dioxide, as follows:

$$6 \text{ NaHCO}_3 + \text{Al}_2(\text{SO}_4)_3 \rightarrow 3 \text{ Na}_2\text{SO}_4 + 2 \text{ Al(OH)}_3 + 6 \text{ CO}_2$$

Al(OH)_3 together with CO_2 forms the foam. This is suitable for fire involving oil. This is more expensive than soda-ash-water type but is more versatile and is used for both A and B classes of fire. Foam-spray extinguishers are not recommended for fires involving electricity but are safer than water-type, if applied inadvertently.

CO$_2$ (carbon dioxide) type extinguisher   Carbon dioxide is in gaseous state under normal pressure and temperature but can be in liquid state even at normal range of atmospheric temperatures if the pressure is sufficiently high. In this type of extinguisher, CO_2 is liquefied under a pressure of about 236 kgf/cm^2 (about 235 times the normal atmospheric pressure) and stored in the extinguisher cylinder. If this liquid is allowed to come out to atmospheric pressure conditions, it suddenly expands and evaporates to gaseous state. The required latent heat of evaporation is taken from the liquid itself, causing a huge drop in temperature to about $-78°C$. This creates a cooling effect on the surroundings. A discharge valve at the top of the extinguisher is opened by simply rotating the wheel when the gas comes out through a horn-shaped nozzle with gradually expanding area. This extinguisher works by

- Cooling the materials
- Blanketing the space, thereby reducing oxygen supply
- Inhibiting combustion

This is suitable for B and E classes of fire, but not on chip pan fires. Recharging should be done immediately after use. A few lines of caution may be that the atmosphere may become deficient of oxygen and the discharge horn may become dangerously cold to touch.

Dry chemical powder (DCP) type extinguisher   In this, sodium bi-carbonate (NaHCO_3) is used as the dry chemical powder. It is fluidized (suspension of fine solid particles in a gas or air called carrier gas) by using carrier gases such as nitrogen (N_2) and CO_2. The extinguisher, therefore, contains a mixture of NaHCO_3 and N_2 or CO_2 under pressurized condition. Air cannot be used as the carrier gas as it contains oxygen, which helps the fire. When fire occurs, the carrier gas comes out of the outlet nozzle at a high velocity carrying the DCP along with it. Thus, fire is extinguished under totally dry condition and the extinguisher is suitable for fire involving petrol, oil, grease, paint, and especially for class C fire. It can also be applied for A and B classes of fire.

Special powder for class D fire   This is also a type of dry chemical powder specially meant for Class D fire (metal fire). For example, a mixture of trimethyl boroxine (TMB) and methyl alcohol is used as fire suppressant for fire caught on magnesium. TMB decomposes into boric oxide, which forms a coating on the burning magnesium and acts as blanket.

Halon (halogenated hydrocarbon) At high temperatures, halons decompose and release halogen (chlorine, bromine, fluorine, iodine, etc.) atoms, which combine readily with hydrogen atoms, arresting the flame propagation reaction and quenching the flame even when adequate fuel, oxygen, and heat remains in the fire zone. Halons are able to destroy the fire at much lower concentrations than are required by those using the more traditional methods of cooling, oxygen deprivation, or fuel dilution. Normally, Halon 1301 and Halon 1211 are used as fire suppressants.

Halons are very effective on classes A (organic solids), B (flammable liquids and gases), and E (electrical) fires, but they are totally unsuitable for class D (metal) fires, as they will not only produce toxic gas and fail to extinguish the fire, but in some cases pose a risk of explosion. Halons can be used on Class F (kitchen oils and greases) fires, but offer no advantages over specialized foams.

Halon 1211 is typically used in hand-held extinguishers, in which a stream of liquid Halon is directed at a smaller fire by a user. Halon 1301 (bromo tri-fluoro methane) is more usually employed in total flooding systems. In this mode, extinguishment is achieved by cooling and oxygen deprivation at the core of the fire, as well as radical quenching over a larger area. After fire suppression, the halon moves away with the surrounding air, leaving no residue.

It is considered a good practice to avoid all unnecessary exposure to Halon 1301, and to limit exposures to concentrations of 7 per cent and for a duration of below 15 min. In practice, the operators of many Halon 1301 total flooding systems evacuate the space on impending agent discharge. There is also a risk of toxic and irritant products.

Nowadays, except for special application areas, for example, in police, army, etc. use of Halon is banned.

Class F fire extinguisher (wet chemical type) Extinguishers designed for cooking oil fires typically include 'wet chemical', 'dry chemical', or are foam-based with special additives. Wet chemical (potassium acetate, carbonate, or citrate) extinguishes the fire by reacting with the hot burning oil to create a thick soapy heat-resistant crust on top of the cooking oil surface, preventing the flammable vapours from reacting with oxygen. In addition, it also cools the oil below its ignition temperature. The formation of this soapy layer is called 'saponification'. The alkalinity of the extinguishing material quickly reacts with the burning oil to create the soap layer. Some of the 'foam with special additive' extinguishers work by covering the hot burning oil with a thick heat-resistant layer on top of the surface while at the same time cooling the burning oil by converting the extinguishing water into steam (thus absorbing latent heat of evaporation from burning oil) in a controlled way.

Wet chemical (essentially potassium acetate because of its ability to cool and form a crust over the burning oil) fire extinguishers have the capability to fight fires in deep fat fryers of up to 75-litre capacity with a 6-litre wet chemical content. It contains a specially formulated wet chemical, which when applied to the burning liquid cools and emulsifies the oil, extinguishes the flame, and seals the surface preventing reignition. An added benefit is that it is also suitable for freely burning materials (class A) such as wood, paper, and fabrics.

Table 9.6 show the fire compatibility of different types of portable fire extinguishers.

Table 9.6 Fire compatibility table of different types of portable fire extinguishers

Class of fire	Combustible materials	Type of extinguishers
A	Paper, wood, cloth, etc.	Water-soda acid type, foam, DCP
B	Oils, paints, grease	Foam type, CO_2, DCP
C	Gas fire	Halon, DCP
D	Metal fire	DCP (special powder only)
E	Electrical fire	Halon, DCP
F	Cooking oils and fats fire	Wet chemical

Colour code for fire extinguishers

Recent BS EN3 STANDARD requires that at least 95 per cent of the extinguisher's body must be coloured red, allowing only 5 per cent colour-coded zones on the extinguisher. These bands of specific colour-coded zones are furnished below.

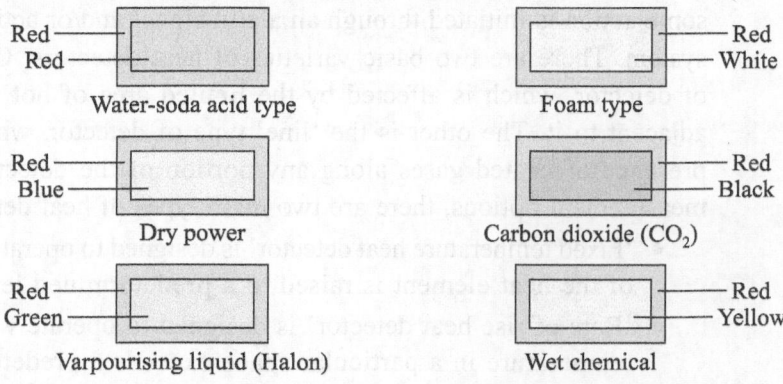

Red — Red
Water-soda acid type

Red — White
Foam type

Red — Blue
Dry power

Red — Black
Carbon dioxide (CO_2)

Red — Green
Varpourising liquid (Halon)

Red — Yellow
Wet chemical

Application areas of firefighting systems

Firefighting system	Application area
Water hose reels	Offices, stores, corridors, kitchen
Sprinkler system	Offices, stores, corridors, kitchen
Hydrant system	Throughout the building
CO_2	Enclosed areas, kitchen
Halon	Computer room, control room
DCP	Universal, all purpose
Water/soda-acid type	General office, reception, front office
Foam	Fuel, oil storage room
Wet chemical	Kitchen room

9.5.2 Fire Detection

Fires are often first detected by human observation. But methods of detection that operate independently of human presence are also very much essential. Fire detectors may respond to any one or other of the manifestations of combustion, such as the generation of heat, smoke, and flames. Various types of fire detectors are available for installation in buildings intended for different occupancies. However, no single detector is suitable for universal application. Great care must, therefore, be exercised while selecting fire detectors. National Building Code, Part IV must be consulted for fire regulations in this regard.

Detector systems

Various fire detector and alarm systems are installed in hotels and other establishments for containment and control of fire. They are discussed as follows.

Heat detectors Heat detectors (Fig. 9.1) are fire alarm systems that work on the principle of sensing heat source by way of measuring temperature and/or rate of rise of temperature. If the rate of temperature rise exceeds set conditions, some action is initiated through an alarm signal and/or activating water sprinkler system. There are two basic varieties of heat detectors. One is the 'point' type of detector, which is affected by the limited area of hot gas layer immediately adjacent to it. The other is the 'line' type of detector, which is sensitive to the presence of heated gases along any portion of the detector line. Based on the measurement options, there are two main types of heat detectors in each variety.

- 'Fixed temperature heat detector' is designed to operate when the temperature of the heat element is raised to a predetermined level.
- 'Rate of rise heat detector' is designed to operate when the rate of rise of temperature in a particular space exceeds a predetermined value. Alarm may start when the rate of temperature rise exceeds 6.7°C to 83°C per minute.

There are more sophisticated heat detectors that combine both the features in one unit.

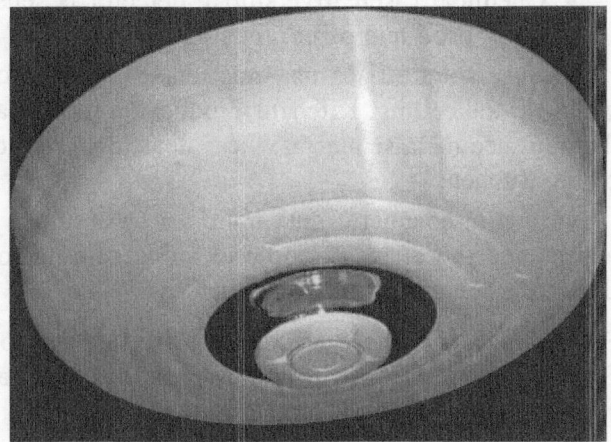

Fig. 9.1 Mechanical heat detector

(*Source:* en.wikipedia.org)

Working of heat detectors A bimetallic thermostat device connects a circuit when a particular set temperature is reached in a location. Thermoelectric detector is a low heat capacity thermocouple device, which creates a voltage based on temperature difference. This detector works on the basis of rate of temperature change. A voltage is developed when the temperature increase at a rate equal or more than, say, 5°C/min. Radiant energy detector use photoelectric cells to detect infrared radiation (thermal radiation from a body due to its temperature) energy radiated by burning materials. Fire warning is given when rapid fluctuations occur in radiation intensities such as due to flames in fire.

Smoke detectors A smoke detector or smoke alarm is a monitoring device that detects smoke and issues an alarm to alert people that there could be a potential fire around. There are three distinct types of smoke detectors. These are discussed below.

- Optical smoke detector or photo-electric detectors employ a light emitter and a receiver. When smoke particles either scatter or obscure transmitted light (visible or near visible), the receiver signal variation triggers action to close a circuit and ring alarm.
- Ionization chamber smoke detector depends upon the effects of the combustion products on ionization currents within the detector. This detector contains a small amount of radioactive material that establishes a flow of ionized air between changed electrodes in the conductor. The presence of

smoke changes this flow of ionized air by changing its conductivity and triggers the detector.

- Chemically-sensitive smoke detector is sensitive to carbon monoxide or other products of combustion.

Because smoke rises up, most detectors are mounted on the ceiling or on a wall near the ceiling. Kitchens produce lot of heat and smoke and hence to avoid the nuisance of false alarms, these detectors are mounted away from the vicinity of the kitchen.

Normal atmospheric dust build-up on a smoke detector can result in 'false' alarms and therefore, periodic cleaning of detectors is required. In addition, a detector's sensitivity also changes over time.

Flame detectors   These are designed to respond to radiation emitted by flames. Since fires, in practice, invariably produce both heat and smoke, detectors which respond to these are accepted as the best general-purpose fire detectors.

Characteristics and selection of fire detectors

Detailed regulations for selection and citing fire detectors are furnished in Appendix B of the National Building Code of India, NBC (SP-71), Part 1V. Here we give some general remarks as adapted from the code.

Heat detectors   Fixed temperature heat detectors are particularly suitable for use where normal ambient air temperature is high and/or may rise and fall rapidly over short periods.

Rate of rise heat detector is particularly suitable where the normal ambient air temperature is low and/or may vary over a wide range slowly. These detectors sense how fast the temperature of a space changes and operate whenever there is an abnormally sharp rise in temperature due to a fire. However, they are liable to give false alarms at ambient temperatures approaching 43°C and should not be used where ambient temperatures approach 43°C. Such fire detectors do not conform to the standard specifications and should not be used alone, but only as supplements to the fixed temperature type fire detectors. They are useful in low ambient temperatures where the temperature remains around 4°C.

In kitchens, boiler houses, furnace kiln rooms, and the like, where height and/ or suddenly rising temperatures are likely, only fixed temperature type of detectors should be used. If ambient temperatures are likely to exceed, its nominal operating temperatures may need to be greater than is permitted in accepted standards (Art. B-8.7.1 of NBC SP71).

Smoke detectors   These detectors are used at places where the ambient temperature is within the limits of 0°C to 38°C. Beyond this temperature range, smoke detectors are not satisfactory.

Optical smoke detector will not respond to the invisible smoke from a clear burning fire but will respond quickly to smoke, which is optically dense. This type of detectors can be used only in free and clean atmospheres. Deposits of atmospheric dust and dirt on the sensitive surface of the photosensitive element and/or the exciter lamp will impair the efficiency of the detector in course of time.

Ionization chamber smoke detectors respond quickly to invisible smoke from a clear burning fire, but may respond slowly to optically-dense smoke. This type of detectors can be used only in free and humidity-controlled atmospheres. False alarms may be caused by smoke and other fumes, dust (including slow accumulations of dust and distributed aerial dusts), fibres, steam, and condensation produced by normal processes, activities, and the environment by vehicle engines and insect infestation. Self-cleaning ovens may cause anionization detector to operate. Very fast air flows, for example, in a warehouse exposed to high wind conditions, could cause some ionization smoke detectors to give false alarms. The presence of explosive gas mixtures can cause explosions. Ionization-type smoke detectors do not give timely alarms or may fail to detect smokes produced by burning materials like polyvinyl chloride (PVC).

Chemically-sensitive smoke detectors have a chemically-coated sensitive element that reacts to the presence of carbon monoxide or other products of combustion present in smoke. Depositing of dust or moisture affects the operation of the sensitive element and can cause the detector to give false or no alarms. These detectors may not be suitable for residential occupancies.

Where there are production processes that produce smoke in a manner which would operate smoke detectors, an alternative detector should be used or an operational routine established to avoid unwanted alarms.

Flame detectors   These operate by the presence of flame. But the fact that combustion is not always accompanied by flame restricts the application of flame. Another factor is that radiation from flames travels in straight lines and a clear line of sight is desirable, although reflected radiation may actuate a detector. Flame detectors are, therefore, used mainly in special applications and to supplement heat and detectors. There are ultraviolet ray (UV) flame detector, infrared (IR) flame detector, and combined UV/IR flame detectors. They are highly accurate, durable, and robust.

9.5.3 Fire Sprinkler System

A fire-sprinkler system consists of a water supply with adequate pressure and flow quantity to a water distribution piping system. Fire sprinklers are connected to this system at the end. They discharge water when a fire has been detected, such as when a predetermined temperature has been reached as sensed by a heat detector fitted with the sprinkler head.

Many lodging establishments now contain sprinkler systems. Besides their contribution to suppressions, these systems also detect fires and trigger an alarm. The alarm is triggered by a blow in the sprinkler piping, which is sensed by piping flow sensors connected to the building alarm. The alarm also identifies the particular flow sensors involved, making it possible to identify the approximate location of the fire.

9.5.4 Total Active Fire Protection System

Total active fire protection systems include the following.

1. Fire fighting system
2. Fire detection system
3. Fire alarm system
4. Fire alarm and control panel
5. Fire hydrant system
6. Monitoring system

9.6 | PASSIVE FIRE PROTECTION

Passive fire protection measures are more concerned with the integrity of building structures as well as the building envelope and compartmentalizing of different zones in the building. Passive fire protection is an all-encompassing fire-safety, concept, which includes the passive measures in fire containment design and in addition, augments the active measures. It is a proactive approach taken at the building design stage and aims at a comprehensive solution to the fire problem. This system does not require power or water to operate in case of fire.

Present situation in India

Passive fire protection in India is in developing stage, concentrated on the passive fire protection for load-bearing structures (beam and columns) using concrete mostly in the industrial segment. In the commercial segment, virtually no passive fire protection concept exists except for some paints and sprays. Various fire tragedies in the recent past led to an awareness of these types of fire hazards.

The scope of application for passive fire protection depends on the following:

- Structural fire protection
- Compartmentalization and consequent containment of fire spread
- Safe and unobstructed escape routes and safe refuge for occupants
- Preserving the function of active fire safety measures
- Safety of fire service personnel

9.7 | FIRE PREVENTION

In spite of all precautions and care, fire does occur. However, by adopting proper safety practices and norms, the frequency of fire occurrence can be brought down significantly and damage significantly reduced.

Sources of ignition and prevention

It is always wise to follow the proverbial saying that 'prevention is better than cure'. This is very important for fire. A few common sources of fire and their possible methods of prevention are given below.

Electrical equipment Electrical defects are generally due to poor maintenance, e.g., short circuits, etc. Preventive maintenance of equipment and regular checking needed

Friction Defects occur due to hot bearings, misaligned or broken parts, jamming of material, etc. Following a regular schedule of inspection, maintenance and lubrication is needed.

Open flames Gas cutting and welding of material may result in fire in an inflammable environment. Follow the rules of welding and gas cutting. Use of open flames near combustibles, such as cotton waste, paper, grease etc., should be strictly avoided.

Smoking Smoking is dangerous near flammable material. Smoke only in permitted areas.

Hot surfaces Exposure of combustibles to furnaces, hot ducts, or glowing elective lamps or irons, etc. is dangerous.

9.8 | CONTROLLING FIRE IN BIG HOTELS AND CASE STUDIES

Many cases of big fires have occurred in hotels in the past. These not only involve death and devastating injury to people but also loss of property, revenue, and reputation of such establishments. Good planning and measures for fire prevention, as outlined below, can mitigate the possibilities of fire.

9.8.1 Fire Prevention and Control in Big Hotels

Fire prevention is clearly linked to maintenance. Different steps and measures to make a hotel more fire-safe are given below.

1. Establishments with on site laundry facilities need to have regular cleaning of hot air ducting for the dryers and frequent removal of covering lint from filter in order to minimize the risk of fire.

2. Sources of many instances of fire are electrical short circuits. Inspection of the building electrical system, especially electrical wiring, helps prevent fires.

3. Kitchen areas pose a great risk of fire due to the presence of heat, flame, and combustibles such as cooking oil, fat, etc. Attention to proper housekeeping practices in the kitchens and regular cleaning of duct work can help prevent fires and reduce their severity if they do occur.

4. Proper ventilation and exhaust hood system are most important for fire safety. Cleaning a hood system is a major maintenance work. Periodic clearing of duct work and fans are also necessary to reduce the risk of fire and avoid a grease dump on the roof or wall area near the fan outlet. The cleaning is done either by in-house maintenance department or by a contract service.

5. Trash storage and disposal should also be designed and positioned with the possibility of fire in mind. Strong combustible trash means the fire is waiting only for a match to start playing havoc.

6. Fire prevention is a key issue during renovation. Renovation may involve nothing more than replacing some interior finishes or it may consist of substantial changes to the building. When replacing interior finishes, it is wise to consider the relative flammability and smoke development potential of the material chosen. Carelessly thrown cigarettes on inflammable material can create a fire. Some major hotel fires have started in this way. For more extensive renovations, in order to demolish and remove positions of the building, contractors often use cutting torches and fuel tanks in the vicinity, which may create a fire risk.

7. All doors that must be passed through to 'Exit' should remain unobstructed and always operable from the inside. Doors leading to stairways at each floor level should be kept closed, not locked. The staircases, corridors, and lift landing should be kept free of any material that can possibly affect a safe and speedy exit.

8. Flammable and combustible materials should be kept in safety containers and should be stored and protected properly.

9.8.2 A Few Things to Remember for Safe and Reliable Fire Control

1. Extinguishers come in dry chemical, foam, carbon dioxide, water, halon, or Class F wet chemical types. They should be labelled by a testing laboratory.

2. There should be a record as to the date of refilling and service of the extinguishers, as suggested by the dealer.

3. Recharge the extinguishers after any use. Some extinguishers are disposable after use. They should be disposed of properly..

4. Extinguishers should be installed away from potential fire hazards and near an escape route.
5. It is very dangerous to use water or an extinguisher labelled as an A type on an electrical fire. It must have the E type rating on it.
6. Extinguishers should be installed beyond the reach of small children.
7. Fight only small fires with the extinguisher.
8. Stay between the fire and the exit. Do not let the fire block the escape path.
9. If one cannot put the fire out within 30 seconds, then fire is out of control.
10. Call the fire department if the fire seems out of control.
11. If one inhales any smoke, immediate medical advice should be taken.

Although there are many extinguishers that work with the directions below, please read the instructions for the extinguisher for any variations.

While fighting fire, the word 'PASS' should be remembered.

P   'pull the pin'

A   'aim low so that the extinguisher nozzle points at the base of the fire'

S   'squeeze the handle, which releases the extinguishing agent'

S   'sweep from side to side'

9.8.3   Fire in Hotels—Case Studies

The direct and indirect losses due to fire in India are estimated to be more than Rs 1200 crore annually, with nearly 20,000 fatal injuries. With rapid rise in industrialization and urbanization, fire accidents and resulting losses are increasing at a much faster rate.

The actual losses due to fire are, in all probability, far higher than what is actually reported. The general cause for the fires and more importantly the fire spread are also often not reported correctly. Short circuit has been identified as the primary cause in almost all the fires, which might not be the actual case. Lack of proper active fire protection systems, lack of regular and proper maintenance, and fake fire approval certificates are few of the other probable causes that costs us very heavily. These loopholes in our system should be addressed properly to reduce occurrence of fire and its spread and extent of damage.

Cases of fire

Siddartha Hotel (Now Vasant Continental), Vasant Vihar, New Delhi, 22 January 1986

- Electric short circuit from false ceiling
- Smoke spread through vertical shafts and ducts

Report

- No battery backup for emergency services. All electrical services including emergency fed from the same electrical circuit
- No smoke extraction system
- Wet risers and hose reels non-operational, due to no electricity supply
- No horizontal or vertical compartmentalization

Casualties

37 dead (35 due to smoke, 1 due to burns, 1 because of fall)

MGM Grand Hotel and Casino, Las Vegas, Nevada, USA, 21 November 1980

- Fire at the deli
- Started at a short circuit in wiring behind a refrigerated pastry display case
- Quickly spread throughout the casino
- Smoke propagated through the egress passageways

Report

- The wiring had been improperly installed
- The automatic fire-alarm system either did not activate or did not sound
- Most guests became aware of the fire situation after they saw or smelled smoke, heard shouting from other occupants, or saw or heard the fire-department rescuers
- Several factors contributed to the resulting devastation
- The level of fuels available within the casino, the building configuration, and the lack of fire-resistant barriers contributed to the rapid growth and development
- The lack of fire extinguishers at the start of the fire allowed the fire to develop to a dangerous level
- The vertical openings within the building enabled smoke to spread with little effort
- The means of egress were compromised by shoddy construction
- Smoke infiltrated into the smoke-proof enclosures and stairwells, as well as through the HVAC system
- Was not equipped with a complete sprinkler system and was not required to be by local law

Casualties

85 dead

The Mason Hotel, San Diego, USA, 17 December 2004

- Fire sparked by a careless smoker tore through the hotel
- Flames began spreading through the three-story hotel

- People escaped the 1345 Fifth Ave. structure through windows and down ladders
- Very thick smoke

Report

- Several maintenance oversights worsened the crisis, most notably the fact that one of two first-story exits was padlocked shut
- Firefighters had to break down the door. But the door should have been unlocked
- Also, several stairwell doors were improperly left open, allowing flames to spread quickly through the top stories of the hotel

Casualities

1 dead

Deep Forest Hotel, Luoyang, Henan Province, China, 30 January 2004

- Improper use of fire by a hotel staff started the fire
- Firefighters in Luoyang rushed to the Shensen group hotel two minutes after they received the report

Report

- The blaze was caused by a security warden at the hotel, who accidentally set his overcoat ablaze with a coal fire
- Firefighters had to break down the door. But the door should have been unlocked
- Also, several stairwell doors were improperly left open, allowing flames to spread quickly through the top stories of the hotel

Casualties

7 dead

Hotel Caledonian, Kristiansand, Norway, 5 September 1986

- Two night porters were on duty and the porter checking security was alerted by the fire alarm. He immediately took the lift to the first storey reception area where he was greeted by flames and smoke in the vicinity of the stairwell
- The fire brigade in Kristiansand did receive a telephone call with the message 'Fire in the reception of Hotel Caledonian' and seconds later the automatic fire message came through
- The brigade responded immediately. On arriving at the hotel three minutes later, the firemen were confronted by a very serious fire
- The emergency staircase led down to the reception area which was in flames

- The two jets initially available proved no match for the intense heat
- The fire spread rapidly and soon involved the first four storeys of the hotel
- Conditions were extremely difficult with heavy smoke pouring from the building

Report

- The hotel was thought to be of fire-resisting construction with good fire protection for guests
- Staff employees were reportedly well trained in fire precautions and fire drills were regularly carried out
- In the event, the fire protection features of the hotel proved inadequate and staff training not very effective
- The hotel, constructed in 1968, complied with the building regulations in force at the time
- It also complied with the Norwegian Hotel Fire Precautions Act 1963, which, among other requirements, ordered that the entire premises be protected by an automatic fire-alarm system
- Unfortunately, the system installed comprised only heat-activated detectors but it did have a direct link to the fire brigade

Casualties

14 dead

Paris-Opera Hotel, Paris, France, 15 April 2005

- The fire broke out around 2:30 a.m. – possibly in a first-floor breakfast room
- The fire started on the first floor and swept through the six-storey building via the stairwell

Report

- There was only one staircase and the fire broke out on the lower floors
- The only main staircase quickly became engulfed in flames and this could be the reason so many people were trapped on the upper floors
- Guests jumped from upper floor windows to escape the flames and choking smoke
- The cause was in all probability accidental and it is unclear how the fire started
- A separate fire exit was not mandatory for older buildings according to regulations

Casualties

21 dead

Qutub Hotel, Katwaria Sarai, New Delhi, 1 August 2008

- The fire broke out in an air conditioner on the second floor
- Two members of the hotel staff fighting the fire were engulfed by smoke while dousing the flames
- Nine fire tenders were pressed into service and the blaze was doused in 40 minutes

Report

- It seems that the fire was caused by a short circuit

Casualties

2 injured

Mandarin Hotel, Bangkok, 5 September 2007

- The fire appeared to have started in a ground-floor cafe under renovation
- It caused heavy damage to the hotel

Report

- The injured guests suffered from smoke inhalation
- Some hotel guests complained that hotel staff were slow in alerting and evacuating them when the fire started after midnight

Casualties

12 injured

Grand Hotel, New Delhi, 26 January 2008

- Fire apparently broke out in the basement kitchen under a Japanese restaurant
- Later spread to the lobby
- A thick cloud of smoke engulfed the hotel
- Forced a complete evacuation of the hotel
- Forty fire tenders were rushed to the spot
- The sprinkler system ran out of water, which is why almost forty fire tenders had to be pressed into service
- The windowpane of a room was smashed to provide an outlet for the smoke

Report

- The fire was suspected to have been caused by a short circuit
- Because of the wooden ceiling and furniture, the fire spread rapidly to the ground and first floors
- Firefighters claimed the blaze could have been brought under control more swiftly had the hotel authorities alerted the Delhi fire service at once

- The hotel staff first tried to douse the flames themselves and called the fire service 20 minutes after the fire first broke out when the situation had gone out of hand
- Precious moments were also lost when a water pipe burst and the fire safety equipment kept at the hotel reportedly malfunctioned

Casualties

2 firemen injured

9.9 | CARE AND MAINTENANCE OF FIRE CONTROL EQUIPMENT

Most countries in the world require regular fire extinguisher and detection system maintenance by a competent person/vendor to operate safely and effectively as a part of fire safety legislation. Lack of maintenance can lead to an extinguisher not discharging when required, or rupturing when pressurized. Deaths have been reported to occur from explosion of corroded extinguishers. Fire detection systems play a vital role in controlling fire, particularly under unmanned condition. They should also be available and fully functional in case of emergency. Proper testing and maintenance schedule must be followed to have all these effective at all time. Comprehensive maintenance programme includes inspection, testing, and maintenance as follows.

Inspection   A visual inspection of a water-based fire protection system or portion thereof to verify that it appears to be in operating condition and is free of physical damage. The value of an inspection lies in the frequency, regularity, and thoroughness with which it is carried out. The frequency will vary from hourly (in severe environments) to monthly, based on the specific needs of the situation.

Testing   It is a procedure to determine the status of a system by conducting physical checks on water-based fire protection systems such as water-flow tests, fire pump tests, alarm tests, and trip tests of dry pipe, deluge, or pre-action valves.

Maintenance   It is the work performed to keep equipment in operable condition by making repairs or replacing, if required. Fire extinguishers should be maintained at regular intervals (at least once a year), or when specifically indicated by an inspection. Maintenance is a 'thorough check' of the extinguisher. It is intended to give maximum reliability for an extinguisher for its efficient and safe working in case of fire.

Inspection, testing, and maintenance service   A service program provided by a qualified contractor or owner's representative in which all components of the systems are inspected and tested at the required times and necessary maintenance is provided. This program includes keeping relevant records and maintaining log of the activities.

As a general rule, there may be three types of maintenance to be carried out for fire extinguishing systems. They are as follows.

- Basic service: All types of extinguishers require a basic inspection annually to check weight, correct pressure, and for any signs of physical damage or corrosion.
- Extended service: Water, wet chemical, foam, and powder extinguishers require a more detailed examination every five years, including a test discharge of the extinguisher output and recharging. Some manufacturers recommend shaking dry chemical extinguishers once a month to prevent the powder from settling/packing.
- Overhaul: CO_2 extinguishers, due to their high operating pressure, are subject to pressure vessel safety legislation and hence the relevant fire extinguishers should be pressure tested (a process called hydrostatic testing) after a number of years to ensure that the cylinder is safe to use. Operating manual, extinguisher label, or the manufacturer is to be consulted to see when such testing schedules are due.

All extinguishers must be recharged immediately after use regardless of how much they were used.

9.9.1 Maintenance and Refilling Schedule of a Few Types of Fire Extinguishers

In this section, statutory regulations regarding maintenance and refilling work of different types of fire extinguishers as specified in Indian Standard (IS 2190-92) are given.

Water-soda-acid type   Refilling is only required when used or once in every two years as per IS-2190-92.

Water-gas cartridge type   Refilling is required when the extinguisher is used or every three years along with hydraulic testing as per IS:2190-1992.

Water-stored pressure type   Refilling is only required when the extinguisher is used or if the extinguisher shows a loss of pressure of more than 10 per cent.

DCP   Refilling is only required when used or as per IS:2190.

Mechanical foam   Refilling is only required when extinguisher is or every three years as per IS:2190-1992.

9.9.2 Detectors and Sensors Testing and Maintenance

Detectors and sensors are very important components of modern fire-control system. They should be kept at the perfect working condition. This necessitates periodic testing and regular maintenance of these devices for their proper functioning.

Heat detector A heat detector works on the principle of sensing temperature and hence needs regular cleaning of the detector surface for correct degree of sensitivity. The steps for regular maintenance include the following:

- Cleaning the surface for dirt by vacuum cleaning/blowing air
- Checking for fungi at electrical contact points and removing them, if any
- Removing oil and grease, particularly for those installed inside kitchen, by spirit/petrol

The maintenance frequency should be at least once in a fortnight.

Modern designs of heat detectors come with an arrangement for continuous checking for the continuity of electric circuit needed for initiating alarm. A small light goes on blinking in the device at a regular interval of 6–8 seconds. So long the light blinks, the system is alive and can be used without any time limit. So, replacement requirement can not go unnoticed.

Smoke alarm Smoke alarms should be tested at least once a month. All smoke alarms come with a test button that can be pushed to check out the entire alarm, including its sensitivity. Sensitivity is indicated by the amount of smoke it needs to set off. If the testing mechanism does not work properly, the alarm should be replaced immediately. Open flame devices, such as a lit match stick, underneath the smoke alarm sensor to test an alarm is not at all advisable. Occasional vacuum cleaning of dust or cobwebs accumulated over the surface of the detectors is also vital for smooth operation of the detector. Comprehensive cleaning and testing of smoke detectors should be done at least twice a year.

Smoke alarms need no maintenance other than changing batteries (in those that have batteries). Batteries need to be changed because they deteriorate over time and need to be replaced at proper time for reliable operation of smoke detectors. This replacement of batteries should be done annually.

Every smoke alarm comes with a homeowner booklet, which describes how to use and take care of that particular alarm.

Smoke alarms have a useful life of about 10 years. At that time they should be replaced, even if they seem to be working alright. This ensures that the alarm shall work when one needs it.

9.9.3 Sprinkler Maintenance

Sprinklers are required to remain at the topmost operating condition. There are extensive inspection, testing, and maintenance schedules for them. A well-maintained fire suppression system can save priceless human lives, millions of rupees in structural damage and property loss, and irreparable loss of reputation for a hotel. A perfectly good sprinkler system can fail in an emergency for reasons as simple as one of the control valves being in the wrong position. This highlights the need for periodic inspections. Listed below are some general steps followed for sprinkler maintenance.

Following are a check list for important components in the fire-control system based on the frequency of supervision needed.

Quarterly checks and maintenance

The following cheeks should be done quarterly.

- Main drain test
- Wet pipe system flow
- Priming level
- Low-pressure alarm
- Pre-action system
- Control valves check
- Check for availability of spare sprinkler heads
- Static pressure check
- Dry pipe system
- Residual pressure check
- Flow alarm
- Checking of deluge system
- Checking nameplates for legibility

Annual checks and maintenance

All the quarterly maintenance activities plus the following should be done annually.

- Maintenance of tanks
- Maintenance of pumps
- Checking for fidelity of water supply
- Maintenance of control valves
- Functional test of all systems
- Functional test of all alarm devices

Summary of minimum inspection, testing, and maintenance for sprinkler

Activity	Frequency
Main drain flow test	Monthly
Inspector's test drain (located at end of building) flow test	Monthly
Test for water flow alarm	Monthly
Pressure gauge calibration	5 years
Complete sprinklers test	20 years

9.10 | REGULATORY REQUIREMENTS

National Building Code of India 2005, published by Bureau of Indian Standards, provides for fire and safety regulations in buildings. Part IV of the code details the fire safety regulations to be incorporated and observed in different categories of buildings. There are some general regulations for all types of buildings, while there are some specific regulations for different categories. Hotels fall under the A-5 and A-6 categories of buildings in the code and any particular specifications for them are very explicitly outlined therein.

9.11 | CONCLUSION

Fire is capable of playing havoc and has done so on countless occasions. Fire has multiple effects in terms of damage being inflicted. Fire results in loss of human and animal life, loss of property, and psychological trauma to many. Hotels are among the worst sufferers in fire accidents. Apart form the damages caused, as mentioned, a hotel has to bear tremendous financial loss in terms of compensation to the affected customers, loss of morale on the part of employees, loss of business till the fire-ravaged facilities are refurbished to their normal operational condition, serious loss of reputation in its hospitality business, etc. Hotels, therefore, should put in maximum effort to ensure a fire-safe stay for the boarders and should not hesitate to make adequate investment on fire prevention and fire safety measures being implemented and properly maintained. Service of professional fire-safety agencies should be hired for the purpose. All the employees should be properly trained so that they adhere to proper safety norms in their regular work and are capable of tackling emergency fire situations.

KEY TERMS

Active fire protection	Active fire protection involves the control of smoke, detection, and communication process signaling the occurrence of a fire outbreak and triggers some sort of counteraction towards extinguishing the fire.
Explosive limit	The explosive limit of a gas or a vapour in air is the limiting concentration of the gas or vapour in air that is needed for ignition and explosion.
Fire	Fire means a rapid oxidation-reduction reaction, resulting in the production of heat and generally of visible light in the form of a luminous flame.

Fire point	The lowest temperature at which burning occurs in presence of an external igniter and continues for at least 5 seconds is called fire point.
Fire pyramid	It is a graphical representation of the four components that should be controlled for suppressing fire shown in the form of a pyramid or tetrahedron.
Fire sprinkler system	A fire-sprinkler system consists of a water supply with adequate pressure and flow quantity to a water distribution piping system where fire sprinklers are connected at the end discharging water when a fire has been detected
Fire triangle	It is a graphical representation of the three components necessary for fire, shown in the form of a triangle.
Flame detectors	These are fire-detector systems that operate by the presence of flame.
Flash point	Flash point is the lowest temperature at which a liquid fuel will give up just enough vapour to form an ignitable mixture with air, which will burn momentarily but will not sustain burning.
Fuel	Fuel may be defined as a substance, which when burnt gives off heat and/or light for useful purpose.
Heat detectors	Heat detectors are fire-alarm systems that work on the principle of sensing heat source by way of measuring temperature and/or rate of rise of temperature.
Ignition temperature	The minimum temperature at which a substance will continue to burn without application of external heat is called ignition temperature.
Passive fire protection	It is an all encompassing fire-safety concept, which includes the passive measures in fire containment design and in addition augment the active measures.
Smoke detector	A smoke detector or smoke alarm is a monitoring device that detects smoke and issues an alarm to alert people that there could be a potential fire around.

REVIEW QUESTIONS

9.1 Explain a fire triangle. How is a fire pyramid different from it?

9.2 Explain the principles of fire suppression.

9.3 Classify different types of fire.

9.4 Prepare a table showing different types of extinguishers used in different types of fire.

9.5 Discuss about various colour codes associated with different fire extinguishers.

9.6 Explain the importance and functions of fire detectors.

9.7 Discuss about the applicability of different types of fire detectors.

9.8 Make a table showing different types of extinguishers against their possible locations of installation.

9.9 Mention period and time after which soda-acid type and mechanical-foam type fire extinguishers should be replaced.

9.10 Explain briefly the maintenance of fire-detector-cum-alarm system.

9.11 Discuss about the testing and maintenance of automatic water-sprinkler system.

9.12 List 10 steps to be taken to make a hotel more fire-safe.

10

Chapter

Refrigeration System

Learning Objectives

In this chapter the reader is introduced to the science, technology, and components of a refrigeration system. The primary objectives of the chapter are to:

- explain the necessity and importance of refrigeration system in the hotel industry
- introduce a few important terminologies related to refrigeration
- discuss the basic cycles of operation of a refrigeration plant
- introduce different types of refrigeration units as employed in the hotel industry

After going through the chapter, the reader should be able to comprehend the essentials of refrigeration system and its utility in the hotel industry.

10.1 | INTRODUCTION

In many situations, we need to maintain the temperature sufficiently lower than the prevailing normal room temperatures. In hospitals, some medicines are to be kept at lower temperatures, much lower than the normal atmospheric temperatures, to preserve their usefulness. In hotels and households, food items, both cooked and uncooked, can be preserved for a long time at low temperatures. Storage of fruits and vegetables makes it possible to add fresh salads to the menu all year round, and to store fish, meat, and other perishable items safely for long periods. Seasonal food items can be bought from the market in bulk quantity at relatively cheaper price, stored in refrigerated conditions, and later used during off season, thus offering the guest both economy and gastronomy satisfaction.

All cooling applications need removal of heat from the space to be cooled. This can be accomplished in many ways. In one system, e.g., in household and commercial refrigerators (reach-in, walk-in refrigerators as employed in kitchens of hotels), a cold liquid (called refrigerant) is circulated through a pipeline in contact with the inside walls of the room to be cooled. The walls and hence the air inside get cooled by transferring their own heat to the cool refrigerant-carrying pipeline, and thus the inside space of the room and the items kept therein get cooled (Fig. 10.1a). In another application, for example, air-conditioning system, air is separately cooled by cold refrigerant and this cold air is then circulated via ducting through openings (louvres or grills) into the room (Fig. 10.1b) to produce the cooling effect inside the room.

The subject of refrigeration is, in principle, related to production and use of this cold refrigerant for different purposes, as discussed. So, we may formally define refrigeration as follows.

Refrigeration is the science of producing and maintaining a temperature condition in a closed space lower than the surrounding ambience by removing heat from the space and transferring that heat to another space or substance, usually the atmospheric air space. In 'cooling', the final temperature of the cooled substance need not be lower than the surroundings. An example of this phenomenon is cooling of a hot room but not below the external surrounding temperature, whereas in refrigeration the final temperature is necessarily lower than the surroundings. An example of this is adding ice to a cup of water.

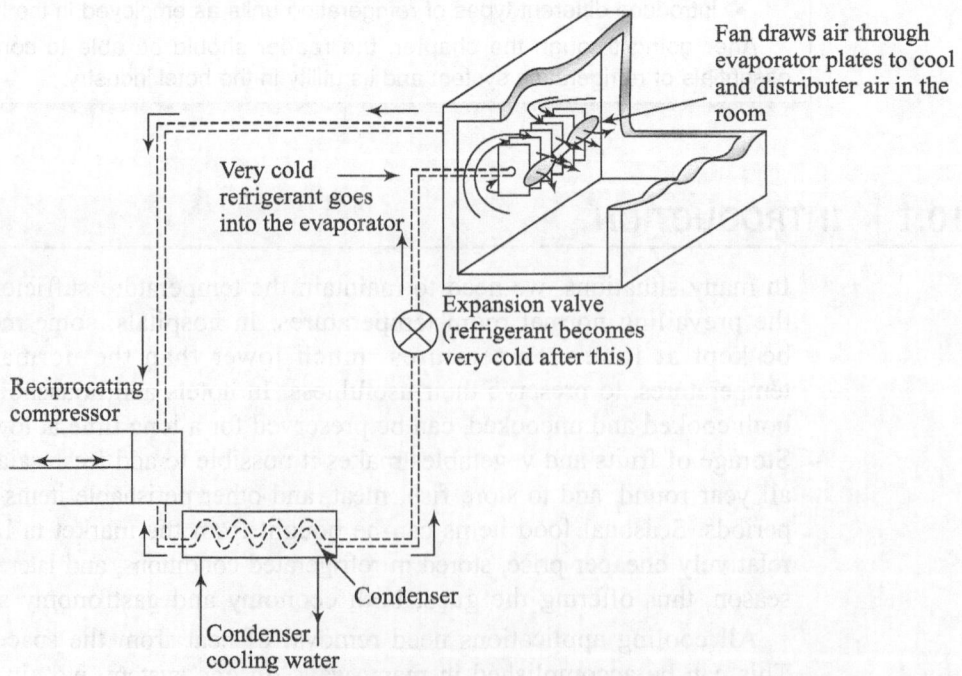

Fig. 10.1 (a) Direct refrigeration of space by keeping evaporator coil of refrigeration unit inside the space

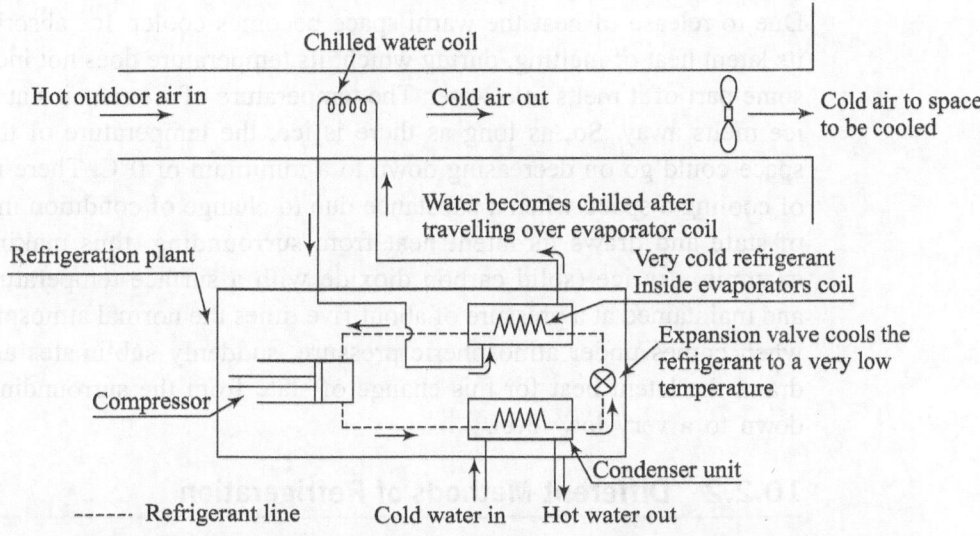

Fig. 10.1 (b) Indirect cooling of air by use of chilled water passing through evaporator coil of refrigeration unit

---

10.2 | PRINCIPLES AND METHODS OF REFRIGERATION

Before we discuss the basic principles governing the science of refrigeration, we enumerate the various uses of refrigeration as follows:

- Preservation of cooked food materials
- Preservation of vegetables, raw meat, chicken, fish, etc.
- Storage of cold water, ice, and ice creams
- Storage of milk and milk products
- Preservation of units, jam, jellies, butter, etc.
- Storage of eggs
- Cooling of food and drink for sale from vending machine

10.2.1 Principles of Refrigeration

Creating a refrigerated space is based on the fact that if we keep a cold substance in a warmer environment, due to the temperature difference, heat will flow from the warmer substance/space, which will slowly cool. But the cold substance will also become less cold in the process; slowly the refrigeration effect will be lost and the space will again become warm. So we have to make sure that the temperature of the cooling substance does not decrease even when it absorbs heat from the surrounding hot space. How can we do that? We all know that if ice is kept in a warm space, heat will flow from the space into the ice due to temperature difference.

Due to release of heat the warm space becomes cooler. Ice absorbs this heat as its latent heat of melting, during which, its temperature does not increase; instead, some part of it melts into water. The temperature of ice remains at 0°C till all the ice melts away. So, as long as there is ice, the temperature of the refrigerated space could go on decreasing down to a minimum of 0°C. There is another way of cooling a space when a substance due to change of condition makes a change of state and draws its latent heat from surrounding, thus making it cool, for example, dry ice (solid carbon dioxide with a surface temperature of –78.5°C and maintained at a pressure of about five times the normal atmospheric pressure) when comes under atmospheric pressure, suddenly sublimates at –78.5°C and draws the latent heat for this change of state from the surrounding which cools down to a very low level.

10.2.2 Different Methods of Refrigeration

There are two broad categories of producing refrigeration effects as used in practice. They are (a) non-cyclic or natural process and (b) cyclic or mechanical/artificial process.

Non-cyclic process

Cooling by melting of ice We all know that many perishable items are kept cool by putting them in an environment of ice. The ice absorbs heat from the space in the form of latent heat of melting and starts melting. Ice melts at 0°C and hence when ice is kept in a space warmer than 0°C, it starts taking away heat from the space, which is thus cooled. This method of cooling a space to very near 0°C is one of the earliest methods of refrigeration. However, we cannot maintain the space cool for a long time as the refrigerant (in this case, ice) gets depleted, as it starts melting down to water. Further, we cannot maintain the temperature of the space below 0°C. Formation of water also poses cleanliness problems. Control of temperature and cooling management becomes exceedingly difficult and cumbersome. So, with ice, we get what is known as a non-cyclic process of refrigeration as the refrigerant is not replenished by recycling the melted ice.

Cooling by sublimation of solid carbon dioxide (also called dry ice) Solid carbon dioxide is made from liquid carbon dioxide under pressure. This solid carbon dioxide is also called dry ice. Dry ice when kept under atmospheric pressure quickly sublimates directly from the solid state to the gaseous state. The latent heat of this sublimation process instantly cools the surroundings. The sublimation temperature of dry ice is –78.50°C. So, one can imagine the cool refrigeration effect it can produce on the surroundings. It is used for flash freezing of many food items such as fish, which requires very fast cooling and a very low temperature for storage. Since there is no liquid state being formed, it is a very clean process. The temperature of the refrigerated space can be much lower than 0°C. However, dry ice is very costly and is not economically viable for large-scale commercial use. Further control of temperature and cooling management also becomes difficult and cumbersome.

Cyclic process

Of greater importance are the methods where the refrigerant is not consumed and discarded but used again and again in what is called the thermodynamic cycle, as explained later. In this system, the refrigerated space can be cooled to any desired temperature and much greater control of operation can be exercised, making the process almost universally accepted as the commercial means of refrigeration.

Let us, in the beginning, define two very important terms associated with refrigeration.

Capacity of refrigeration Capacity of a refrigeration unit is estimated from how quickly it can remove a given amount of heat from a space. For example, a popular capacity index is tons of refrigeration.

If a refrigerator can produce 1 ton of ice at 0°C from 1 ton of cold water at 0°C in 24 hours, we say that the refrigerator is a 1-ton refrigerator. If the refrigerator unit is capable of accomplishing the same task in 12 hours, the unit will be a 2-ton refrigerator. This does not mean that the refrigerator will always convert water at 0°C to ice at 0°C. Actually, it is a measure of the capacity of heat extraction. The following example will make it clear.

Latent heat of ice (i.e., heat needed to be extracted from water at 0°C to make it ice at 0°C) = L = 333.55 kJ/kg ≈ 144 Btu/lb

Let us start with 1 ton of water at 0°C, which is to be converted to ice at 0°C. Let us calculate the heat needed to be taken away from the water to do so.

Mass (m) of 1 ton = 2000 lb = 907 kg

Therefore, heat needed to be extracted = $m \times L$ = (2000 × 144) BTU = 288,000 BTU

Now, let this water be fully converted to ice in 24 hours. Then every hour, heat to be extracted is 288,000/24 = 12,000 Btu. This rate of heat extraction of 12,000 BTU/h or 200 BTU/min is, therefore, the heat extraction capacity for 1-ton refrigeration machine. However, it is not necessary that a 1-ton refrigerator has to make ice only of the amount under these conditions. This is only ice melting heat equivalent of the cooling capacity of the refrigerator.

Therefore, 1-ton refrigeration means extraction of heat at the rate of 200 Btu/min or 211 kJ/min or 3.517 kW power equivalent of refrigeration. If the extraction rate is 880 BTU/min, the capacity of the unit will be 880/200 = 4.4 ton machine. Refrigeration units may

(Contd)

(Contd)

vary in capacity from 1 to 10 tons or even more. Let us make a simple calculation for determining the capacity of a freezer for a specific task.

Suppose, a freezer needs to freeze 30 kg of chicken (with 74 per cent water) at 30°C down to –4°C. The freezing point of the chicken is –2.8°C and specific heat of chicken before freezing is 3.32 kJ/kg/s and that after freezing is 1.77 kJ/kg/s. The latent heat of the chicken can be taken as 247 kJ/kg. This cooling is to be done within 15 min to prevent any bacterial growth to other food items kept there. With these data, we can proceed to work out the capacity requirement of the refrigerator as far as the preservation of chicken is concerned.

Total heat needed to be extracted consists of three stages as (i) sensible cooling from 30°C to –2.8°C, the freezing point plus (ii) extraction of latent heat for converting chicken with water and other liquid components to their frozen state plus (iii) extraction of heat for sensible cooling from –2.8°C to –4°C.

Total amount of heat being extracted = $30 \times 3.32 \times (30 + 2.8) + 30 \times 247 + 30 \times 31.77 \times (-2.8 + 4) = 10740.6$ kJ of energy. Now, this amount of heat energy is to be extracted in 15 min. So extraction rate is 10740.6/15 kJ/min. or 716 kJ/min. We have seen that extraction rate of 211 kJ/min means 1 ton capacity machine. So, for the present job, 716/211 or 3.39 ton machine or a 4-ton machine is needed.

Another way of specifying a refrigerator capacity is in terms of volume of internal space within the refrigerator. The capacity of a refrigerator is expressed in either litres or cubic feet (USA). For a combined fridge-freezer, this volume may be further split into that of fridge volume and freezer volume. Typically, the volume of a combined unit may be 250 litres split into to 100 litres for the freezer and 250 litres for the refrigerator, although these values are highly variable. It is quite apparent that a higher volume capacity will also mean, in general, a higher tonnage of refrigeration.

Refrigerant   Refrigerant is the substance used for heat transfer in a refrigeration system. It picks up heat evaporating at a low temperature and pressure and gives up heat by condensing at a higher temperature and pressure and thus works in a cycle.

10.3 | REFRIGERATION CYCLES AND SYSTEM

We discuss different refrigeration cycles and systems in this section. We also introduce a term called coefficient of performance (COP) for comparing between different units.

10.3.1   Coefficient of Performance (COP)

A refrigeration engine is based on energy input and output quantities for the system (Fig. 10.2). The refrigerating machine has to be provided with some work

(W) from outside. The unit will absorb heat (Q_1) from the hot objects inside and reject some heat (Q_2) to the outside atmosphere. The performance of a refrigerator is defined in terms of what is called coefficient of performance (COP) and is calculated as

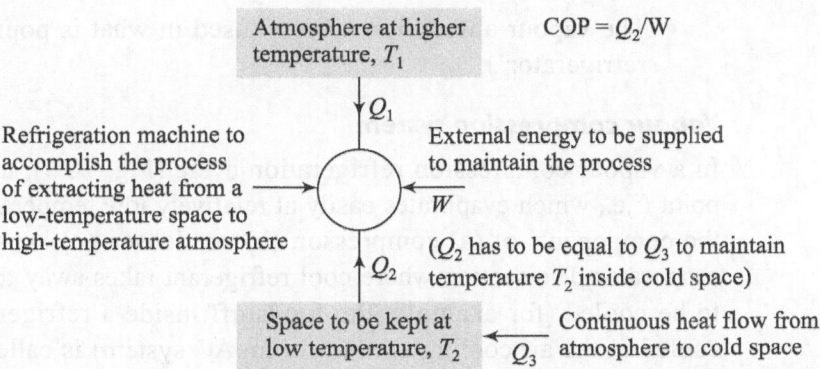

$$COP = Q_2/W$$

Fig. 10.2 Basic refrigeration engine

COP = Heat absorbed from objects to be cooled/work supplied from outside
 = Q_1/W.

This work can be 'mechanical work' supplied by a compressor in one type of refrigeration system (vapour compression cycle) or 'thermal work' of generator supplied by heat energy (vapour absorption cycle) in the other, as discussed later in the chapter. Although actual COP depends on a number of factors, good systems have COP around four to five.

It should be remembered that that actual work input to the refrigeration system is less than the power input to the motor and can be found out if efficiencies of the motor and compressor are known as $W_{\text{cycle}} = W_{\text{motor}} \times$ motor efficiency $\times$ compressor efficiency. The selection of proper W depends upon the requirement. If performances of two refrigerator units having same COP are to be compared, it is always logical to compare them by taking motor input as the work input because that power means the money one has to spend for running the unit.

A refrigerator extracts heat from foodstuff at a capacity of 1.5 ton of refrigeration. The electric motor driving the compressor of the refrigeration unit consumes power of 1.4 kW. Determine the COP of the unit.

Heat extracted by a 1 ton machine is 211 kJ/min or 3.517 kJ/s. Hence heat extraction (Q_1) in one second by 1.5 ton machine is $1.5 \times 3.517 = 5.275$ kJ. Input power is 1.2 kW or 1.2kJ/s or work (W) consumed in 1 s is 1.2 kJ. Hence, by definition,

$$COP = Q_1/W = 5.275/1.2 = 4.39$$

10.3.2 Refrigeration Cycles and Systems—Working Principles

Two different basic systems of refrigeration are employed, which are

- The vapour compression system (used in what is popularly known as 'electric refrigerator')
- The vapour absorption system (used in what is popularly known as 'gas refrigerator')

Vapour compression system

In a vapour compression refrigeration cycle (Fig. 10.3), usually, a low boiling point (i.e., which evaporates easily at relatively low temperatures) passes through the components of (a) compressor, (b) condenser, (c) expansion valve, and (d) evaporator. The section where cool refrigerant takes away the heat of the objects to be cooled (for example, the foodstuff inside a refrigerator or the air to be cooled in the air cooling section for an AC system) is called evaporator. At this section, the refrigerant cools the objects and itself gets heated up. This heat absorbed by the refrigerant is Q_1 as shown is Fig. 10.2.

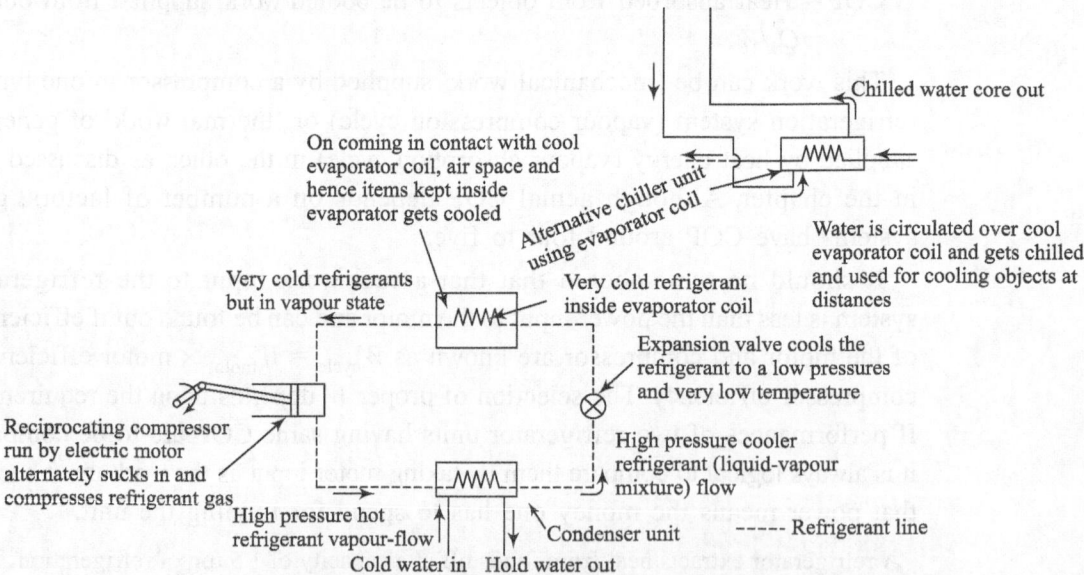

Fig. 10.3 Scheme and components fo a refrigeration plants

The flow of the refrigerant is continuous through the system and undergoes various conditions of pressure and temperature in course of its journey through the components. We start with the condition before the compressor. As already stated, the refrigerant after the coming out of the evaporator section gets heated up but this heat is latent heat of evaporation of the liquid refrigerant. Therefore, the temperature of the refrigerant does not change but it only undergoes a change of state, i.e., from liquid refrigerant to vapour (sometimes a little quantity of liquid remains). This vapour refrigerant then passes through the compressor (thus receiving the work, W,) and is compressed to a high pressure and temperature vapour state. It subsequently passes through a condenser coil or tubes where it rejects heat either to surrounding air (as in a domestic refrigerator and we feel the heat at the outer wall of the refrigerator) or to cooling water (as in a large commercial and industrial refrigerators). It is condensed to relatively cold liquid state but still at a high pressure. This heat going out from the hot refrigerant in the condenser is the heat rejection is Q_2, as shown in Fig. 10.2. This high-pressure but relatively cold refrigerant then passes through a valve called expansion valve (actually a restricted passage in the flow pipeline), which serves to cool down the refrigerant to a very low temperature as well as reducing the pressure of it. The cool refrigerant is now ready for absorbing heat from objects while passing through evaporator coils and the cycle continues. This heat is obtained from the interior of the insulated cabinet via the walls of the evaporator.

Expansion valve can be adjusted to control the temperature and pressure to which the refrigerant comes out of the valve into the evaporator. Thus by controlling this valve knob, we can make the refrigerated space under 'Hi', 'Lo', and 'Med' or any other temperature conditions. Now after a few cycles of operation, the temperature of the interior space becomes as desired and any further operation would make it still cooler and the compressor will unnecessarily work and consume energy from the driving electric motor. To avoid this continuous running, a thermostat control senses the interior temperature and stops the electric motor running the compressor and this stops the refrigerant flow. When the interior temperature rises above certain value due to continuous generation of heat from food staff inside and leakage of heat from hot outside atmosphere through refrigerator door and walls (this heat is again the Q_1 which is absorbed by the refrigerant in the evaporator section continuously during running) the thermostat again causes the motor to run.

In brief, the warmth from the interior of the insulated cabinet is (a) converted into the latent heat of evaporation of the cold liquid refrigerant, (b) carried to the outer side of the cabinet by the refrigerant, and (c) dissipated into the surrounding air/water in the condenser. A compressor is needed to maintain these basic operations in a cycle.

Vapour absorption system

Like a vapour compression refrigeration system, a vapour absorption refrigeration system also has the basic processes of (a) compression (raising the pressure and temperature of the refrigerant, (b) heat rejection in the condenser where the refrigerant gets cooled to a liquid state but is still at a high pressure, (c) expansion to a very low temperature and pressure, usually during its passage through a restricted passage called expansion valve or otherwise, and (d) heat absorption from the space/hot staff/air called evaporator. However, in contrast to a vapour compression system, a vapour absorption system has a generator-absorber (thermal compressor) assembly instead of the mechanical compressor. In the generator, heat is supplied either by steam or some other heat source. Hence, vapour absorption system is essentially a heat-operated unit in which the refrigerant is alternately absorbed and released from an absorbent, as explained later.

There are two variants of this type of refrigeration system. They are

1. Water-ammonia system with ammonia as the refrigerant

2. Lithium bromide water system with water as the refrigerant

Water-ammonia system   Figure 10.4 Illustrates the basic ammonia vapour absorption. Ammonia is the refrigerant and water is the absorbent. Water is used as the absorbent because ammonia is vigorously absorbed in water. Thus, low-pressure ammonia vapour from the evaporator (freezer chamber, etc.) comes out of it and mixes with dilute solution of ammonia (i.e., rich in water, hence, also called weak solution of ammonia) in the absorber. It is readily absorbed releasing a large quantity of heat to the cooling water and the temperature of ammonia solution remains almost constant. This solution, now rich in ammonia (hence called strong or rich solution), is pumped to the generator where heat is supplied from external source (for example, steam, gas flame, etc.). Since ammonia is much more volatile than water, it vapourizes and is given off from ammonia-water mixture at a high temperature and pressure. Water containing a little left-over ammonia (weak solution) returns to the absorber and dissolves ammonia coming from the evaporator, as explained earlier. Ammonia vapour leaving the generator enters the condenser where it rejects heat and becomes liquid. Liquid ammonia still at quite high temperature is passed through expansion valve, which causes to cool the ammonia vapour to a low temperature (now itself ready to cool other objects/space). This cool refrigerant then passes through the space called evaporator, absorbs heat from the evaporator and the cycle repeats. In some systems, liquid ammonia from the condenser drops into the evaporator filled with hydrogen, which has a similar effect in a vacuum and causes the ammonia to evaporate. The process of evaporation absorbs heat from the refrigerator cabinet.

The ammonia vapour being heavy sinks to the bottom of the evaporator from where it passes to the absorber, Here it is absorbed by the water that has returned (after losing its ammonia) from the generator, as already explained. The strong solution of ammonia in water is formed, and this is conveyed through a pipe to the generator the working cycle begins again. Practical systems contain arrangement called analyzer-rectifier combination for trapping any carry-over water in the ammonia vapour.

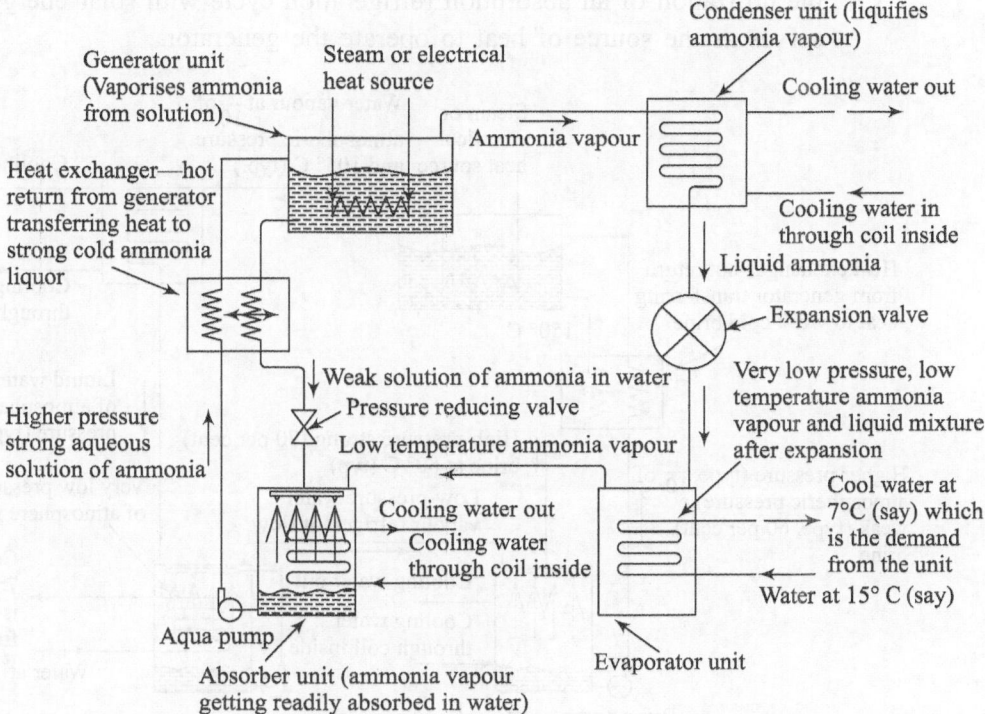

Fig. 10.4 Ammonia vapour absorption refrigeration system

Lithium bromide water system In service industry, the other system, i.e., lithium bromide water system is popular. In many applications such as in air-conditioning system to cool air, the refrigerant need not be cooled to below 0°C. In such applications, water is used as a refrigerant that is absorbed in lithium bromide solution (brine), which has a strong affinity for water. The basic operations are similar to ammonia absorption system. Fig. 10.5 Illustrates the basic operations of such a system

Solar absorption system

Solar energy conversion, having attained a mature status, is harnessed very effectively for heating the generator in vapour absorption system. We discuss, in brief, the solar refrigeration system.

The demand for refrigeration is generally greatest at the time of maximum solar radiation. Thus, the seasonal variation of solar insolation (incident radiation on earth) matches very well with the needs of refrigeration by the use of high concentrating solar collectors. Flat-plate solar collectors are commonly used in solar space heating. However, only a few practical methods are available in flat-plate solar-operated cooling processes. One of the most promising schemes is the utilization of an absorption refrigeration cycle with solar energy. Solar heat serves as the source of heat to operate the generator.

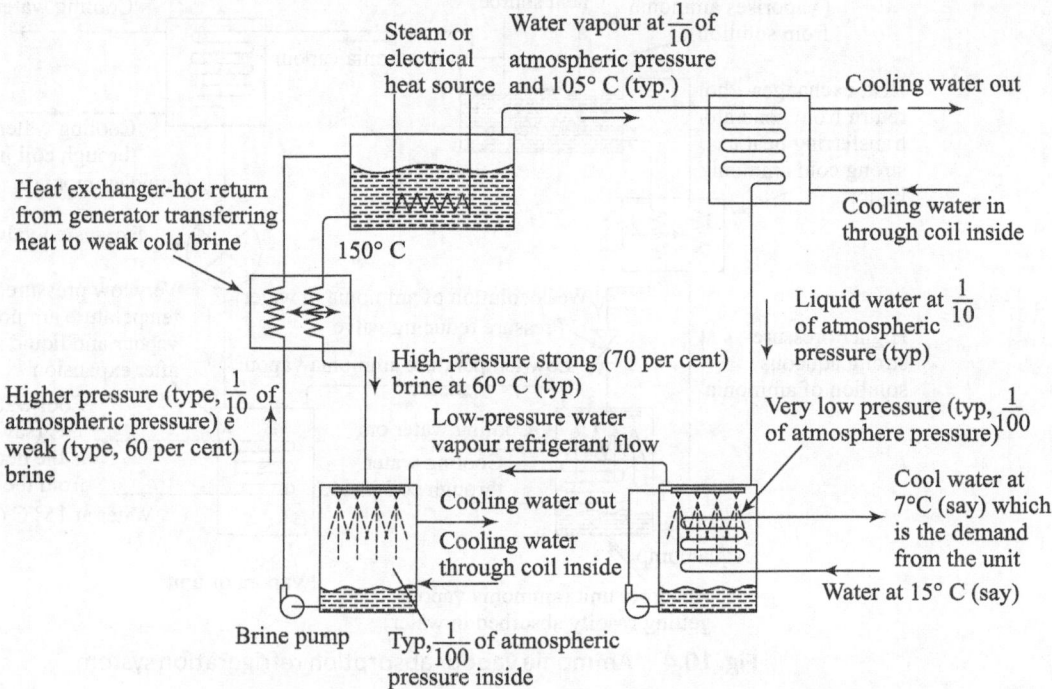

Fig. 10.5 Lithium bromide vapour absorption refrigeration system

A solar-operated absorption refrigeration system is made of a solar collector and a vapour absorption refrigeration cycle. The system differs from a customary fossil-fuel-fired unit in that the energy supplied to the generator is directly from the solar collector. The high-temperature thermal energy available to the generator in this system makes it thermodynamically more efficient. However, like any system, solar absorption refrigeration system has also its advantages and disadvantages as listed below.

Advantages

- Perfect matching of refrigeration demand pattern with solar energy availability pattern
- Renewable nature of solar energy makes the systems sustainable
- Very low running cost
- Combined heating and cooling is possible

Disadvantages
- Not yet a fully developed technology and there is need for R&D
- Commercialization is not fully developed
- Entails high initial cost

10.3.3 Comparison of Vapour Compression and Vapour Absorption System

Both the systems of refrigeration are widely used in the industry. However, they have their specific areas of application and relative advantages and disadvantages. Table 10.1 provides a comparison.

Table 10.1 Chart for comparing vapour compression and vapour absorption systems of refrigeration

Comparison attributes	Vapour compression system	Vapour absorption system
Grade of energy	Needs high grade electrical energy of electric motor which drives the compressor	Requires 'low grade' energy in the form of heat
Operational hazard	Noisy and entails vibration	Very silent operation
Maintenance	Rotating and reciprocating mechanical parts require involved maintenance	Requires much less maintenance
Efficiency	High and hence it forms majority of refrigeration systems	Much less efficient. Economical only when large amount of waste heat (steam or fuel gas from boilers etc.) is available.
Space requirement	Less space for the same capacity	Because of requirement of energy efficiency it requires large space.
Use of solar energy	Electric power from solar energy for running the cycle will be cost prohibitive	Using solar energy for heating in the generator is feasible

10.3 | REFRIGERANTS—PROPERTIES AND NOMENCLATURE

Desirable properties/characteristics of a good refrigerant are listed below.
- Non-flammable
- Non-explosive
- Non-toxic under all conditions
- Easily detectable in leaks
- Stable during all stages of operation

- Non-corrosive on metals
- Fairly high latent heat
- Low condensing pressure
- Evaporating at pressure near normal atmospheric pressure at the refrigeration temperature
- No effect on lubricating oils
- Boiling point at pressure after expansion lower then the temp of the space to be cooled at the pressure

Most of the widely used refrigerants are a group of halogenated (chlorinated or fluorinated) hydrocarbons under various proprietary names of Freon, Genetron, Isotron, Frigen, and Mafron (used in India). These are either methane (CH_4)- or ethane (C_2H_6)-based. The hydrogen atoms are systematically replaced by chlorine and/or fluorine atoms. Methane-based compounds are designated by two-digit numbers where the first digit minus one is the number of hydrogen atoms and the second digit indicates the number of fluorine atoms. We call these substances CFC (chlorofluorocarbons) compounds when there is no hydrogen and has pronounced atmospheric ozone layer depletion properties and are being replaced by HCFC (hydrochlorofluorocarbon) and HFC (hydrofluorocarbon) compounds. R-21 or refrigerant-21 means that the number of hydrogen atoms is one (2 minus 1). This indicates that remaining three atoms of hydrogen in methane are replaced. The number of fluorine atoms that replaces hydrogen is one. So, the rest of the hydrogen atoms (two in number) in methane are replaced by two chlorine atoms, and hence the chemical formula of R-21 becomes $CHCl_2F$. A very commonly used domestic refrigerant is R-22. If the compound is ethane-based, a three-digit number system is used. The first digit is always 1 so that it is differentiated from methane-based system. The rest of the numbering nomenclature is the same as in a methane-based system.

While the halogenated hydrocarbon refrigerants are in popular use because of their many attractive features such as low toxicity, low reactivity, and low flammability, many of them are accused of being responsible for depletion of ozone layer (due to emission during leakage) and are selectively banned, as indicated earlier, for use as refrigerants.

There are many other refrigerants in commercial use. The important among them are ammonia, sulphur dioxide, carbon dioxide, water, etc. Ammonia is used both in vapour compression and absorption system. Though cheap, it is toxic and inflammable and proper care must be taken while handling and using ammonia.

However, small refrigeration units come as pre-charged, while installation of bigger units is accomplished by people of the manufacturer who charge them with refrigerants as specified in the design and as such, users have very little to do about this selection. However, recharging (to compensate for leakage of gas out of the refrigeration system) is usually done by engineering department of the user establishments. All that is required is to keep a supply of compressed refrigerant in store to replenish any leakage. Special equipment and chemical analysers are necessary to locate a leak.

10.4 | TYPES OF REFRIGERATING UNITS IN HOTEL INDUSTRY

The usual categories of refrigerating units employed in the hotel industry are
1. Reach-in units
2. Counter-top or table-top units
3. Walk-in units
4. Special-purpose units

Although counter-top units are essentially reach-in type of refrigerators but much less in size, they are usually differently categorized because of their extensive use and very special type of use on serving counters.

In industry, the term cooler is used to denote units cooling down to 0°C. Units that can cool to sub-zero conditions are known as freezers. Each of the aforesaid categories can have both cooler and freezer sections in them. Some refrigerators are now divided into four zones to store different types of food. They are as follows:

- Subzero (upto −18°C) — Freezer section
- 0°C — Meat preservation section
- 5°C — Refrigerator
- 10°C — Vegetables section

10.4.1 Reach-in Refrigeration System

A majority of these units are pre-designed by the manufacturer and are available as a standard range of products except for minor changes to suit the actual conditions of the user. As such, they are efficient having minimal maintenance and a service life of 10–20 years.

Coolers

Capacities of such units are usually expressed in terms of the internal space volume (cubic metre) or in mass (kg) of material being stored using an average density of the materials. A typical conversion factor of 480.6 kg per cu m of space may be used for capacity specification. Another factor for comparison of performance/cost is energy consumption. Typical units run for 16–18 hours a day. The annual operating cost in each case has to be carefully worked out and an inference is drawn. Another factor that influences the energy bill is the duration of operation during a day. Typical interior temperature range will be 4.4°C to 7°C. Food coolers should not exceed the average temperature setting of 4.4°C except for very short period of time during opening of the cooler or putting in warm food items. Setting the evaporator temperature (i.e., cooling temperature) lower means longer duration of running of the compressor and higher electric bill. Frequent on-off of the compressor motor also decreases life of the components.

The cooling coils or evaporator coils carrying the cool refrigerant are usually placed on the walls of the cooling space. Most of the food items do not come in direct contact with them. They get cooled by what is known as natural circulation of air inside the chamber, which comes in contact with the cooling coil, gets cooled, becomes heavy, comes down, and in turn cools the items inside. Quite clearly, if a fan is used inside, the air can be circulated inside (forced circulation) more easily and uniformly enhancing cooling efficiency. Of course, in the process, some of the food items lose some moisture. Forced circulation will, thus, allow a higher evaporator temperature and a consequent lower energy requirement.

Reach-in freezers

Many of the foregoing discussion for coolers also hold good, in principle, for freezers. However, while the capacity for them is also expressed as cubic metre of volume or kg of mass, the conversion factor from volume capacity to mass capacity is 720.9 kg per cubic metre for freezers. This factor is applicable only when the freezer is loaded with already frozen material. Cooling items to the frozen condition needs removal of substantial heat and could be a deciding factor for many situations while comparing specifications/performance of similar machines for purchase decision.

The freezing rate of normal foodstuff could also be an important factor for freezers. Some units have a hot-gas defrosting system, thus reducing energy cost but also increasing the possibility of large temperature differences within the chamber. Close control of temperature is critical for freezers as this could affect food quality, storage life of food, and loss of moisture.

Many hotel establishments have now installed many reach-in coolers and freezers at local work places, thus reducing the movement of employees and material to and from the centrally-located walk-in refrigerator.

10.4.2 Counter-top Units

Small counter-top refrigerators are frequently used to facilitate service. They are usually coolers and have features similar to those of reach-in coolers but of much smaller size usually up to a size of 100 litres only.

10.4.3 Walk-in Refrigeration System

As the name suggests, food operators can directly move into such units. They are much bigger in size and height than walk-in refrigerators. A typical size would be 4 × 5 × 3 m, or even more, or as less as having a floor area of 2 sq m. As such, they are normally custom-built. Manufacturers have usually a large range of components, which in combination allow them to build these units covering a wide range of capacity and other requirements. Many designers prefer a cooler room with a separate freezer chamber as a single unit instead of a separate walk-in cooler and a freezer. There is some energy-saving in this arrangement. However, in order to walk into the freezer chamber, one has to walk through the cooler chamber and the door in between the cooler and the freezer will reduce the actual capacity of the cooler space.

The number, type, and combinations of different refrigeration units can be best decided by professional engineers of the manufacturer and the hotel engineering division in consultation with the kitchen staff.

10.4.4 Special-purpose Units

Apart from the units discussed, there are many special purpose cooling units widely employed in food service industries. Some of them are listed below.

Ice-cream conservator and frozen food conservator They are usually 'well type' and open vertically through a lid. This minimizes heat loss as the cool and heavy air inside cannot escape upward. They are usually maintained at temperatures between −23°C and −18°C.

Wet fish cabinets They contain deep galvanized drawers in which fish can be embedded in ice. They keep a moist storage conditions necessary for maintaining the quality of fish. They are usually maintained at temperatures between 0°C and −2°C.

Display cabinets In many hotels, there is a refrigerated display cabinet kept in the dining for the customer to choose varieties of meats for grilling. They may be of (a) fully enclosed showcase type, (b) enclosed display slab type, and (c) open slab display type. Open slab display type has an easier access but least protection.

Bottle coolers They are used in dining and more usually in bars. They are constructed in two ways. One is open refrigerated trough type and the other is glass front cabinet type with hinged door or sliding doors. The open trough type offers easy access and is commonly used for quick service in bars.

Ice makers Earlier ice was made in ice-making plants in large metal containers. It was quite a difficult task to break the large chunks of ice into small pieces suitable for use. Modern ice-making machines need only to be connected to a cold water supply line. In one design, cold water is allowed to flow over an inclined refrigerated plate where the flowing water changes to a slab of ice to a set thickness. This slab of ice goes over a wire-net, which is electrically heated and the ice melts along those hot lines of the wire mesh and falls through the square holes of the mesh as separate small cubes of ice.

10.5 | CONCLUSION

In the end, we find that refrigeration systems are critical systems for the hotel industry, where they used for various purposes starting from long-term storage of perishable items such as fish, meat, vegetables, etc. to smart small units of bottle coolers and ice-makers. There is a wide choice of systems available for every specific need. The selection and setting of refrigeration systems is a professional job and professional help must be sought for this purpose and the hotel engineer must also discuss the requirements with the end users such as hotel chefs and food and beverages (F&B) manager for finalization of such projects. Operation and maintenance of large systems are done by skilled humanpower either in-house or through a contractual system.

KEY TERMS

Cooling	It is the lowering of the temperature of a substance or space but not necessarily below ambient temperature.
Coefficient of performance (COP)	This is an efficiency index of a refrigeration system and is defined as the ratio of heat extraction to the work needed to be supplied to the system.
Cooler	It is a refrigerator used to cool items down to 0°C.
Evaporator	It is the unit in a refrigeration plant where the cold refrigerant is passed through coils through a space where the things to be cooled are kept.
Freezer	It is a refrigerator used to cool items to sub-zero temperatures.
One ton of refrigeration	This is the capacity of a refrigeration plant capable of producing one ton of ice from water at 0°C in 24 hours and equivalent to a heat extraction rate of 200 Btu/min or 211 kJ/min.
Refrigeration	Refrigeration is the science of producing and maintaining a temperature condition in a closed space lower than the surrounding ambience by removing heat from the space and

transferring that heat to another space or substance, usually the atmospheric air space

Refrigerant   A substance used in the refrigerant plant, which alternately get heated and compressed and then cooled to low temperature producing the refrigeration effect.

REVIEW QUESTIONS

10.1 Discuss the importance of a refrigeration system in the hotel industry.

10.2 Calculate the heat needed to be extracted by a 2-ton refrigeration plant. If COP of the plant is 4, find the power needed to be provided to the compressor for the purpose. [1.75 kW]

10.3 Discuss the principles and various methods of refrigeration.

10.4 Explain, in brief, vapour compression refrigeration system.

10.5 Explain, in brief, vapour absorption refrigeration system.

10.6 With the help of a neat sketch, discuss a solar energy absorption refrigeration system, citing its relative merits and demerits.

10.7 Compare vapour compression and vapour absorption refrigeration systems.

10.8 Discuss the desirable properties of refrigerants.

10.9 Discuss walk-in and reach-in refrigeration units in relation to the hotel industry.

10.10 List the special types of refrigeration units commonly used in the hotel industry.

11 Ventilation and Air Conditioning

Chapter

Learning Objectives

Ventilation and air conditioning are very important for healthy and comfortable living conditions and proper storage of things.

The objectives of the present chapter are to make the reader:

- understand the basic conditions of body comfort
- familiar with the terms related to ventilation and air conditioning
- gain an understanding of various processes of ventilation and air conditioning
- know the different systems of ventilation and air conditioning with particular reference to hotel industry

After reading the chapter, the reader should have a general idea about ventilation and air conditioning and should be able to answer questions on basic terminologies, processes, and systems related to the field.

11.1 | INTRODUCTION

It is well known that if a closed space is occupied by human beings and/or contains organic matter as in a kitchen or store room, the air inside, if not properly circulated and replaced by fresh outside air, soon becomes hot, humid, and stale. Staying inside will become very uncomfortable and the organic matter would also deteriorate in quality under such conditions. The purpose of ventilation is to provide controlled air to such spaces to prevent unpleasant conditions of internal air. Proper ventilation can provide a very comfortable interior environment to such spaces. Ventilation is, thus, necessary for public areas, guest rooms, banquet halls, kitchen, and other stores. However, it is to be borne in mind that a substantial amount of energy is expended to move a large volume of air through various units in the ventilation route for the purpose.

Ventilation is, thus, about helping air to circulate inside rooms and public areas in the hotel. It allows moisture and airborne pollutants to escape, and fresh, clean air to be drawn into the desired space. Well-designed ventilation provides cooling in summer. In winter, it lets stale air out to keep warmth in. There could be two basic mechanisms for ventilation, namely (a) natural circulation (also called passive ventilation) as air flows naturally through doors and windows and (b) mechanical ventilation using fans to drive air in and fans to extract air out in the ventilation system of a given space.

Building code practices specify ventilation rate requirement for different types of space. There are many procedures adopted for determining the quantity of ventilation that air needs. They could be on a (a) room-volume basis, (b) floor-area basis, (c) occupant-based method, and (d) equipment-based method.

Apart from the requirement of good air ventilation, many spaces require circulation of conditioned air free from dirt and have the desired level of temperature and humidity conditions. In these cases, air-conditioning system is integrated with ventilation system in a building. In the following sections, we will discuss first about ventilation system and then air conditioning, as generally practised in hotels and other related establishments.

11.2 | CONDITIONS OF COMFORT AND VENTILATION SYSTEM

The feeling of comfort for persons inhabiting a space depends largely on the following factors.

- Amount of oxygen available for easy breathing
- Speed and distribution of air passing around them
- Quantity of dirt, smoke, and odour in the room
- Humidity or amount of moisture in the air
- Temperature

The purpose and design of a good ventilation system are to provide optimal conditions of comfort. Not all spaces in an establishment require the same degree of control for all the factors of comfort. For example, a kitchen may not have close control on temperature, odour, and smoke, while a banquet hall may have very close control of almost all the factors of comfort. Ventilation, thus, has to perform the following functions for complete control of the factors of comfort:

1. Supply of fresh air to
   (a) Replenish oxygen in the space
   (b) Remove smoke and odour

(c) Carry away the heat given off by human body, organic matter, and heat appliances in a kitchen
2. Proper distribution of air at different spaces inside an establishment through sound design of ducting
3. Control of temperature by air cooling/heating arrangement
4. Control of humidity as a part of air-conditioning system
5. Control of dirt by using filters in the air intake system

11.2.1 Types of Mechanical Ventilation Systems

There are three basic types of mechanical ventilation systems. These are as follows:

The balanced system It uses a fan to supply air to the interior and another one to remove air from the interior. In many ventilation systems, a lot of energy is saved by a design where a part of the return air from various rooms (already cold/hot and relatively free from dirt albeit a little less fresh) is mixed with fresh air intake (Fig. 11.1).

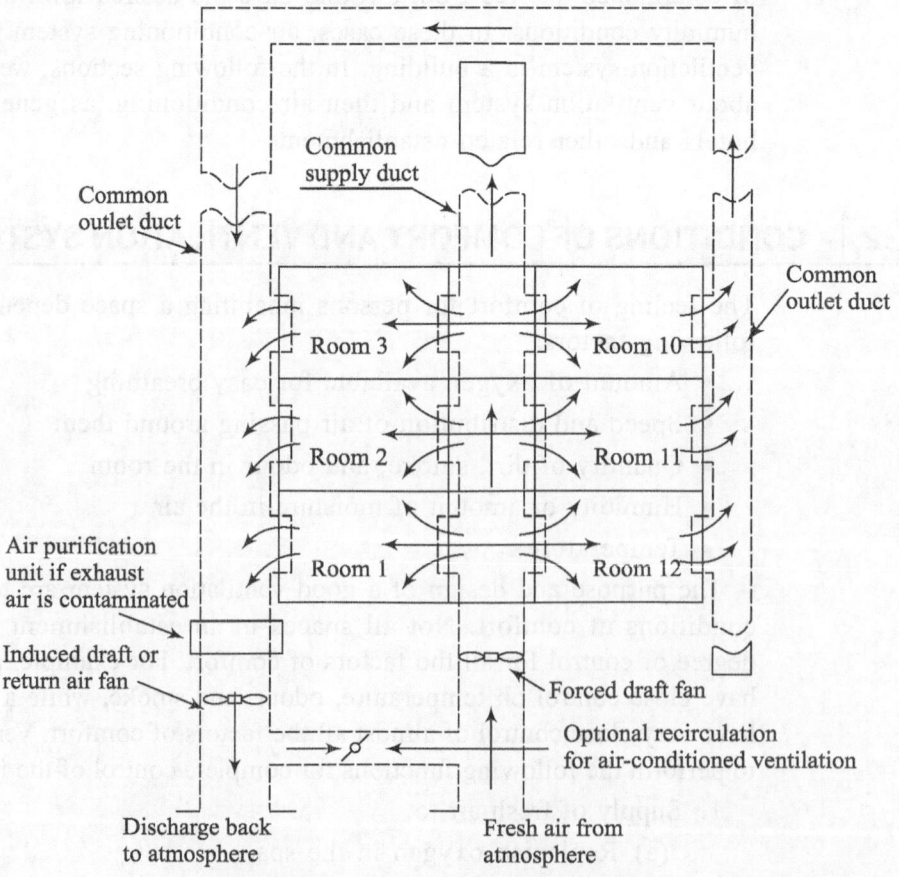

Fig. 11.1 Balanced ventilation system

The supply-only system   It uses one or more fans called 'forced draught' fan to supply air to the building interior, relying on exfiltration outflow through doors, crevices, etc. to remove air from the building. The air inlet is generally at a high level with louvers that are adjusted to prevent downward draught of cold air.

The exhaust-only system   It uses one or more fans called 'induced draught' fans to draw air out of the building interior, relying on infiltration or inflow through doors, crevices, etc. to provide the make-up air supply required.

11.3 | VENTILATION AIR QUANTITY REQUIREMENTS

Depending on the exact nature of space to be ventilated, building codes and ventilation engineering handbooks provide the standards for requirements of fresh air in the space. Fresh air requirement to replenish is only 20 per cent of the air needed to remove body odour. So, the amount of fresh air needed is usually based on the requirement to remove odour and the heat generated inside the room. Kitchen requirement usually would not be guided by the requirement of providing an odour-free environment. As already mentioned, different procedures adopted for determining the quantity of ventilation air needs are as follows:

1. Room-volume basis
2. Floor-area basis
3. Occupant-based method, and
4. Equipment-based method

11.3.1   Room-volume Basis

As the name suggests, room-volume basis specifies air requirements for 1 cu m of room volume. Codes usually specify the requirements in terms of number of air changes per hour. One air change is equal to the introduction of fresh air of volume equal to the volume of the room. Thus, if the size of the room is 5 m × 6 m × 3 m (volume being 90 cu m) and if the code specifies that the particular room needs 20 air changes per hour, it would mean that 20 × 90 cu m of air has to be introduced into the room every hour. Some codes specify this requirement not on the basis of gross volume of room space but on the free air space available therein. The number of air changes per hour required for the room depends upon the type of activity being carried on in the room. Table 11.1 provides a typical standard requirement of air changes for different types of room activities.

Table 11.1 Typical recommended number of air changes per hour

S.No.	Type of room	No. of air changes/hour
1.	Kitchen	20–40
2.	Washing-up rooms	10
3.	Restaurants, large meeting rooms, and public rooms	8
4.	Smoking rooms	3
5.	Bed rooms	10
6.	Internal lavatories, W.C.S., and bathrooms	1
7.	Offices	6
8.	Laundries	3
9.	Boiler house	10

11.3.2 Floor-area Ventilation Method

The second common procedure involves calculation of fresh air requirements on the basis of gross floor area. A common figure could be 3 cu m of fresh air per hour per square meter of floor area.

11.3.3 Occupant-based Method

In this method, the fresh air requirement is specified on the basis of volume of fresh air needed per hour per person. The volume of fresh air needed per person per hour depends on the air space available for each person. Clearly, if free air space available is more per person, the fresh air needed is less. This requirement further depends upon whether or not smoking is permitted in the space. For example, a space where 3 cu m of air space is available per person, the fresh air requirement is 60 cu m per hour per person for non-smoking zone (82 cu m for smoking zone), whereas a space where each person has 12 cu m of air space available inside the room, the fresh air requirement drops to 22 cu m per hour per person for non-smoking zone (30 cu m for smoking zone). The minimum requirement as specified in different building codes range from 18 cu m per hour per person to 21.5 cu m per hour per person.

11.3.4 Equipment-based Method

In spaces such as kitchen, stores, machine room, boiler, engine room, and laundries, factors such as the number, size, power consumption, safety requirements, heat potential of the equipment as also the moisture given off by the equipment could be the appropriate basis of calculation of fresh air requirement. The requirement could be specified either per unit of equipment or per net sq m occupied by the equipment. This figure would again vary depending on the type of equipment based on the factors as mentioned.

11.4 | SCHEMES OF VENTILATION SYSTEM

There are many schemes of operations and related installation layout for ventilation system. They can be classified as follows:

1. Central ventilation system with complete central control
2. Central ventilation system with separate heating/cooling control for each room/space
3. Separate ventilation system for each room
4. Zonal ventilation system

11.4.1 Central Ventilation System with Complete Central Control

A central ventilation system includes all sorts of air condition controls such as heating, cooling, humidity control, and dust control (Fig. 11.2). However, a particular estate may not decide to install all the conditioning elements in its ventilation system. As shown in the figure, a part of the return air is fed back into the inlet circuit. This arrangement economizes on the power requirement as the return air is relatively cleaner and cooler/hotter, as the case may be, than fresh air and to put it in simple words, it is to a large extent already conditioned except for the presence of exhaled carbon dioxide and odour.

The important features of central ventilation systems are as follows:

- There is continuous air (cold/hot, etc.) movement through all the rooms whether or not the rooms are occupied. Even if only one room is occupied, the entire system must be operated to provide ventilation to that room.
- Only one set of equipment for each functions (for example, one fan, one cooler/one heater, one dehumidifier/one humidifier, one air filter, etc.) is required for the purpose.

An analysis of these features indicates that though the capital cost of a central ventilation system may be less, its running cost tends to be high, and it lacks selective control of conditions for different spaces.

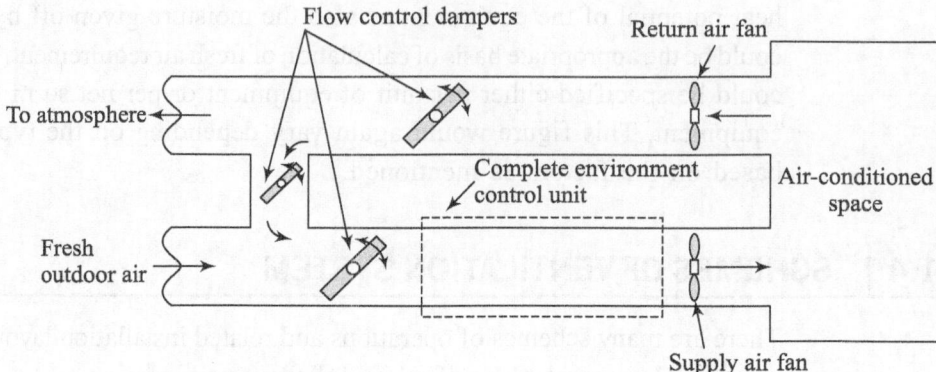

Fig. 11.2 Central ventilation system with complete central control

11.4.2 Central Ventilation System with Separate Heating/Cooling Control for each Room/Space

This system introduces control on selective cooling/heating of rooms while retaining the essential features of a central ventilation system (Fig. 11.3). Although whenever a room is not occupied, the particular heating/cooling unit can be put off, it still circulates fresh air inside unoccupied rooms causing energy and money wastage.

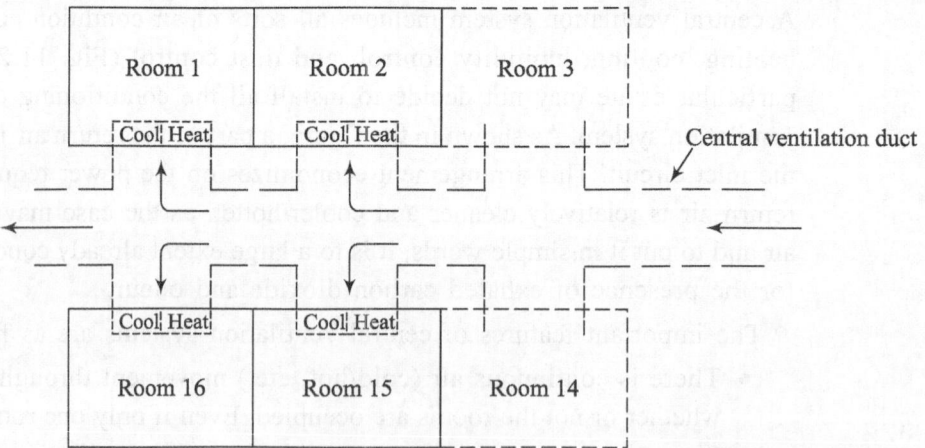

Fig. 11.3 Central ventilation system with separate heating/cooling control for each room/space

11.4.3 Separate Ventilation System for each Room

This system is opposite to the concept of central ventilation system and provides for local fan and other components for ventilation for each room/space (Fig. 11.4). The major advantage is that each room/space can be closely controlled for ventilation requirements. However, the total cost of components is liable to go up as also the cost of carrying large inventory goes up.

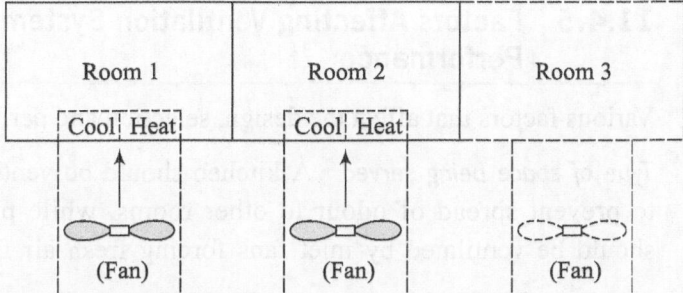

Fig. 11.4 Separate ventilation system with fan for each room/space

11.4.4 Zonal Ventilation System

This is a compromise between local and central systems of ventilation. Separate independent ventilations units are provided for all the rooms/spaces in a particular zone (Fig. 11.5). A zone could be a floor, or a wing, or a building depending on the size, importance of the rooms/space, and level of control desired. If there are some guest rooms still available in zone I while rooms in zone II are all vacant (and hence ventilation system there not put on operation), a new guest should be encouraged to take room in zone I instead of zone II to economize on costly energy.

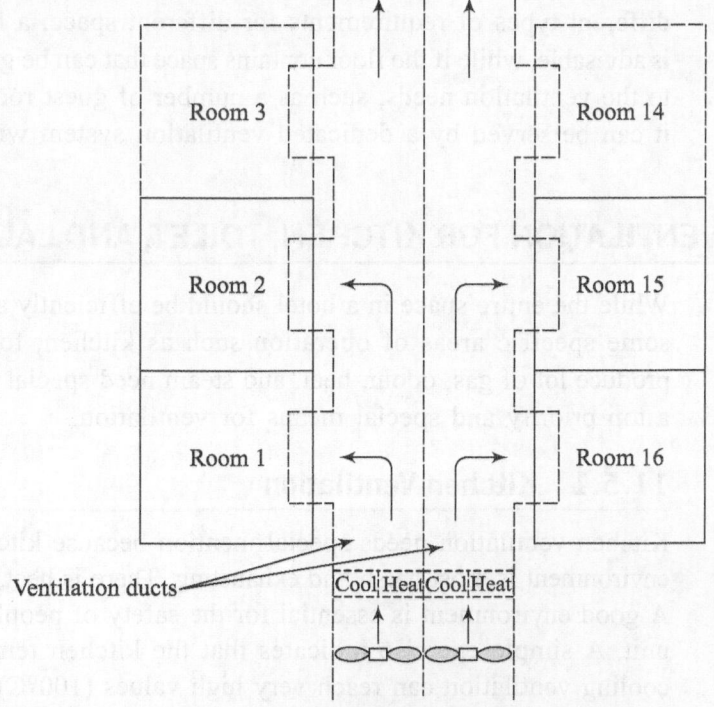

Fig. 11.5 Zonal ventilation system with heating/cooling control for each zone

11.4.5 Factors Affecting Ventilation System Design and Performance

Various factors that affect the design, selection and performance are the following:

Type of space being served A kitchen should be ventilated by an extraction fan to prevent spread of odour to other rooms, while public rooms, store house, should be ventilated by inlet fans forcing fresh air free of any odour.

Number of probable occupants Bedrooms can have less air changes per hour than a kitchen or a boiler house.

Amount of dust and pollutants in free air This affects the efficiency of cleaning of the filter units in a ventilation system.

Type of gas being ventilated out Exhaust outlets from areas that may be contaminated by noxious dust, fumes, steam, gases, odour or other contaminants harmful to people should be placed above the roof level. The discharge to the atmosphere should be located as far as possible usually not less than 8 m from any operable window, door, and/or outdoor intake for a fan.

Types of space to be ventilated on a particular floor If a particular floor has all different types of requirements for different space, a locally controlled system is advisable, while if the floor contains space that can be grouped together according to the ventilation needs, such as a number of guest rooms on a particular floor, it can be served by a dedicated ventilation system with common control.

11.5 | VENTILATION FOR KITCHEN, TOILET, AND LAUNDRY

While the entire space in a hotel should be efficiently and adequately ventilated, some specific areas of operation such as kitchen, toilet, and laundry, which produce lot of gas, odour, heat, and steam need special mention, as they demand a top priority and special means for ventilation.

11.5.1 Kitchen Ventilation

Kitchen ventilation needs special mention because kitchen is a place where the environment is most harsh and exhausting. There is heat, odour, smoke, and noise. A good environment is essential for the safety of people and productivity of the unit. A simple exercise indicates that the kitchen temperature without proper cooling ventilation can reach very high values (100°C) within a short period of operation while working with standard electric cooking appliances. However, the cold air circulated through the dining room can be fed into the kitchen for cooling it. In addition to the main ventilation system working throughout the kitchen, an additional exhaust ventilation subsystem is needed for kitchen appliances that generate heat and smoke. Examples of such systems are hood or canopy type exhaust systems installed in individual or group of such appliances (Fig. 11.6). All the ventilator hoods are connected to a common ventilation duct and exhausted by means of exhaust fans. They must incorporate grease filters in them for trapping grease coming out with the smoke during cooking. In other arrangements, ceiling grills are placed at regular intervals over the ceiling and connected to the ventilation exhaust system.

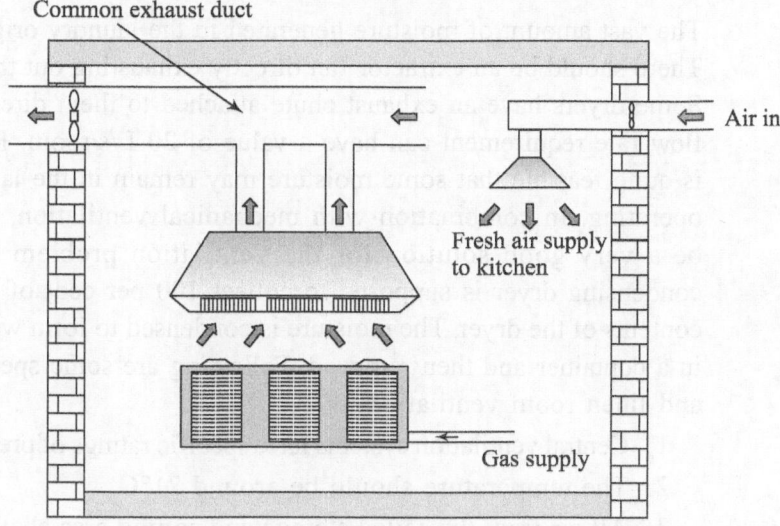

Fig. 11.6 Kitchen ventilation system with hood

Because of its relatively foul nature, kitchen exhaust is not recirculated into the ventilation system and is exhausted directly out of the building.

Good conditions are provided by the entry of filtered and slightly warmed air to avoid cold draught inside the kitchen. Pastry rooms and larder should be ventilated cool, while service points should be kept a little warm to avoid rapid cooling of items to be served.

11.5.2 Toilet Room Ventilation

Toilet rooms produce odour, pungent chemical gas, and steam. They require a good ventilation system to maintain hygienic and comfortable conditions. Following are some recommendations for proper ventilation of toilets:

1. Exhaust air may be recirculated through a central ventilation system that is provided with final filters with some specified rating. Alternatively, all air shall be exhausted directly to the outdoors.

2. Minimum exhaust ventilation rate should be 1.5 cubic feet (cft) per min per square foot (0.45 cu m per min per square m) of floor area, but in any case, the total flow rate should not be less than 50 cft per min (1.4 cu m/min).

3. Make-up air supply for exhaust requirements may be provided from a mechanical ventilation system or by transfer from adjacent areas.

All these specifications apply equally for closets, linen, and trash chute rooms as well, except that, for them, all air should be exhausted directly to the outdoors.

11.5.3 Laundry Ventilation

The vast amount of moisture generated in the laundry originates from the dryer. There should be an extractor fan directly exhausting out the steam to outside air. Some dryers have an exhaust chute attached to them directly. The minimum air flow rate requirement can have a value of 20 L/s/room. However, at this rate it is quite feasible that some moisture may remain in the laundry when a dryer is operating. In combination with mechanical ventilation, condensing dryer can be a very good solution for the ventilation problem of laundry rooms. A condensing dryer is supposed to collect 100 per cent of the moisture from the contents of the dryer. The moisture is condensed to form water, which is collected in a container and then removed. Following are some specifications for laundry and linen room ventilations.

1. Central ventilation systems have specific ratings of pre-filters and final filters.

2. The temperature should be around 21°C.

3. All air from the soiled storage and sorting area shall be exhausted directly to outdoors.

4. Air from the clean area may be recirculated within the laundry complex, but must pass through a lint screen or trap before returning to the air handling unit.

5. While circulation and ventilation rates may vary according to requirement, sufficient fresh air should be supplied to make up for exhaust.

6. Minimum circulation of unconditioned air at summer conditions in India should be 0.9 cu m/min per square m of floor area or 15 air changes per hour, whichever is larger. The figures might change subject to actual atmospheric conditions at a particular place and the service load.

11.6 | PRINCIPLES OF AIR CONDITIONING

Air conditioning, in the most general sense of the term, means any kind of conditioning of air. This includes circulating air in a room, filtering air, cooling or heating air, humidifying or dehumidifying air. Complete air conditioning would mean all such processes being carried on air in a ventilation system. Normally, when we say the space is air conditioned, we mean the air is cooled. Cooling of air can be through evaporative cooling, for example, when air flows over a bucket of ice-cold water, the water gets evaporated due to air convection and draws its latent heat required for evaporation from air, which in turn gets cooled air or through heat exchange with refrigerated water from a refrigeration plant. Again, usually air is cooled in air-conditioning units by using refrigerated water and hence a refrigeration plant is necessary for such air-conditioning systems. Before we understand the utility of all the components therein, we recall that temperature and humidity are two very important conditions for feeling of comfort. As such, we define and explain a few terms for the basic understanding of the science of air conditioning.

11.6.1 Dry Bulb Temperature

This is the temperature of air as measured by a thermometer. The thermometer must be put in open air but not directly under the sun or other objects radiating heat so that the thermometer records the temperature on the basis of heat gained from the air only. When the dry bulb temperature of air exceeds our body temperature (about 37°C), we start receiving heat from air and the condition becomes very tormenting for the body. This is to be kept in mind that when we go out we always receive radiation from the sun and other bodies and start feeling a sense of hotness much before the dry bulb temperature of the air exceeds our body temperature. In any case, dry bulb temperature would dictate, at a gross level, the feeling of hotness in human beings.

11.6.2 Dew Point Temperature

Atmospheric air is never fully dry. It always contains water in vapour form. If the air is now slowly cooled down, it will reach a temperature when the water vapour can no longer be in vapour state but would form water droplets and come out of air. This temperature at which the vapours become water droplets is called the dew point. Thus, dew point may be formally defined as the temperature at which the water vapour in air, upon cooling, would condense. Higher moisture content in air would mean a higher dew point. So when a person is exposed to air with low dew point, it means that the air has low moisture content and can take more moisture. Our body constantly generates moisture and if it does not evaporate we start feeling hot and humid. Under the condition of low dew point, the body moisture quickly evaporates into air and we feel a sense of coolness because the moisture on our skin takes its latent heat of evaporation from our body. However, if the dry bulb temperature is low, the air quickly attains a state of saturation when it can no longer absorb moisture from the body and we start feeling uncomfortable again.

11.6.3 Wet Bulb Temperature

From the foregoing discussion, it is clear that there should be a relationship between dew point, dry bulb temperature, and degree of human comfort. Wet bulb temperature is a measure of relationship between dew point and dry bulb temperature (Fig. 11.7). If a stream of air is passed across a water-soaked wick, water evaporates taking its latent heat required for evaporation from the wick and thus the temperature of the wick is reduced. Air in contact with the wick thus gets cooled. At one point, equilibrium is reached and there will be no further cooling of air. The temperature of the wick/air at that condition denotes the wet bulb temperature. When the wet bulb temperature is low, one feels cold and uncomfortable. As this temperature goes up, one starts feeling comfortable and still higher value would lead to a warm and uncomfortable environment. However, wet bulb temperature along with either dry-bulb temperature or dew point would indicate the degree of human comfort. However, it is quite difficult to formulate the kind of correlation needed for ready understanding of the degree of comfort obtained from a description of these temperatures. We need to identify a single quantity, which would incorporate the combined effects of these temperatures and would indicate the degree of comfort.

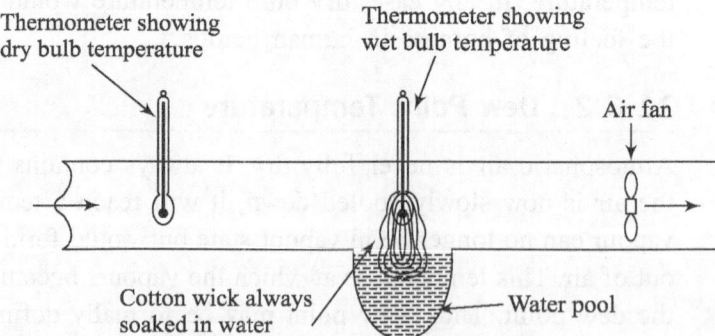

Fig. 11.7 Definition sketch of dry bulb temperature (DBT) and wet bulb temperature (WBT)

11.6.4 Specific Humidity (or Humidity Ratio)

It is defined as the total quantity of water vapour (moisture) contained in moist air and is expressed as kg of moisture per kg of dry air. As an example, suppose under certain conditions of temperature and pressure, the specific humidity, W, is 0.02. It means that the air has 0.02 kg of moisture in every kg of dry air or 1.02 kg of the air contains 0.02 kg of moisture. The maximum specific humidity at 30°C under normal atmospheric pressure (1.013 bar) can be found from appropriate table (called psychometric chart) as 0.0273. This means that 1 kg of dry air at 30°C under 1.013 bar pressure may contain a maximum of 0.0273 kg of moisture. If, under this condition, air can hold no more moisture and no evaporation of sweat from our body is possible, and even if the atmospheric temperature is not very high, we feel tremendous discomfort due to lack of cooling effect our body gets due to evaporation of sweat from our body. Thus, specific humidity for the situation can range from 0 to 0.0273. Now let us take up a situation when the temperature of air is 40°C. From the appropriate table, we find that the maximum moisture this air can hold is 0.05 kg/kg of dry air. Now suppose this air contains 0.0273 kg of moisture per kg of dry air or the specific humidity is 0.0273. In the earlier case also, when temperature is 30°C, suppose that the specific humidity is 0.0273. When the air is hotter, this value of specific humidity is less oppressive for the body because the moisture content has not reached its limit of 0.05 but for the less hot air, this value is the maximum holding limit of moisture in air and as explained, we shall feel very uncomfortable. Thus, we observe that specific humidity or the absolute quantity of moisture is not the true indicator of comfort conditions. We shall see how relative humidity is defined to more explicitly indicate the comfort condition of ambient air.

11.6.5 Relative Humidity (RH)

This is a very important property in the science of air conditioning and can be defined as the ratio of the actual moisture content of a given volume of air at a particular temperature to the maximum amount of vapour it can hold at that temperature and volume when it will be saturated with water vapour and can no longer absorb any further moisture. Thus, when we say that the relative humidity of air is 0.8 or 80 per cent, it means that air can still absorb 20 per cent more of moisture compared to what it holds now. Thus, air with 98 per cent (say) relative humidity can absorb very little quantity of moisture and becomes fully saturated and the environment becomes very uncomfortable with human beings as there is be very little removal of moisture from our skin and we start sweating even while sitting in a room. On the other hand, air with relative humidity of 30 per cent (say) would mean very quick evaporation of moisture from our body and we start feeling a sense of drying up of skin. Both these conditions are not comfortable. A relative humidity of 45 per cent to 60 per cent, depending on the dry bulb temperature, is considered to be an optimum range for comfort.

Thus, we find that only temperature is not the sole criterion of comfort. A good low temperature (say 25°C) with a high humidity ratio could be much more uncomfortable than a higher temperature with low humidity ratio. If any two conditions of air are given, a specially prepared thermodynamic chart, called the psychometric chart, provides information about all other properties described, in addition to the total heat content of air. Air-conditioning system controls both these parameters by means of various processes. This will be clear from the following examples.

Atmospheric air is at 26°C with a relative humidity of 80 per cent. It is desired to get conditioned air at 19°C with relative humidity of 60 per cent inside the dining hall of a hotel. We work out the steps for the conversion as follows.

Maximum possible moisture content in air at 26°C is 0.0186 kg of vapour per kg of dry air (from psychometric chart). Now, RH = actual moisture content/maximum possible moisture content. So, actual vapour content in air is 0.8 × 0.0186 = 0.015 kg of moisture per kg of dry air. Now, we have to find out the total amount of moisture that must be retained in air in order to achieve the conditions of air as required, i.e., 19°C with 60 per cent R.H. We again refer to the chart and observe that the maximum quantity of moisture that air at 19°C can hold is 0.0138 kg of moisture per kg of dry air. If we want air to have a relative humidity of 60 per cent, it should than only contain 0.6 × 0.0138 = 0.0083 kg of moisture per kg of dry air. But originally, we had 0.015 kg of moisture/kg of dry air, so we must get rid of (0.015 − 0.0083) kg or 0.0067 kg of air. If we now start cooling air, the maximum moisture holding capacity will come down as slowly as moisture will start condensing out of the vapour state. Therefore, our strategy will be to cool the air up to that point when it can no longer hold moisture more than 0.0083 kg of moisture. From psychometric chart, one can find that air at 11.5°C air has a maximum moisture carrying capacity of 0.0084 kg. So we cool the air from 26°C to 11.5°C so that the desired amount of moisture comes out and then heats the air up to 19°C. The steps of air conditioning in this example are, therefore, as follows:

1. Cool the 26°C air to 11.5°C by some refrigerated water or other cooling means (cooling and dehumidification).
2. Heat the air then to 19°C to get the desired conditions of air at 19°C with 60% R.H. (only heating).

As shown in the above example problem, air is needed to be cooled and dehumidified during summer time. In winter, the problem is one of heating the air and adding moisture. Figure 8.11(a) shows cooling, dehumidification, and heating process. This is because the moisture content will remain the same if we only heat the air but its moisture-carrying capacity becomes higher since temperature is now higher due to heating. This would mean a relatively lower RH (we recall that RH is the actual moisture present divided by maximum moisture carrying capability of air at the prevailing conditions of temperature and pressure), thus.adding to discomfort with a feeling of dryness. So, along with heating, moisture must be added to raise the RH. Figure 11.8(b) illustrates this process. Table 11.2 furnishes the values of temperature and RH in combination for generally acceptable levels of human comfort in different seasons.

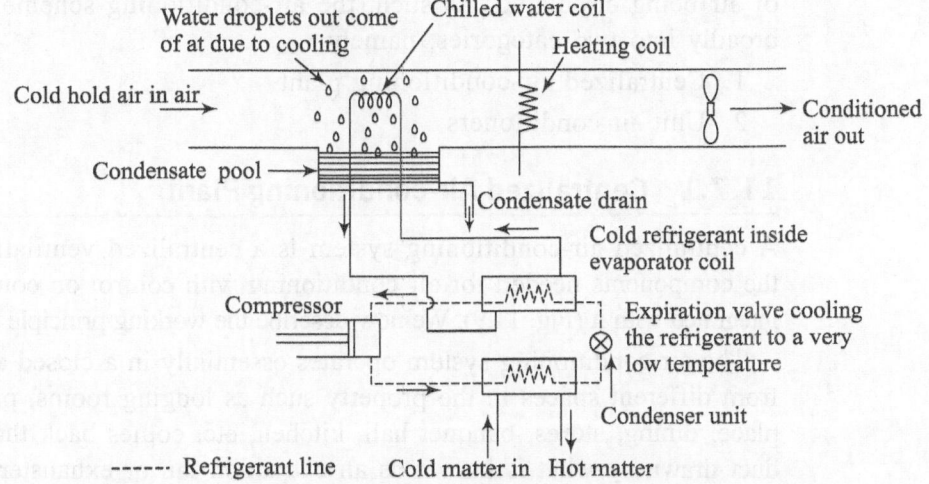

Fig. 11.8 (a) Cooling, dehumidification, and heating process

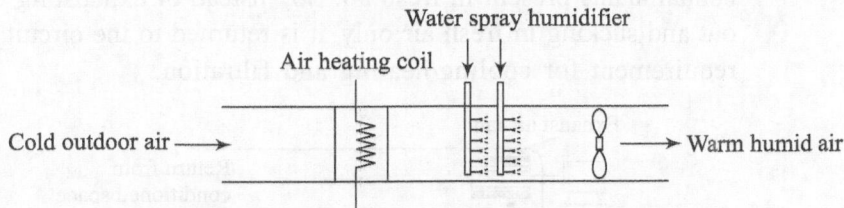

Fig. 11.8 (b) Heating and humidification process

Table 11.2 Temperature and humidity level combinations for comfort in different seasons

Winter						
Temperature	15.5	16	16.5–20.5	21	22.2	23
Relative humidity(%)	70	50–70	30–70	30–60	30–40	30
Summer						
Temperature	17.5	18.5	20–23	25		
Relative humidity (%)	70	50–70	30–70	30		

For the majority of people, 20°C during winter and 21°C during summer are comfortable temperatures.

11.7 | TYPES OF AIR-CONDITIONING PLANTS

As already explained in connection with ventilation system, a fully centralized air-conditioning plant will have all the controls for physical condition and quality of air being circulated. As such, the air-conditioning scheme can be divided broadly into two categories, namely:

1. Centralized air-conditioning plant
2. Unit air conditioners

11.7.1 Centralized Air-conditioning Plant

A centralized air-conditioning system is a centralized ventilation system with the components needed for air conditioning with control on comfort parameters integrated with it (Fig. 11.9). We now describe the working principle of such a system.

The air-conditioning system operates essentially in a closed cycle. Return air from different spaces in the property such as lodging rooms, public circulation place, dining, stores, banquet hall, kitchen, etc. comes back through the return duct drawn by what is known as an extraction fan or exhauster. This return air although a little stale is still cold/hot and relatively free from dirt and other contaminants present in fresh air. So, instead of exhausting the entire return gas out and sucking in fresh air only, it is returned to the circuit to minimize energy requirement for cooling/heating and filtration.

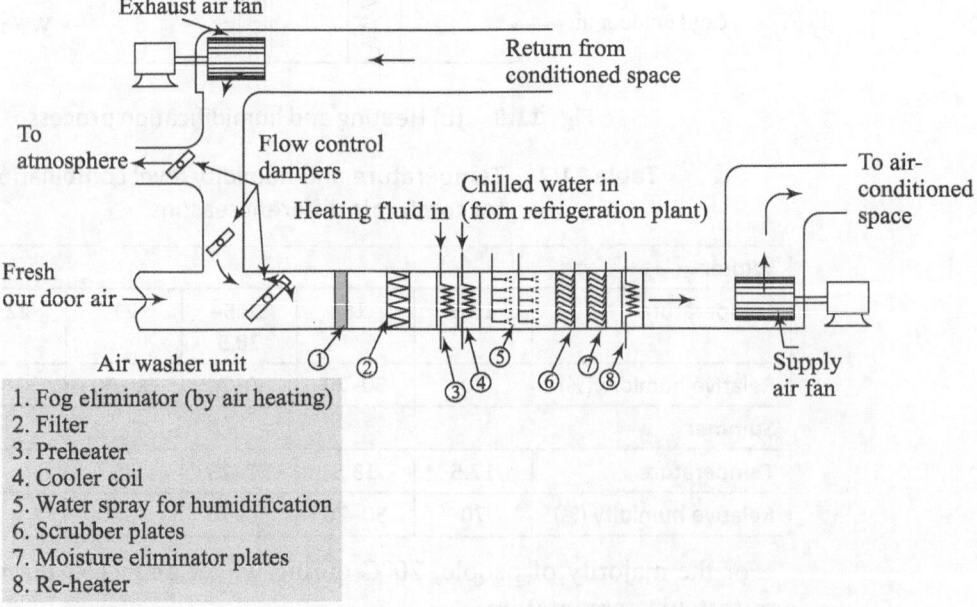

1. Fog eliminator (by air heating)
2. Filter
3. Preheater
4. Cooler coil
5. Water spray for humidification
6. Scrubber plates
7. Moisture eliminator plates
8. Re-heater

Fig. 11.9 Central ventilation system with complete air-conditioning control

However, since this air lacks freshness (as there will be odour, carbon dioxide etc.), a part of this air is released to atmosphere and rest is fed back to the system. On its way back to the system, fresh air from the atmosphere joins it to make up for what has been released earlier and to improve the freshness quality of the return air. The fresh air inlet has louvers/grilles fitted in it for uniform mixing with return air. This combined stream of air then passes through a fog eliminator, which is essentially an air heater. After the fog eliminator, air passes through a filter unit, which arrests suspended particles and dirt present in the air. We have seen in earlier examples how air is heated/cooled and then sprayed with water and then again heated to control humidity and temperature conditions in a space. To accomplish this, air is then passed through a pre-heater/cooling coil and then sprayed with water in air washers. In many cases, instead of separate section for pre-heating/cooling, water spray is warmed/cooled. Pre-heated/warm water spray warms the air and increases humidity, while refrigerated cooling coil/chilled water spray cools the air and decreases humidity. Chilled water is obtained employing a refrigeration plant. An additional dirt arrestor called scrubber makes air further free from dirt. Traditionally, the term 'scrubber' refers to pollution-control devices that use liquid to wash unwanted pollutants including suspended solid particles from a gas stream. During the process of water spraying, excess small water particles tend to hang on to the air and flow along with it. To arrest these water particles, air is passed through a water eliminator section fitted with zigzag plates where heavier water particles cannot negotiate the bends and separates out of air and is drained away. An additional mist eliminator section may be provided where air is forced to pass through specially designed bed of vanes and non-metallic pads that trap fine water particles but allows air to flow through. But unlike filters, which hold particles indefinitely, mist eliminators coalesce (merge) fine droplets and allow the liquid to drain away. Air flows vertically upwards through the packed bed. At this stage, the air has the desired degree of cleanliness and moisture content (in vapour state) in it. Next, it passes through a re-heater to bring the cool air to comfortable temperature. Beyond this point, a fan sucks in the air and delivers to various spaces through an air conditioning duct and thus the cycle continues.

Components of air-conditioning system

Given below is a list of typical major components used in a centralized air-conditioning and ventilation system. In actual applications, the components may change depending upon requirements.

Fresh air inlet It admits fresh air through a louver/grille situated on an outside wall very much away from kitchen, boiler house and water closet (WC) outlets.

Filters It is employed to arrest suspended particulate matter such as dirt and dust from inlet air.

Air washers This is the air-conditioning section which comprises pre-heater coil, refrigerating coil, and banks of water sprays. The functions of air washers are to control humidity and clean air.

Scrubber plates These are a series of zigzag plates with a stream of water running down them and are used to catch the dirt-laden water droplets coming out from water spray section in air washer.

Water eliminator plates These are a series of dry zigzag plates to remove any remaining free moisture in the air stream.

Mist eliminator It is a metallic/non-metallic bed through which air is flown upwards. Fine mist is trapped in the bed and fall down as they coalesce to bigger size.

Air reheater This is composed of heating coils to bring the air up to the desirable room temperature, if necessary.

Inlet fan It is a fan for drawing air through the inlet, ducting and forcing it into the various spaces in the hotel/building.

Fresh air inlet/louvres These are fitted at the inlet of the fresh air for uniform entry of fresh air and hence smooth mixing with return air before entry to the air-conditioning section.

Room inlet grilles One or more grilles in each room for the controlled inflow of air.

Room extraction grilles One or more grilles in each room for the outflow of air-one to discharge outside into the atmosphere and another leading back to the AC plant inlet for the re-circulation, as explained earlier.

Extraction fan It is the fan for extracting air in the return path of the air ducting coming back from rooms and spaces.

Central extraction damper In case where complete central control is exercised a part of the return air from all spaces is discharged to the atmosphere through a control damper valve in the delivery side of the extractor fan.

Ducting Ducting is a term used to denote the passage for flow of air from the AC plant through various passages in the building opening into the closed space to be cooled and return to the inlet ducting in the AC plant. This also includes the ducting for exhaust to atmosphere.

In some terminologies, scrubber plates and water eliminator plates are included in air washer.

Controls in AC systems

Air-conditioning control essentially involves control of temperature and humidity. These controls are effected through thermostats and humidistats, which sense the temperature and humidity of air, respectively and send control signals for corrective actions. Corrective actions include controlling (a) the temperature of air heaters, (b) the temperature of water spray, (c) the operation of refrigeration plant, and (d) the amount of recirculated air (by controlling the damper opening). Different rooms and spaces may require different conditions of air. For example, bedroom and banquet halls should have different requirements of air quality and conditions. The number of occupants should also be a factor in determining the individual requirements. Control can be achieved by any of the following three methods.

1. There is complete central control where the central unit does everything and the whole building receives air at the same uniform temperature and humidity conditions

2. Instead of installing the components of air washer and eliminator in the central plant, these units are located at each room so that the conditions can be controlled very locally.

3. Instead of locating the air-washer components and eliminator for each room, spaces with similar requirements are grouped and separate controls with one set of components consisting of fan, filter, and cooling/heating coil is installed for each such zone.

11.7.2 Unit Air Conditioners

These are small self-contained, cabinet-style units (Fig. 11.10). The units are made small enough to fit into a standard window frame. Almost all the components of a central air-conditioning plant and the refrigeration plant are incorporated within the cabinet. The refrigeration plant (small enough to be housed within the box) provides the cool refrigerant for directly cooling the air.

Fig. 11.10   View of a window-type unit air conditioner

(*Source:* hometips.com)

Working of unit air conditioners

We recall that a vapour compression refrigeration system needs a (a) compressor to pressurize and heat the refrigerant, (b) a condenser to cool the hot refrigerant gas, (c) an expansion valve (a constriction in the pipeline carrying the refrigerant, which lowers the pressure and also the temperature of the refrigerant to a very cool state, (d) an evaporator coil through which the cool refrigerant passes and surrounding which hot room air circulates to get cool, (e) appropriate refrigerant, and (f) pipeline for circulating the refrigerant through all the devices as mentioned. The air-conditioning section consists of a blower (a fan which can develop higher pressure), which sucks in return air from the room through inlet louvers/grilles over the cool evaporator coil and forces it over the same coil finally back to the room again through delivery louvers/grilles. Figure 11.11(a) and 11.11(b) illustrates the working of room air conditioners.

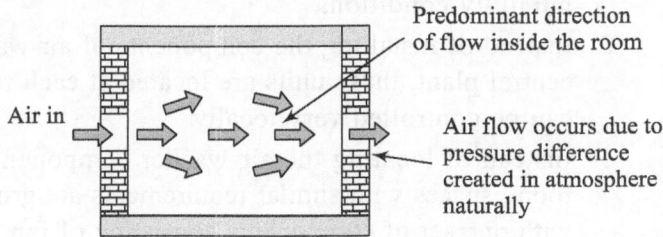

Fig. 11.11 (a) Wind ventilation

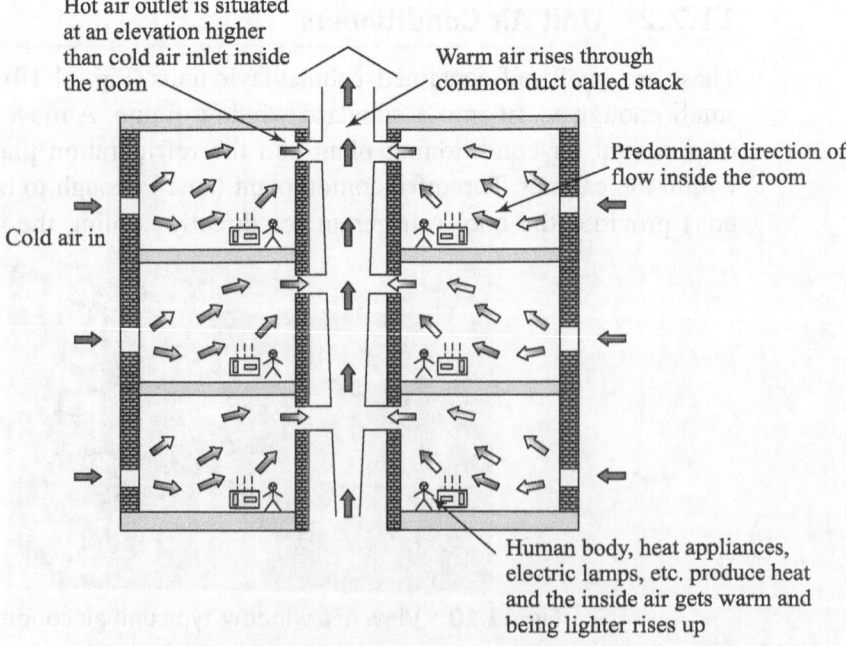

Fig. 11.11 (b) Stack ventilation in a multistoried building

The orientation of delivery louvres can be adjusted for individual comfort. A motor drives the blower on one side and through the same shaft drives a fan on the other side. This fan draws atmospheric air to flow over condenser coils in the refrigerator section for lowering the temperature of the hot gas from the compressor flowing through the condenser coil. A separate motor drives the compressor. The room air-conditioning box has two compartments usually separated by a metal plate. The outer compartment is the part outside the wall and contains the condenser, compressor, fan, and fan motor. The compressor has an in-built electric motor in it. The inner compartment extending into the room contains the inlet and outlet grilles, blower, expansion valve, and evaporator cooling coils. In one variety, fresh air is also inducted in and mixed with room return air for improving freshness of air. In still other design, water sprays are used for humidity control. Thus, in higher models both temperature and humidity can be controlled.

From the foregoing description, we can now list the components we normally encounter in a unit-type room air conditioner.

Refrigeration section

- A compressor for raising the temperature and pressure of the refrigerant
- A hot coil or condenser (on the outside) to reduce the temperature of the high-pressure refrigerant
- An expansion valve to reduce the pressure and chill the refrigerant
- A chilled coil or the evaporator coil (on the inside) to carry the chilled refrigerant for cooling the air, in turn
- Tube for carrying refrigerant

Air-conditioning section

- Chilling coil fan/blower to suck in return air from the room and force it over the chilling (or evaporator) coil
- Air inlet/outlet grilles to the unit from room for re-circulating the room air in a stream-lined manner
- Fresh air inlet grilles (for some models only) for admitting fresh air into the system

Other components/devices

- A fan for drawing atmospheric air to blow over hot condenser coil to lower the temperature of hot gas from the compressor
- A drive motor for running the condenser fan as well as chilling the coil blower in the air-conditioning section
- A control unit to control temperature and humidity

Models without a fresh air inlet operate entirely on recirculated air and are suitable only for rooms with small fresh air requirement. If we recollect refrigeration system, we will notice that the first four components comprise a basic refrigeration system. The cold refrigerant passes through the chilling coil over which flows the warm air from outside (in case fresh air is admitted) and the relatively cooler return air from inside the room forced over the chilling coil by a fan. Another fan forces atmospheric air to cool the hot refrigerant from the compressor flowing through the hot or condenser coil. These two fans are driven by a single drive motor.

Split-system units

The development of split air-conditioning units has further made the installation of room air conditioners more flexible. In this system (Fig. 11.12), the cold side of the unit (such as expansion valve, chilling coil, etc.) is physically separated from the hot side (such as hot coil or condenser coil). The refrigerant flows through a long pipeline connecting the hot side and the cold side. The cold side is placed inside the room and the hot side outside the building where there is no problem of condensation of water (during passage of atmospheric air over hot coil, moisture condenses out), which had been a problem in the old single-unit compact version for rooms having all walls inside the building. The unit consists of a long, spiral coil shaped like a cylinder. Inside the coil is a fan, to blow air through the coil, along with a weather-resistant compressor and some control logic. This approach has evolved over the years because of its low cost, and also because it normally results in reduced noise inside the house (at the expense of increased noise outside the house). Other than the fact that the hot and cold sides are split apart and the capacity is higher (making the coils and compressor larger), there is no difference between a split-system and a window air conditioner.

Thus, the main advantages of split-type air conditioners are

- Low cost for relatively larger size units
- Absence of noise (due to placing the compressor out in the hot side). Hence the split-type units have evolved very popular in recent years

However, in larger buildings and particularly in multi-storied buildings, the split-system approach begins to run into problems. Either running the pipe between the condenser and the air handler exceeds distance limitations (runs that are too long start to cause lubrication difficulties in the compressor), or the amount of duct work and the length of duct becomes unmanageable.

In warehouses, large business offices, malls, big department stores, and other big buildings, the condensing unit normally lives on the roof and can be quite massive. Alternatively, there may be many smaller units on the roof, each attached inside to a small air handler that cools a specific zone in the building.

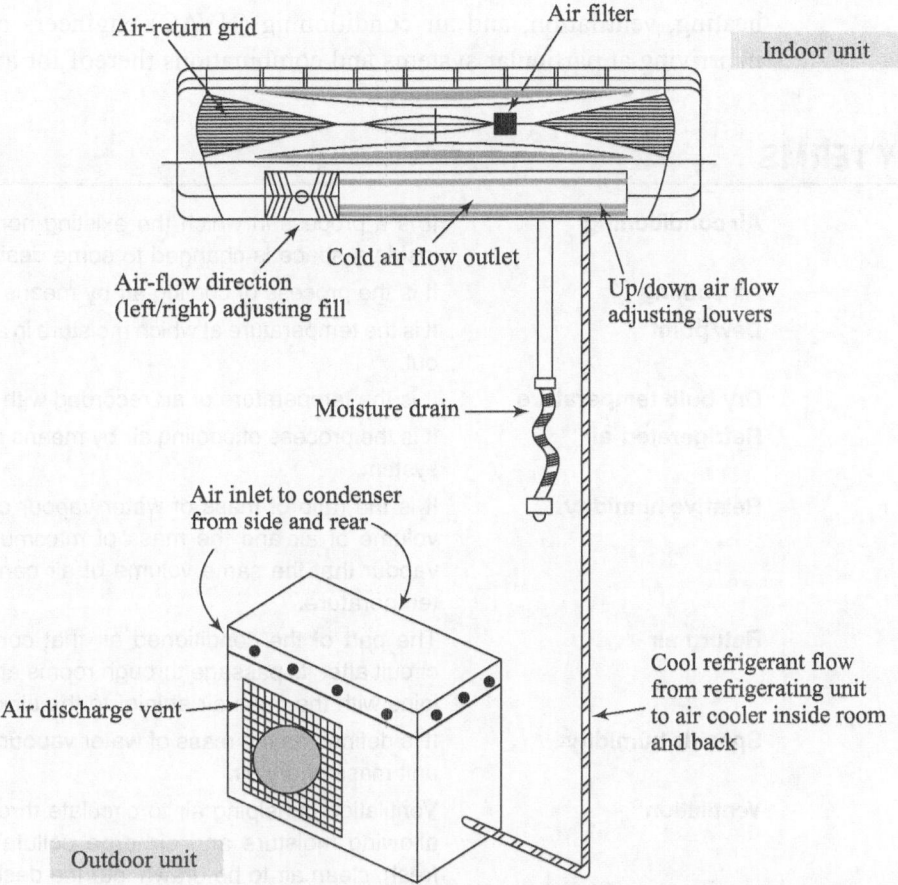

Air-return grid

Air filter

Indoor unit

Cold air flow outlet

Air-flow direction
(left/right) adjusting fill

Up/down air flow
adjusting louvers

Moisture drain

Air inlet to condenser
from side and rear

Cool refrigerant flow
from refrigerating unit
to air cooler inside room
and back

Air discharge vent

Outdoor unit

Fig. 11.12 Schematic arrangement of a split unit air conditioner

11.8 | CONCLUSION

We have already outlined the relative merits and demerits of central air conditioning and local or unit air-conditioning system in previous chapters on ventilation systems. It can be restated that unit/window air conditioners may be more economical in capital cost than a central air-conditioning plants when only a small number of window air conditioners are required. They are also useful for isolated rooms where it is very difficult to connect the room with the central ducting system. The self-containing and factory-assembled nature of these units mean that they can be installed quickly with relatively little on-site labour requirement. The noise of compressor and fan in old window-type air conditioners is also eliminated in the split-type units. As such, both unit type and central system are widely used in their specific areas of applications. Complete control of temperature, humidity, and dust require elaborate air-conditioning system and will be very costly. Therefore, heating, ventilation, and air conditioning (HVAC) engineers must be consulted in arriving at particular systems and combinations thereof for a specific property.

KEY TERMS

Air conditioning	It is a process in which the existing normal condition of air inside a space is changed to some desired level
Air cooling	It is the process of cooling air by means of evaporation.
Dew point	It is the temperature at which moisture in air starts condensing out.
Dry bulb temperature	It is the temperature of air recorded with a dry bulb.
Refrigerated air	It is the process of cooling air by means of refrigerated water system.
Relative humidity	It is the ratio of mass of water vapour contained in a given volume of air and the mass of maximum amount of water vapour that the same volume of air can retain at the same temperature.
Return air	The part of the conditioned air that comes back to the air circuit after its passage through rooms and other spaces and joins with the fresh air at inlet to the air-conditioning system
Specific humidity	It is defined as the mass of water vapour contained in air per unit mass of dry air.
Ventilation	Ventilation is helping air to circulate through a given space, allowing moisture and airborne pollutants to escape, and fresh, clean air to be drawn into the desired space.
Wet bulb temperature	It is the temperature of air recorded with the bulb enveloped by a moist cotton wick.

REVIEW QUESTIONS

11.1 Discuss the various physical parameters affecting comfort in a space.

11.2 What are the two basic mechanisms of ventilation and describe, in brief, any one of them.

11.3 Draw neat sketches showing central, room, and zone ventilation systems.

11.4 Discuss the relative merits and demerits of central, room, and zone ventilation systems.

11.5 List the factors affecting ventilation.

11.6 Discuss, in brief, kitchen and laundry ventilation systems.

11.7 Draw a neat sketch of a central air-conditioning system indicating the main components on it.

11.8 Discuss, with the help of a neat sketch, a window air-conditioning system.

11.9 Discuss unit air-conditioning system.

11.10 Explain the working of a split-type window air conditioner, stating its relative merits.

12 Chapter
Maintenance Management

Learning Objectives

This chapter discusses the concepts of maintenance engineering and management. After reading this chapter, the students should be able to:

- understand the importance of proper functioning of plant and equipment in the hotel industry
- realize the impact of equipment operational level on customer satisfaction and profitability
- appreciate the importance of maintenance department and to understand the role played by it
- know the different types of maintenance generally followed in hotels and their economics
- know the structure of the maintenance department
- have a fair idea about contract maintenance and tendering process

After reading the chapter, the students should be able to understand the importance of proper maintenance as a vital service for continuation of essential service and production in lodging and catering industry.

12.1 INTRODUCTION

Engineering system in a hotel property consists of numerous mechanical and electrical machines, pipelines, electrical lines, electronic control equipment, cooking appliances, and buildings, among others, which are subject to deterioration due to wear and tear. As such, proper and methodical maintenance procedures must be adopted to keep the service level of the equipment and quality of the property at the top-notch level. As the hotel industry, is increasingly using conventional and sophisticated engineering systems, reliable and uninterrupted functioning of all these systems have assumed vital importance. Therefore, their maintenance programmes and their management by expert professional engineers have become critically important in the hotel industry. An engineering/maintenance department looks after these aspects in addition to carrying out other functions such as procurement, replacement, and installation of engineering components and systems, maintenance of safety standards and practice, as well as accounting and costing of maintenance department. However, in spite of the best efforts and planning for maintenance, failures of systems are bound to occur because of many unpredictable reasons. Proper maintenance practice reduces the downtime of the plant and keeps the overall system running by adopting alternative operational routes in case of failure, unless it is of a very serious nature.

12.2 | ROLE AND IMPORTANCE OF MAINTENANCE DEPARTMENT

We have already noted that keeping the engineering facilities in good operating conditions is vital for the business of a hotel. In hotel industry, non-availability of a facility does not only result in a loss of revenue due to underutilization of the facilities but also leads to a loss of reputation if the guest does not get the facility in addition to the loss of future business potential. Hotel industry invests a huge sum of money on engineering equipment and cannot afford to throw away the capital resources unless they have been utilized to the fullest extent possible. Maintenance department aims at keeping all the engineering facilities in the best operating condition. However, maintenance work can continue for a particular piece of equipment till the average annual operating cost remains low. Upgradation of technology is also an important criterion in this regard. It is under this context that an efficient maintenance department is critical. Although the term 'maintenance department' is still used, it is more justified to term it as the 'engineering department' because apart from maintaining equipment, the department does a lot of other engineering functions. It is a full-fledged department like any other department in the hotel industry and is headed by the chief engineer. However, analyses have revealed that, in the past, many hotel managers had very little idea about the importance of engineering systems and were almost oblivious of the existence of an engineering/maintenance department. It was looked upon as a necessary evil. However, hotel managers have now realized the complexity and importance of engineering systems and their maintenance. The chief engineer as the departmental head has to perform the following functions:

- Repair and construction of buildings and roads, maintenance and installation of plumbing and sanitation work, sewage treatment plant, etc. under a civil engineer
- Repair, maintenance, and operation of boilers, diesel generating sets, refrigeration and air-conditioning plant and equipment, swimming pool, water-treatment plant, pumps and compressors, kitchen equipment, etc. under a mechanical engineer
- Repair, maintenance, and installation of equipment, such as electric transformer and substations, all electrical motors, all electrical wiring in the building, audio-visual equipment, etc., under an electrical engineer
- Maintenance and control of inventory of spare parts and consumables in the engineering stores through a stores officer
- Establishment of equipment replacement policy and initiating the process of purchase of new equipment through appropriate processes such as tendering, etc.
- Vendor development for supply of spare parts, equipment, and contract maintenance
- Maintainnace of an engineering office for administrative work and processing/preserving documents
- Negotiation with employees' union on matters of mutual interest
- Look after the recruitment, training, and proper placement of human power
- Coordination with other departments of the establishment

Thus, the complexity and enormity of the job and the associated responsibility of the chief engineer cannot be overemphasized. A lapse on the part of some department might affect some individual guests but failure of an engineering system, such as an air-conditioning/ventilation system, will play havoc for the entire establishment. In some cases, such failures could be catastrophic for the business of the hotel and the importance of the maintenance department has to be seen in this light. Figure 12.1 illustrates the organizational chart of a typical engineering/maintenance department of a big hotel. In small hotels, of course, the department may not be so elaborate.

12.3 | TYPES OF MAINTENANCE PROCEDURES IN HOSPITALITY INDUSTRY

Modern machines and equipment are designed and manufactured with a very high degree of reliability. But even then, failures are common occurrences and lend themselves to different maintenance philosophies for consideration in

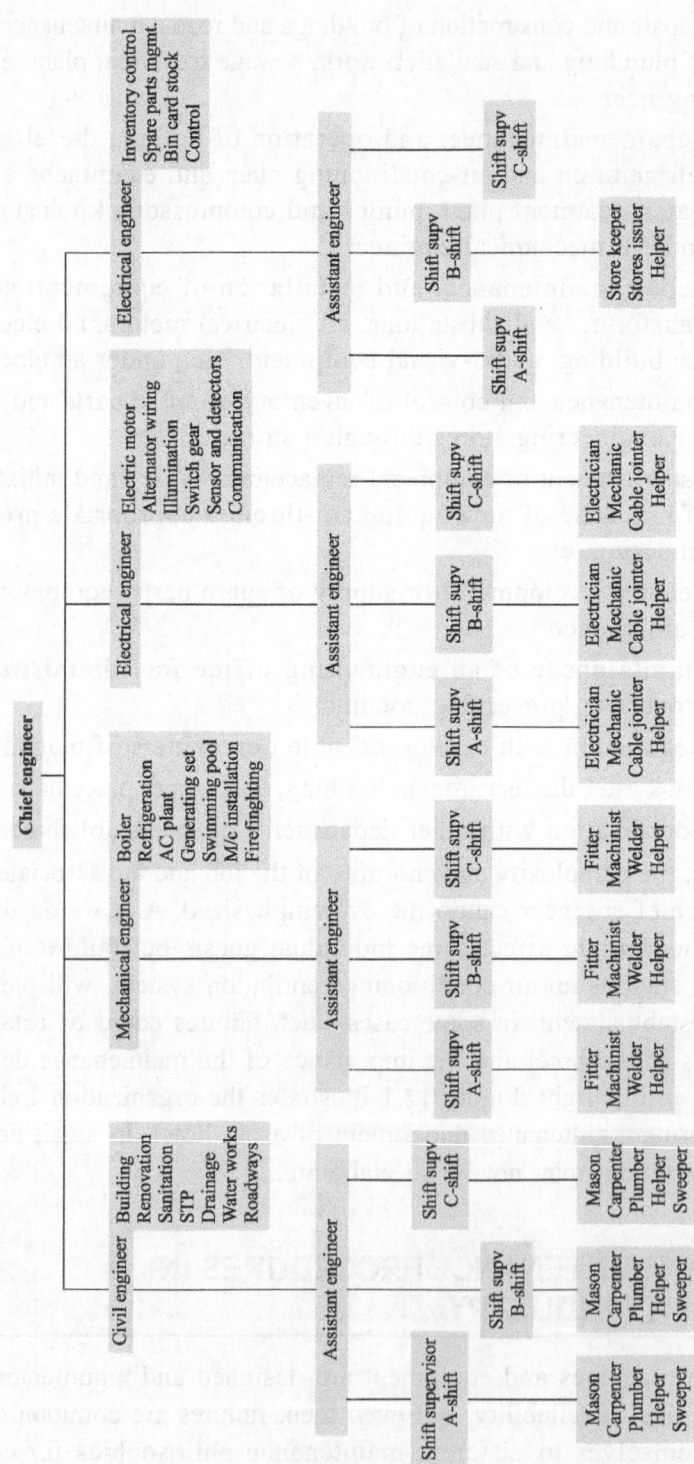

Fig. 12.1 A typical organizational chart of engineering/maintenance department of a big hotel

standard maintenance programmes in an organization. We shall discuss the various maintenance practices in relation to Fig. 12.2.

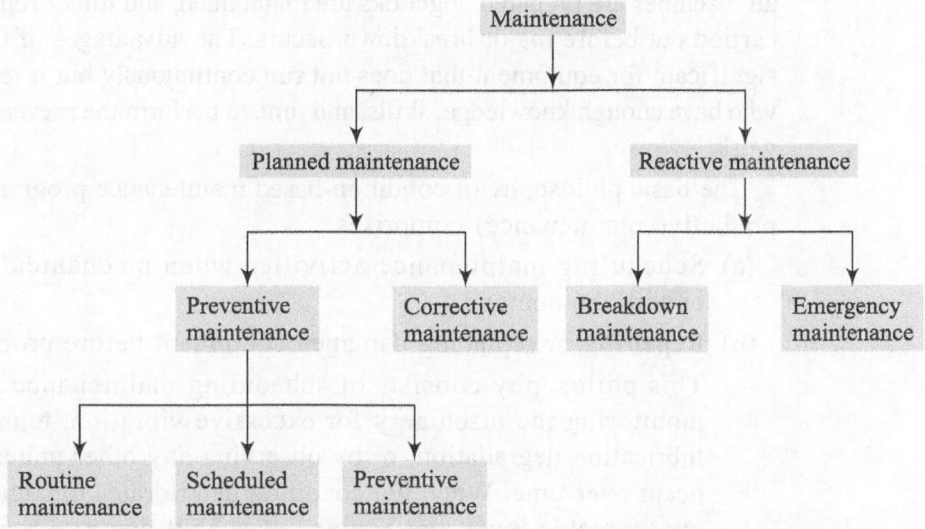

Fig. 12.2 Flow chart for different types of maintenance practice

12.3.1 Planned Maintenance

As the name suggests, in this programme, equipment will be maintained throughout its working life, before a complete breakdown requiring replacement/major repair. Planned maintenance can be of two categories. They are preventive maintenance and corrective maintenance.

Preventive maintenance Preventive maintenance can be defined as the maintenance actions carried out on a time-based and condition-based schedule that detect, preclude, or mitigate degradation of a component through controlling degradation to an acceptable level.

Essentially, it is based on the principle of 'prevention is better than cure'. While it can definitely not prevent failure, it can extend the working life of the machines. Also, through constant inspection and checking, it can predict failures so that alternative facilities can be kept ready to avoid complete stoppage of service.

The basic philosophy of time-based preventive maintenance comprises

- Scheduling maintenance activities at predetermined time intervals
- Repairing or replacing damaged equipment before obvious problems occur

These are achieved through (a) routine maintenance and (b) scheduled maintenance programmes. Routine maintenance programmes are cleaning the machines, regular lubrication of moving parts such as door hinges, bearings, etc.

Scheduled maintenance programmes are prepared and followed using experience and following manufacturers' maintenance manuals. Maintenance schedule of all machines are prepared, logbooks are maintained, and minor repairs are routinely carried out before major breakdown occurs. The advantages of this approach are significant for equipment that does not run continuously but it requires personnel who have enough knowledge, skills, and time to perform the preventive maintenance work.

The basic philosophy of condition-based maintenance programme (also called predictive maintenance) comprises

(a) Scheduling maintenance activities when mechanical or operational conditions demand

(b) Repairing or replacing damaged equipment before problems occur

This philosophy consists of scheduling maintenance by periodically monitoring the machinery for excessive vibration, temperature, and/or lubrication degradation, or by observing any other unhealthy trends that occur over time. When the condition deteriorates to some predetermined unacceptable level, the equipment is shut down to repair or damaged components replaced so as to prevent occurrence of a more costly failure.

Thus, the basic preventive maintenance programme consists of the following:

(a) Regular periodic inspection of machinery, utilities, and buildings. The schedule is determined based on experience of the maintenance engineer and the recommendations of the manufacturer.

(b) Preparation of information database regarding nature and possible causes of breakdown of different equipment so that they can be later analysed and corrective maintenance action can be taken to ensure that they do not become repetitive.

Corrective maintenance   Corrective maintenance is also sometimes called running maintenance. This is a maintenance procedure where minor repair work is carried out on machines, which is initiated by a formal work order from the machine user department. This work order can clearly or vaguely identify the problem; the maintenance crew inspects the machine and does repair work to run the machine usually in a short time. Corrective maintenance, therefore, attempts to meet the known needs in an orderly and timely manner consistent with the requirement of the property.

12.3.2   Reactive Maintenance (Breakdown Maintenance or Run-to-failure Maintenance and Emergency Maintenance)

As the name suggests, reactive maintenance is a maintenance action as a reaction to failure of equipment/building civil work. While breakdown maintenance is a strategy, emergency breakdown is responding to emergency situations where failure occurs, in spite of other modes of maintenance being carried out on them.

Breakdown maintenance is a maintenance strategy based on the philosophy of

- Allowing the machinery to run up to the point of failure
- Repairing or replacing damaged equipment when obvious problems occur

Breakdown maintenance implies restoration of a facility to almost its original conditions for which it is designed either by major repair or by total replacement. Many modern equipment, such as modern cars, washing machines, vacuum cleaners, dish washers, mincers, mixers, etc., are made with a high degree of sophistication and reliability. With such a high degree of reliability, it often proves very advantageous from cost point of view to allow them to run till they fail. This is also true for sophisticated electronic control systems, such as sensor etc., which will be either very costly to maintain or the skill needed to maintain will be very high to justify the items when they totally fail. To add to the list still more, items such as electric bulb are replaced when they fail, although a failure pattern may be established for them to have a group replacement policy where some bulbs in the group may be otherwise still working. In breakdown maintenance, there cannot be any delay in repairing/replacing the components, otherwise the entire operation will come to a halt.

The advantages of this approach are that it saves a lot of regular maintenance work in terms of labour and spares inventory. However, if the service is to be immediately restored with the same machine, cost of overtime, and emergency spares may be very high. It, therefore, works well if equipment shutdowns do not affect service and if labour and material costs do not matter, in case of emergency repair.

Emergency maintenance is carried out when a piece of equipment suddenly fails, but usually the maintenance can be delayed such as emergency repair of a guest room, keeping the room out of service for some time but it could result in a high level of general inconvenience. It is also an expensive way of maintenance.

12.3.3 Preventive Maintenance versus Reactive Systems of Maintenance

While preventive maintenance is not the only maintenance programme that is practised in industry, it does offer several advantages over a purely breakdown programme. By performing the preventive maintenance according to manufacturer's instructions, the life of the equipment may be achieved closer to design. Preventive maintenance (for example, lubrication, filter change, etc.) will generally run the equipment more efficiently, resulting in energy savings. It cannot fully exclude the probability of catastrophic failures, but the frequency of such failures in the property is expected to be reduced. Reactive programme reduces labour and spares cost but if proper inventory is not available at the time of breakdown and if the equipment is very critical in nature, it might entail additional cost, which could far outweigh the benefits. Following are the advantages and disadvantages of these two basic maintenance practices.

Advantages of preventive maintenance

The advantages of preventive maintenance are as follows:

- Is cost-effective in many capital intensive equipment
- Increases operational life of equipment
- Reduces the number of equipment and/or service failures in a property
- Provides flexibility for the adjustment of maintenance schedule
- Operates at design conditions
- Gives maximum return on capital investment
- Results in an estimated 12 per cent to 18 per cent cost savings over that found in a reactive maintenance programme
- Improves worker morale by reducing idle time of workers
- Provides safety from hazards to personnel and property by reducing the possibility of sudden catastrophic breakdown such as in boilers, compressors, transformers, etc.

Disadvantages of preventive maintenance

The disadvantages of preventive maintenance are as follows:

- It cannot eliminate complete failures
- It is more labour- and time-intensive
- It requires extra facilities and may lead to poor/underutilization of facilities
- Maintenance activities that are really not required, are often carried out
- It is not economical for small property which use cheaper and non-critical equipment
- Savings potential is not readily seen by management, which usually focuses on running costs

Advantages of reactive maintenance

The advantages of reactive maintenance are as follows:

- Has lower costs due to not regularly attending and maintaining the facility
- Requires fewer staff to implant the scheme

Disadvantages of reactive maintenence

The disadvantages of reactive maintenance are as follows:

- Increases costs due to unplanned downtime of equipment
- May involve prolonged downtime if proper humanpower and components are not readily available
- Increases labour costs, especially if overtime is needed
- May increase costs associated with repair or replacement of equipment on an urgent basis
- May result in possible damage to associated equipment (for example, failure of an electric transformer may cause some damage to other electric equipment connected down the line)

Additionally, we may define two more types of maintenance practices specifically coined for the hospitality industry. They may be described as follows:

Routine maintenance

This includes activities that pertain to the general upkeep of the property and recur on a regular basis (daily or weekly). Typical examples are cleaning the floor, sweeping the carpets, cleaning the accessible windows, polishing the door handles, cleaning the guest rooms, replacing defective lamps, etc. These do not require any formal work order or skilled personnel for execution.

Guest room maintenance

Guest-room maintenance has a special meaning in the hotel industry. While some include it in the category of preventive maintenance, others call it corrective or running maintenance. The most important aspect of the hotel industry is the level of comfort and ambience provided to the guest in his/her room. For the guest, the condition of the room is the most visible thing he/she can see of the property as soon as he/she steps in and assesses the quality of the property based on this perception. Hence, guest-room maintenance should have the top-notch level of maintenance programme. The condition and proper operation of furniture, fixture and equipment such as television, audio system, air conditioner, etc., the appearance of the walls and ceilings, conditions of carpet, window louvres, toilet fittings, supply of water in the toilet, door hinges, and the overall cleanliness are all included in the maintenance programme of guest rooms.

12.4 | MAINTENANCE MANAGEMENT POLICY

In order to get the best mileage out of a car and to prevent breakdown on road, as far as possible, the car needs proper care and maintenance and so do the mechanical, electrical, and civil systems in a hospitality system. Failure of service of such systems could have far-reaching consequences as far as guest service reputation is concerned. Unfortunately, since maintenance activities do not produce revenues directly, they are often considered not so important an activity and people are inclined to think that equipment should always run smoothly. Whenever an equipment fails, the management thinks that since there is a maintenance department, it should not have failed and wants the maintenance crew to set it right almost instantaneously. People in the maintenance business often find it a thankless job.

Management should realize that expenses on maintenance is more than compensated by reducing downtime and saving loss of reputation. Hence, this activity is critical for revenue generation and continuation of business. In the light of the above, role of the property management may be enumerated as the following.

1. Establishing a policy regarding the quality of mechanical equipment, materials, and systems with respect to performance, economic life, operation, service, and repairs. The cost of maintenance and breakdown can be reduced by properly planning production (e.g., food preparation) and operation of mechanical and electrical facilities. For example, replacing/repairing an equipment should not involve shifting or removing of other facilities, which could even be a very heavy machine.

2. Management must decide as part of its policy:
   • Whether to adopt a preventive maintenance programme or reactive maintenance (breakdown maintenance) programme or a combination of both, which is very common
   • Whether to do maintenance work in-house or hire contract service and if so to what extent and in which areas
   • Whether to have an operating and maintenance manual prepared in-house, if there is no such from manufacturer and if so, how comprehensive this manual should be

3. Developing a maintenance and operating programme and providing a qualified and well-managed staff for the purpose

4. Management must establish a proper equipment replacement

12.5 | ELEMENTS OF MAINTENANCE PROGRAMMES

All maintenance programmes are composed to some degree of the following:

1. Management maintenance policy
2. Inventories and records of systems, equipment down to the component level and controls that include drawings with specifications, equipment installation, service, and maintenance manuals and spare list

3. Procedures and schedules, which are the action part of programme relating to the following:
   (a) Operating instructions that include
      - Starting and stopping procedures, sequences
      - Adjustments and regulations
      - Seasonal changeover
      - Logging and recording procedures
   (b) Inspection involving
      - Equipment to be inspected
      - Areas/functions to be inspected
      - Inspection methods and procedures
      - Frequency of inspection
      - Interpretation and evaluation of observations
      - Recording and reporting
   (c) Service and repair norms related to
      - Frequency of scheduled service
      - Procedures for scheduled service
      - Repair procedures
      - Recording and reporting
4. System of records at a central database file which should involve
   (a) Inventory data
   (b) Operating, inspecting, and servicing instructions, procedures, and schedules
   (c) Records and reports on work performed, time expended, and parts and supplies used
   (d) Logs of operations performed and operating data
5. Operating and maintenance manuals providing vital central reference of organized information and instructions

To carry out the highly specialized and important job of maintenance, the engineering department of maintenance must be organized with proper manpower in a well-managed structural form, as shown in Fig 12.1.

12.5.1 Programmes and Tools for Preventive Maintenance

Preventive maintenance programmes rely on two vital elements of activity, namely

1. Inspection   2. Lubrication

Inspection

Regular inspection of machines and equipment at given intervals is vital for the success of preventive maintenance programmes. Regular checks are likely to detect faults in equipment at an early stage so that proper repair work and replacement can be undertaken and sudden breakdown of vital equipment can be avoided. Inspection schedules are prepared for all the equipment under purview. The frequency of inspection for each individual machine and equipment are decided on the basis of experience and manufacturers' recommendations, if any. The inspection could be weekly, fortnightly, monthly, or quarterly. Figure 12.3 illustrates a typical inspection master schedule for a piece of equipment.

Experienced maintenance staff, usually in the rank of chargeman/foreman, are deputed for inspection who check the machine/equipment, etc., according to the master schedule of the particular machine/equipment. If he finds any fault or abnormality in operation of the machine, he will put the findings in the remarks column. The report on faults will be finally transferred to the machine record card (MR Card) or history card. The maintenance engineer goes through the inspection report on a regular basis and if necessary, will take a shut down from concerned operational department and carry out repair/replacement work. Figure 12.4 illustrates a typical MR Card.

Lubrication

Proper and regular lubrication is essential for smooth and trouble-free operation of machines having rotating and/or sliding parts. Lubrication machine/equipment at the right time and quantity is a vital element for the success of preventive maintenance programme. A complete lubrication master chart is prepared for the year. This should indicate the machine and the parts to be lubricated, frequency of lubrication, and type of lubricants to be used for the particular part. The frequency of lubrication depends upon the rotational and sliding speed of the parts, presence of atmospheric dust and dirt, nature of the materials of the parts to be lubricated, etc. Normally, a semi-skilled worker is deputed for routine lubrication in the beginning of the general shift of duty (typically 8 a.m. to 4 p.m.). But the maintenance engineer must ensure that the right kind of lubricant is used and the master schedule chart is filled up after each lubrication session. Figure 12.5 shows a typical lubrication master schedule.

12.5.2 Work Order

In day-to-day running maintenance and breakdown maintenance, the maintenance department receives work order from the operations department

Name of the Hotel—XYZ																																
Equipment —Raw Water Pump Code: 009	Month—September, 2009 Name of the Inspector—ABC																															
Sl. No.	Parts to be inspected	Frequency of inspection	1	2	3	4	5	6	7	8	9	10	11	12	13	14	15	16	17	18	19	20	21	22	23	24	25	26	27	28	29	30	1
1	Impeller	Monthly																																
2	Bearing	Fortnightly																																
3.	Gland	Daily																																
4.	Shaft sleeve	Monthly																																
5.	Impeller ring	Monthly																																
6	Electric motor	Monthly																																

Put ✓ mark in the box if it is checked   Remarks:

Signature:

Fig. 12.3 Inspection master schedule

Name of the Hotel—XYZ

Name of the Unit: Air-conditioning Plant

Code: 110

Year of Installation—15.3.2005

Year—2009

Sl. No.	Name of the Parts	1	2	3	4	5	6	7	8	9	10	11	......	50	51	52
1	Heating/cooling unit	M*/2/1									M					M
2	Scrubber unit			B 16/1			M								M	
3.	Inlet air fan															
4.	Exhaust fan					M										
5.	Ducting					M								M		
6.	Electric motor				B 25/1					M						M
7.	Filters and water spray nozzles	M/5/1							M						M	

M—Preventive maintenance   B—Breakdown maintenance

Report: Electric motor bearing no. 5340 failed and replaced during breakdown maintenance. P.M. not planned properly.

Scrubber plate failed and replaced

Filters cleaned on P.M. for smooth operation

Authorized Signatory: XYZ

* M/2/1 means preventive maintenance done on 2 January

Fig. 12.4   Machine record (MR) card or history card

Name of the Hotel—XYZ														
Name of the Unit : Boiler Plant														
Code: 105								Year—2009						
Year of Installation—15.3.2005														
Sl. No.	Parts to be Lubricated	Frequency	Name of Lubricant	1	2	3	4	5	6	7	8	9	50	51	52
1	Gear box	Month	Shell Tellus										....		
2	I.D. fan bearing	Fortnight	Grease										....		
3.	F.D. fan bearing	Fortnight	Grease												
4.	Boiler feed pump bearing/safety valve	Week	Grease												
5.	Water-treatment plant, pump, and motor	Month	Grease and Lub. oil												
6	Conveyor roller and bearing	Month	Grease												
	M—Preventive Maintenance B—Breakdown Maintenance														

Put ✓ mark in the box if it is checked

Remarks:

Signature : XYZ

Fig. 12.5 Lubrication master schedule

related to the particular machine/equipment that needs attention. The work order briefly states the nature of the problem encountered while operating and must be signed by an authorized person. The maintenance engineer first reviews the nature of the problem and decides whether to send people immediately to attend the problem or it can wait for some time. He/she then decides the appropriate persons who can attend the job, e.g., if it is problem of electric motor, an electrician has to be sent or if it is a problem of pump, a mechanical fitter has to be sent. Figure 12.6 shows a typical work order format.

Name of Hotel: XYZ

Maintenance Request

Time: 8:10 A.M.
Reported by: XYZ   Date: 15-10-2009
Location/Equipment: Diesel Generating Power House — Cooling Water Pump
Problem—Pump is making excessive vibration. Bearing housing getting hot
Job assigned to: ABC

Date of completion: 15-10-2009   Time taken: 2 hr. 25 minutes
Completed by: ABC   Remarks: Bearing damaged

Fig. 12.6 Sample work order issued by operations department to maintenance department

12.6 | CONTRACT MAINTENANCE

Quite often, a part of the maintenance activities in a property is off-loaded to an external agency through contract under suitable terms and conditions. Before we discuss various aspects of contract maintenance, let us define contract.

Contract

A contract is an agreement in legal form between the two parties (the owner of the property and the maintenance contractor in our case) wherein the terms and conditions of the work under contract are clearly written and adherence to which is binding on both the parties once the contract is signed. The contract, thus, invariably follows a proposal from one party and its acceptance by the other.

Many of the present-day establishments with varieties of engineering equipment and systems and building, premises to maintain, have a mix of maintenance by in-house maintenance personnel and outside contractual maintenance. The decision to perform a particular maintenance activity in-house or through contract is usually controlled by the maintenance engineering manager of the property.

12.6.1 Possible Circumstances for Awarding Contract Maintenance

While contract maintenance may cover all possible types of maintenance including routine, preventive, and corrective maintenance, some specific areas are enumerated below.

- Contract maintenance may handle special needs of a property involving special equipment or when such needs are infrequent in nature. Examples are maintenance of electronic control systems, sensors including heat/smoke detectors in the fire alarm system, lifts, charging of fire extinguishers, to name a few.
- Contract maintenance may be called upon for routine services such as janitorial services for all sorts of building interior cleaning, although in a lodging industry, guest room and other private places are cleaned by in-house staff.
- Ground maintenance and special maintenance needs of pest control activities (needing special licence on the part of the contractor) are generally awarded to contract maintenance.
- In many properties, operation and maintenance of water treatment, sewage treatment and swimming pool maintenance are left to contract service owing to their very special character.
- At small establishments, air-conditioning and refrigeration units are also maintained through contract service.
- Special equipment are often needed for cleaning kitchen ducting, grease trap, grease filters, etc., and their maintenance is also often contracted out.

However, the role of the engineering manager does not surely end with deciding the areas of maintenance to be put under contract maintenance and choosing the appropriate contractor with adequate technical know-how and skilled manpower to do that. The contract has to be carefully constructed to unambiguously define the definite areas of work to perform. The manager has to closely monitor and control the reliability and quality of maintenance being carried out by them. Periodic review is to be made to ascertain the need of any change in the nature of the contract including a possible change in the contractor. The access to the property in areas other than the area of work should be restricted for the employees of the contractor. It is also the duty of the engineering manager to see that the contractors ensure required safety standards for their employees and they are covered by appropriate insurance policies.

12.6.2 Advantages and Disadvantages of Contract Maintenance

There are potential advantages and disadvantages of contract maintenance that the engineering manager must weigh and assess in order to take a decision regarding whether a particular maintenance job is to be contracted out or done in-house.

Potential advantages of contract maintenance include the following:

- Reduction of total labour cost
- Reduction of cost of carrying inventory and special maintenance equipment
- Use of latest techniques and methods of maintenance
- Saving in administrative time on the part of the engineers when they can concentrate more on planning, etc.
- Structure and infrastructural facilities of the in-house maintenance department need not change due to introduction of new equipment with sophisticated technology
- Flexibility to meet emergencies
- Reduction of labour problem
- Need to recruit and train employees is eliminated

Potential disadvantages of contract maintenance include the following:

- Labour cost is not really saved unless the actual number of in-house maintenance staff is reduced
- Revision of contract fees in subsequent years often leads to escalation of total cost if not properly negotiated and controlled
- Often the specialized nature of the maintenance would lead to high charges for such contracts leading to a overall increase in maintenance cost compared to if it were maintained in-house
- At the time of bidding, lack of initiative on the part of the engineering manager would often lead to a non-competitive price quote by a handful of contractors
- Reduction in the number of specialized in-house maintenance crew may result in their unavailability in case of emergency requirement for some other facility
- The engineering manager usually does not get any information feedback regarding the status of the equipment under contract maintenance, except in case when it completely breaks down
- Loss of control over contractor employees, particularly in respect of security, attitude, and sense of belongingness although it has been observed that in-house employees also suffer from these attitudes for further their gains in form of working on overtime or due to absence of proper work culture.

12.6.3   Types of Contract for Maintenance Services

Depending on the requirement for a particular service, maintenance contract usually can be of the following types.

Piece-work or piece-rate contract   In this type of contract, the contractor has to quote separately for each item of work. For example, if a contractor has to maintain a set kitchen ovens, replacement/repair of each items such as burner, valves, and gas pipelines could have different rates for maintenance.

Lump sum rate contract   In this type, the contractor has to quote for maintenance of the equipment as a whole, such as maintenance of all the ovens and burners in the kitchen or maintenance of all the heating equipment in the kitchen to make the scope still broader. This type of contract is usually awarded for a specific one-time job on some specific equipment or group of machines or group of work.

Rate contract   In this type, the contractor agrees to supply items or labour at a fixed rate for a certain period.

Service contract   This type of contract is usually made for a year-round routine, usually involving preventive and scheduled maintenance of a plant/system, and the contractor is needed to carry out these activities for a specific period of time. Maintenance contract of engines, air-conditioning plant, swimming pool, water-treatment plants, etc., if they are contracted out, usually follow this type of contract for maintenance.

12.7 | TENDERING PRACTICE

As we have already seen, many jobs in an establishment are carried out by external agencies. This is true for the hotel industry as well. Some typical jobs may include premises cleaning and janitorial work, security, maintenance and operation of swimming pool, maintenance of many engineering systems such as audio-visual equipment, air-conditioning and refrigeration plants, etc. Purchase of articles in bulk, such as sanitary, laundry, linen, fuel, etc., is also made from identified external agencies. For every organization, there are definite laid down procedures by which these external agencies are identified and the job/supply contract is awarded. One of the most commonly used practices is what is known as 'tendering'. The term 'tender' for the present function is formally defined in many ways. Some documented definitions of tender are as follows:*

* *Sources*: www.thefreedictionary.com and www.en.wikipedia.org

1. A formal offer to supply specified goods or services at a stated cost or rate
2. A written offer to contract goods or services at a specified cost or rate; a bid
3. Tendering of goods and/or services at the best possible total cost of ownership, in the right quantity and quality, at the right time, in the right place for the direct benefit or use of governments, corporations, or local authorities, generally via a contract.

12.7.1   Contract Types for Tendering

Procurement of different categories shall be effected by the following methods of tendering, namely

- Piece-work contract
- Lump-sum contract
- Fixed-rate contract
- Turnkey contract
- Multistage contracting including pre-qualification and two-cover system

While the first three types of contracting have already been discussed in the previous article, turnkey contract usually relates to execution of a job from scratch to finishing such as making a building from green field to a stage ready for possession or supplying, installing, and commissioning of a water-treatment plant, etc. Multistage contracting refers to awarding contract for a job segregated by different stages of execution of the job. This means that the contractor will have order for execution of a job subject to satisfactory completion of different stages. The contract awarding authority keeps the right to cancel the contract at any particular specified stage of the job. For example, instead of awarding the whole contract of making a building, the contract may be made in stages where assessment for the job may be made after completion of each stage. The stages of contract could be laying the foundation, erecting the superstructure, and completing the brickwork, etc. After any particular stage, the contract may be terminated on mutually agreed terms. Pre-qualification would relate to the credential of the party to undertake the job both in terms of technical competence and financial health, while two-cover system relates to the technical ability as well as price quotation competitiveness of the party that are evaluated for the purpose of awarding the contract at the time of tendering. This is discussed in more details later in this section.

The tender-inviting authority shall decide the most suitable method of tendering to be followed in each case having regard to the category, size, and complexity of the procurement/job.

12.7.2   Different Steps of Tendering Process

The different steps involved in the tendering process can be listed as follows:

1. Framing of a tender committee
2. Deciding on the type of contract
3. Preparation of tender document
4. Circulating notice for obtaining tenders
5. Submission of tenders by interested party
6. Evaluation of tenders
7. Acceptance and finalization of tender

We discuss each of the steps in some details as follows:

Framing of a tender committee

A tender committee is formed usually in case the tender value is high, typically more than Rs 15,000 or so, as decided by the organization as its policy. For example, a tender committee should consist of at least one member from the accounts department, one from the concerned executive department such as engineering department or food and beverage (F&B) department, who first raises the requisition for the job/item, and another from a department dealing with such contract matters.

Deciding on the type of contract to be followed (as outlined earlier in section 12.7.1)

Preparation of tender document

This is one of the most important activities in the tendering process where technical, legal, and financial aspects are to be analysed carefully and documented. These are then released for circulation either through publication in the newspaper as notice inviting tender (NIT) or in the website or by other means given below or in combination, as decided by the management.

Circulating notice for obtaining tenders

The following are amongst the different methods of obtaining tenders that may be adopted:

(a) By public advertisement: Works are generally outsourced by competitive bidding by resorting to the system of advertised open tenders. This is the usually adopted method. 'Open tender' comes under this category. Notice inviting tender is given by publishing in national dailies and/or by uploading on website of the organization. It contains important details such as description of work, completion period, date of opening, date of closing, cost of tender form, earnest money, approximate estimated cost of tendered work, minimum eligibility criteria, etc.

(b) By direct invitation to a limited number of firms/contractors: These are called 'limited tenders'.

(c) By invitation to one firm/contractor only: These are called 'single tenders'.

E-tendering   Tenderers may download the tender documents from the designed website free of cost, without any hindrance, for open tenders of value exceeding Rs10 lakh.

Submission of tenders by interested party

The tenderers have to submit their offer in the way described in NIT alongwith an 'earnest money deposit' usually not exceeding 1 per cent of the total tender amount through instruments as specified (such as banker's cheque, demand draft, etc.)

Quite often, NIT has a clause of pre-qualification with a two-cover submission. This is done to assess the credential and technological capability of the bidder for the job at hand.

Two-cover System   It is a procedure under which the tenderers are required to simultaneously submit two separate sealed covers, one containing the earnest money deposit and the details of their credentials and technical capability to undertake the tender and the second cover containing the price quotation. The first cover is opened first and the second cover will be opened only if the tenderer is found qualified to execute the tender. So, for such system, the tenderer has to submit tenders in two different envelopes.

Evaluation of tenders

After the last date of submission is over and on the date of opening the tender, the tenders are opened by the competent authority. If there is no two-cover tender, all the tenders are opened for evaluation. In case it is a two-cover bidding, the following steps are taken for assessment of the tenders being submitted as already briefed.

(a) Prequalification procedures: The tender inviting authority shall for reasons to be recorded in writing provide for pre-qualification of tenders on the basis of

- Experience and past performance in the execution of similar contracts
- Capabilities of the tenderer with respect to personnel, equipment, and construction or manufacturing facilities
- Financial status and capacity

(b) Only the bids of pre-qualified bidders are considered for price bidding financial evaluation. A statement of comparative rates and other important tender conditions is prepared and tabulated for all the tenders received. This comparative statement is prepared by the engineering department and verified by the accounts department.

This tabulated statement is then put to a tender committee, constituted for the purpose, for consideration without any screening by any other official. Along with the tabulation statement, a briefing note duly authenticated is also submitted, indicating the details as per engineering specifications and code.

Determination of the lowest evaluated price Out of the tenders found to be acceptable after the initial examination, the tenderer who has bid the lowest evaluated price in accordance with the evaluation criteria or the tenderer scoring the highest on the evaluation criteria specified, as the case may be, is identified.

The tender committee holds its meetings and after considerations of tenders, records its minutes making appropriate recommendations for acceptance, rejection, or discharge of the tender or for negotiations.

Acceptance and finalization of tender

The tender committee puts up the minutes of the committee to the accepting authority. The accepting authority may accept, reject, or modify the recommendations of the tender committee. After a particular bidder is accepted for awarding the contract, an agreement for execution of the job/supply of items is signed and the successful bidder has to deposit an amount as notified in the tender schedule. The amount is kept as a security deposit so that the contractor fulfils the terms and conditions of the contract and carries out the work satisfactorily. In case the contractor fails to fulfill the terms of the contract, the whole or part of this deposit can be forfeited by the department. This amount does not normally exceed 10% of the contract value or as stipulated in the agreement.

12.8 | CONCLUSION

In the end, it should be clear that maintenance of engineering and other facilities is one of the most critical activities in a hotel. It is both very complex and elaborate in the present era of highly developed technology with its newer areas of application in hotel industry. Although it is not a revenue-generating activity directly, such as F&B or accommodation, proper and regular maintenance definitely saves a lot of money by reducing downtime of machines and equipment. Further, by maintaining uninterrupted service of the facilities, it ensures guest satisfaction and high staff morale and, thus, contributes immensely to the ultimate objective of good business for a hotelier.

KEY TERMS

Breakdown maintenance   It is the maintenance activity that is carried out after the system fails.

Contract maintenance	It is the maintenance activity that pertains to maintenance of systems by outside agencies.
Guest-room maintenance	It is specialized maintenance pertaining to guest-room facilities only.
Maintenance	It is a planned scheduled set of activities carried out on some property to keep it operational and fit for discharging the service for the period it is designed to do so.
Preventive maintenance	It is pre-emptive maintenance that includes inspection and monitoring of the health of the system.
Routine maintenance	It is the maintenance activities which pertain to the general upkeep of the property and recur on a regular basis.
Reactive maintenance	This is a broad category of maintenance procedures, where the maintenance department responds to any fault in the system to the point of failure of continuous service.
Tender	A formal offer to supply specified goods or services at a stated cost or rate.

REVIEW QUESTIONS

12.1 Explain the importance of proper maintenance in the hotel industry.

12.2 Discuss the role and importance of an engineering/maintenance department in the hotel industry.

12.3 List the duties and responsibilities of the chief engineer in a hotel showing the typical organizational chart of the engineering department of a large hotel.

12.4 Discuss, in brief, the various type of maintenance methods.

12.5 Discuss preventive maintenance and reactive maintenance, indicating their relative advantages and disadvantages.

12.6 Discuss the role of management related to maintenance policies.

12.7 Explain contract maintenance, citing possible areas in the hotel industry where they can be employed.

12.8 Discuss the potential advantages and disadvantages of contract maintenance.

12.9 Discuss the role of the engineering manager in respect of contract maintenance.

12.10 State different types of contract maintenance.

12.11 List the stages of a typical tendering process.

12.12 Explain pre-qualification and two-cover system in connection with tendering practice.

13
Chapter

Equipment Replacement Policy

Learning Objectives

Plant and machinery and other service equipment are vital for the hotel industry. Students must know the importance of proper operation and maintenance vis-à-vis the operational cost and the need for replacement of such equipment after a certain period of time. They should be able to understand the different methodologies and their need in respect of replacement of equipment, in order to appreciate the activities of the maintenance department.

The broad objectives of this chapter are to:

- sensitize the students about the need for replacement of equipment
- make them understand the various circumstances under which equipment are replaced
- discuss the various economic factors that govern the decisions regarding replacement
- explain the concept of equipment replacement through solved analytical problems on replacement

13.1 | INTRODUCTION

Plant and equipment are an absolute necessity for successful operations in any industry. We have also seen that proper maintenance is of vital importance for reliable functioning of these systems and is very important for the hotel industry, where loss of service to customers can prove expensive for the business. We have seen that by proper maintenance, the life of the equipment can be prolonged but, after some time, it becomes economically non-viable and the question of replacing the equipment arises. Sometimes, the technology may also become obsolete. The problem is to decide when such replacements become economically meaningful, because the investment made on the old machines has to be realized to the maximum extent possible. Thus, a very important decision, as a substantial part of the investment in a hotel, is made in procuring and operating plant and machinery.

13.2 | CAUSES FOR EQUIPMENT REPLACEMENT

The need for replacement comes from the very fact that no equipment can last for an indefinite period in operation. It is theoretically possible to extend the life of an equipment by taking appropriate maintenance, after a few years by replacing spares. But in the process, maintenance cost and energy cost would soar up so high that it will be irrational to continue with the equipment. Equipment and engineering systems need replacement when any of the following or combinations thereof occurs in respect of a facility:

Replacement because of inadequacy The current level of service may demand a higher capacity/more number of equipment for a particular function.

Replacement because of excessive maintenance activities As the equipment grows older, it requires frequent maintenance to keep it in running condition. It becomes no more viable to maintain because spare parts are very costly. Further, frequent maintenance involves man-hours and possible deprivation of manpower for other areas of maintenance.

Replacement because of technological obsolescence With the rapid development of technology with improved engineering, materials of construction, and control, better equipment become available during the life cycle of old equipment. They are superior in functions energy-efficient and user-friendly, and are also less costly in many situations.

Replacement because of decreasing efficiency Over the period of usage, an equipment, because of wear and tear and because maintenance cannot always, put back all the components to their original conditions in every respect, loses its efficiency, indicating a higher energy cost.

An equipment may have to be replaced if any one or a combination of the causes arise during the operating life of the equipment. The important question is when to replace a particular piece of equipment. This needs a quantification regarding the current condition of the equipment. While the first cause of replacement need, due to demand of increased capacity, does not require any further consideration, other situations can be translated in monetary terms for determining the timing for replacement. Before we proceed any further for establishing different types of quantitative basis for equipment replacement, we distinguish between two basic categories of machines and equipment for the purpose. They are as follows:

1. Equipment which gradually lead to deterioration in performance and failure due to wear and tear and material deterioration due to fatigue, high operating temperature, vibration, corrosion, etc. Examples are compressors, engines, motors, pumps, boilers etc. With the passage of time, they tend to become more maintenance prone and less efficient.

2. Equipment which fail suddenly almost without any warning: The pattern of failures of electric lamps, electric circuit breakers, sealed refrigeration units, etc. fall under this category. Before failure, they continue to work quite satisfactorily and efficiently. These pieces of equipment are replaced as a whole without any maintenance.

The criteria for deciding when to replace the equipment are different for these two categories of equipment.

13.3 | EQUIPMENT REPLACEMENT POLICY

As outlined above, the two basic categories of equipment are replaced using essentially two different approaches. However, for all of them economic factors are the major considerations. Other factors such as shop floor management and human relations may also play vital roles in particular cases.

13.3.1 Equipment Replacement Policy for Equipment that Deteriorate Gradually

The quantification of the causes of replacement of equipment under this category can be made in two ways. They are

1. Average annual cost method
2. Net present value method

Average annual cost method

The causes for replacement of a piece of equipment, as outlined earlier, can be translated to economic terms in the form of average annual cost. Since capital cost is a sunk fund for which investment is already made, it is not considered for the purpose of calculating average annual cost. The objective is to minimize annual cost. The relevant components of cost for calculating annual cost are (a) value lost due to the use of the machine (depreciation), (b) operating cost, and (c) maintenance cost. The average annual cost first decreases with time during the early part of the life of the machine and then starts increasing after a period of time. The year at which this annual cost becomes minimum can be considered as the time when the equipment should be replaced in course of its operating cycle. Now, let us solve a problem of equipment replacement illustrating the concept of average annual cost.

Initial cost of a dishwasher is Rs 1,20,000. Due to heavy depreciation during the first year, the resale value after one year becomes Rs 80,000 only. Depreciation rate thereafter becomes 20 per cent of the resale value of the previous year. This means that the depreciation (i.e., loss of resale value) at the end of second year is 20 per cent of Rs 80,000, i.e., Rs 16,000, at the end of third year it is 20 per cent of Rs 64,000, i.e., Rs 12,800, and so on. Its operating cost, including maintenance cost and cost due to loss of efficiency (because of higher energy being consumed by the equipment, less amount of service is performed due to downtime, etc.), is Rs 5000 in the first year and then increasing by Rs 3000 every year thereafter. We proceed to determine the timing when it is appropriate to replace the equipment. We use the principle of minimum average annual cost.

Table 13.1 shows the calculation table for the above example.

Table 13.1 Calculation table for average annual cost

Year	Value at year end	Depreciation (loss of value) A	Operating cost during the year B	Total cost in the year C=A+B	Cumulative total cost D	Average cost per year/no of years
0	1,20,000					
1	80,000	40,000	5000	45,000	45,000	45,000
2	64,000	16,000 (20% of 80,000)	8000	24,000	69,000	34,500
3	51,200	12,800 (20% of 64,000)	11,000	23,800	92,800	30,933
4	40,960	10,240	14,000	24,240	117,040	29,260
5	32,768	8192	17,000	25,192	142,232	28,446
6	26,214	6553	21,000	27,553	169,785	28,297
7	20,971	5242	24,000	29,242	199,027	28,432
8	16,777	4194	27,000	31,194	230,221	28,777

From the tabulated values, we observe that the average annual cost decreases up to the end of the sixth year and then again increases as mentioned earlier. Hence, the equipment should be replaced after six years.

Net present value (NPV) method

In this method, the cost expected to be incurred and revenue expected to be received in future are discounted through a standard rate down to their present worth. It works on the consideration that Rs 100 obtained after five years will not be as valuable as Rs 100 today. The present value of such an amount may, thus, be calculated as the amount which if deposited in a bank today would produce Rs 100 after five years. It may be Rs 90 or Rs 80 or any value depending on the rate of interest paid by the bank to its customer. We, thus, say that Rs 90 or so is the present value of Rs 100 received after five years. We use the term 'rate of return' or 'discounting rate' in place of the bank interest rate. This method of what is called 'discounting' a sum receivable/payable in future is very useful to compare an earning/expense in future with the investment made today, in order to compare between competing investment options.

Discounting Discounting cash flow is a means of evaluating an investment by calculating the effective annual return on investment (ROI) throughout its life. Discounted cash flow, thus, takes into account the fact that sooner a return is obtained the more worthy it is.

Net present value In this method, a rate of return is applied to the cash flow each year to find the discounted values and such discounted values are calculated for return in each year and added together to find the present value equivalent. It is then subtracted from the capital cost to get the net present value (NPV). If NPV is positive the investment decision is economically justified. The following formula can be used to calculate the NPV of an investment.

$$P = E_t/(1 + r)^t + S/(1 + r)^t - C$$

where P is the net present value of the investment

E_t is the net earning during the t-th year

r is the rate of interest per annum (discounting rate); also called the rate of return on investment

S is the salvage value after the t-th year,

$1/(1 + r)^t$ is the discounting factor for a return made during the t-th year, and

C is the cost of an equipment when purchased or any investment made in the beginning of the period

The first term is the present value of the net earning made during any t-th year and is found out by multiplying the net earning by the discounting factor directly from table or by calculating as given in the above equation. The second term is the present value of the salvage value and the third term is the initial cost.

As long as $P = 0$, the invest proposal is economically viable. Higher the positive value better is the opportunity. For given values of E, r, and C, P and S will be different for different N values. Hence, a trial approach will find a value for N for which the present value, P, will be maximum and the corresponding value of N should correspond to the economic life of the equipment after which it should be replaced. Standard tables in the books of accountancy or financial management furnish the discounting factor after *t*-th year for different rates of return. Let us work out an example using the present value method for evaluating the economic life of a piece of equipment as follows:

A hotel has a sophisticated washing machine for the boarders. The machine is also subject to a high degree of obsolescence. In spite of regular maintenance, the quality of the output deteriorates fast and adversely affects revenue. It is required to find the time when the machine is due for replacement. The following data given in Table 13.2 are available from the hotel's past records.

Table 13.2 Statement of earnings, expenses, and salvage values

Year	1	2	3	4	5	6
Yearly gross earnings from sales	20,000	15,000	12,000	8000	6500	4100
Yearly expenses	1500	2000	2500	4500	3800	4000
Net earning	18,500	13,000	9500	3500	2700	100
Salvage value at the end of the year	10,000	6000	4000	2000	1000	0

The initial cost of the machine is Rs 25,000 and the expected rate of return for company investment (or discounting factor) is 8 per cent, find the economic life of the washing machine. To solve the problem we calculate the NPV of net earnings for different years, as shown in Table 13.3.

Table 13.3 Calculation table for NPV

End of year, t	Net earning E_t (A)	Discounting factor $1/(1+r)^t$ (B)	Present value of net earning $C = A \times B$	Cumulative present value of net earning (D)	Present value of salvage after t-th year (F)	Cumulative present value $G = F + D$	Net present value G–initial cost
1	18,500	0.9259	17129	17129	9259	26388	1388
2	13,000	0.8573	11145	28274	5144	33418	8418
3	9500	0.7938	7541	35815	3175	38990	13990
4	3500	0.7350	2572	38387	1470	39857	14857
5	2700	0.6806	183	40224	680	40904	15904
6	100	0.6302	63	40287	0	40287	15287

Note: Net earnings are calculated by subtracting expenses from gross earnings as shown in Table 13.2. There will be slight change in the values computed if standard table values are used for 8 per cent discount rate as the figures there are furnished as rounded off.

We observe that the NPV starts decreasing after 5 years. Hence, the equipment may be eligible for replacement after every 5 years.

In these days of rapid technological advancements, equipment and machines having an apparently high initial cost may prove economical because of high degrees of operational efficiency and reliability.

13.3.2 Equipment Replacement Policy for Equipment that Fail Suddenly

There are certain categories of equipment that normally do not require maintenance activities while in service. During their lifetime, they operate and give service more or less to the design level and then fail suddenly. The expected time of failure can only be predicted from probability distribution of their failure that is constructed from their failure history. Classical examples of such equipment can be electric lamps, electric-circuit breakers, monoblock pumps, sealed compressor units in a refrigeration system, sensors, etc. The average life of such equipment can be estimated using the theory of probability. For example, when we say that the average life of a particular type of electric bulb is 1000 hours, we mean that at least 50 per cent of the bulbs installed at a point of time will still survive after 1000 hours. Manufacturer's warranty period is based on this estimation of average life.

Group replacement

Continuing with our example of the electric bulb, if one particular bulb in a guest room fails, we need to send an electrician, who collects the tools and starts from the maintenance shop, carries the ladder, removes and replaces the defective bulb, comes back to the shop, and reports the job being done to the superior. It can be said that if the price of one bulb is Rs 10, the cost of replacement (cost of electrician) would be Rs 20. In such situations, it may be economical to replace other bulbs also in the room. This could greatly lower the average cost of replacement per lamp. This is more so important when we consider the inconvenience to the guest due to failure of bulbs. This principle of replacing the entire group is known as group replacement.

The advantages of group replacement are as follows:

- Reduced replacement cost per unit
- Prevention is better than cure principle safeguards the goodwill of the property
- Discounts are available when units are purchased in lot for group replacement.

Quite frequently, a combination of replacing individual unit, when it fails, and a periodic group replacement of all the items (old and the replaced ones) at the end of a period, determined in a way as explained later, often proves economical.

Using the rules of probability, this time interval of group replacement can be found out. Here, we would like to introduce the concept of probability.

Probability is the fraction of times an event (sometimes called favourable event or event of interest) occurs as an outcome of a large number of occasions where other outcomes are also possible. For example, when we make an experiment of tossing a coin a large of times (may be a few thousand times), we would see that the fraction of times head occurs is half and the fraction of times tail occurs is half. We, thus, say that probability of occurrence of the event of head is 0.5 and that of tail is 0.5. The sum of the probability of all possible events in any particular experimentation is always unity.

We now proceed to work out a problem for estimating the optimum period for group replacement for the equipment that fail suddenly.

The following failure rates (also called mortality rate) have been recorded for a type of electric bulb.

Month	1	2	3	4	5
Percent failing by end of the month	10	20	40	15	15
Cumulative percent failing by end of the month	10	30	70	85	100

There are 1000 bulbs in use. It costs Rs 10 to replace an individual defective bulb. If all bulbs are replaced simultaneously, the cost becomes Rs 3 per bulb. It is proposed to employ a policy of combination of individual replacement whenever a particular unit fails and group replacement after a specific period of time. We are to find out the optimum interval of time for group replacement.

From the given data, we observe that 10 per cent of the original number of bulbs fail in the first month. So, the probability that one bulb out of the total number of bulbs fitted initially will fail within one month is 0.1. Similarly, from the given data, the percentage of bulbs failing in the second month is 20 per cent (30 per cent–10 per cent), i.e., the probability that one bulb fails in the second month is 0.2 and the corresponding values for subsequent months are 0.4, 0.15, and 0.15. We also check that they add up to 1. The probability is clearly the difference between the proportion alive at the beginning of a particular month and the proportion alive at the end of the same month. We can tabulate this value as given in Table 13.4.

Table 13.4 Probability table for failure

Week (i)	1	2	3	4	5
Probability that a new bulb fitted at time zero would fail during i-th week (P_i)	0.1	0.2	0.4	0.15	0.15

(Contd)

(*Contd*)

Note: $P_1 = 0.1$, $P_2 = 0.2$, $P_3 = 0.4$, $P_4 = 0.15$ and $P_5 = 0.15$

If we start with 1000 bulbs, we can say that of these, $0.1 \times 1000 = 100$ bulbs are likely to fail in the first month, 0.2×1000 or 200 bulbs are likely to fail in the second month, and so on.

Let n_i denote the number of failures (and hence the number of replacements) at end of i-th month.

Hence, $n_0 = 1000$ (initial installation can be taken as the replacement at zero time)

Therefore, we can proceed to find them in successive months as follows:

$n_1 = n_0 \times P_1 = 1000 \times 0.1 = 100$ (this is number of bulbs that are replaced at the end of first year under individual replacement policy)

n_2 = number of failures in the second year corresponding to old population + failure out of the new population of 100 bulbs that have been replaced in the first year and failing in the second year

$$= n_0 \times P_2 + n_1 \times P_1 = 1000 \times 0.2 + 100 \times 0.1 = 210$$

Similarly,

$n_3 = n_0 \times P_3 + n_1 \times P_2 + n_2 \times P_1 = 1000 \times 0.4 + 100 \times 0.2 + 210 \times 0.1 = 441$

$n_4 = n_0 \times P_4 + n_1 \times P_3 + n_2 \times P_2 + n_3 \times P_1 = 1000 \times 0.15 + 100 \times 0.4 + 210 \times 0.2 + 441 \times 0.1 = 276$

$n_5 = n_0 \times P_5 + n_1 \times P_4 + n_2 \times P_3 + n_3 \times P_2 + n_4 \times P_1 = 1000 \times 0.15 + 100 \times 0.15 + 210 \times 0.4 + 441 \times 0.2 + 276 \times 0.1 = 365$

$n_6 = n_1 \times P_5 + n_2 \times P_4 + n_3 \times P_3 + n_4 \times P_2 + n_5 \times P_1 = 100 \times 0.15 + 210 \times 0.15 \times 441 \times 0.4 + 276 \times 0.2 + 365 \times 0.1 = 282$

$n_7 = n_2 \times P_5 + n_3 \times P_4 + n_4 \times P_3 + n_5 \times P_2 + n_6 \times P_1 = 210 \times 0.15 + 441 \times 0.15 + 276 \times 0.4 + 365 \times 0.2 + 282 \times 0.1 = 309$

We observe that the number of failures first increases till the end of the third month, then decreases, and then again increases. This oscillatory pattern continues with a tendency to settle down to some steady state value and yield what we call an average number of failures per month. However, fortunately, we do not have to take this laborious route to find the average number of failures per month. There is a well known theorem in probability with which we can very easily find the expected life of a bulb from the failure data, as in Table 13.4. and then calculate the average number of failures per month.

The probability that a bulb will fail in the first month is 0.1, the probability that it will fail in the second month is 0.2, and so on. By the application of probability theory, the expected life of a bulb $= 1 \times 0.1 + 2 \times 0.2 + 3 \times 0.4 + 4 \times 0.15 + 5 \times 0.15 = 3.15$ months

The average expected number of failures per month per bulb installed will, therefore, be the reciprocal of the expected life of a bulb and is equal to 1/3.15 bulbs. For 1000 bulbs, it will be $1000/3.15 = 317.46 = 318$ bulbs.

(*Contd*)

(Contd)

Hence, if we choose to adopt a policy of replacement of defective bulbs individually, the cost incurred will be Rs $10 \times 318 = $ Rs 3180

Now, we can proceed to find the alternative policy of combined individual replacement and group replacement which means that the replacement for individual defective bulbs only will continue for some period of time and then all the bulbs will be replaced simultaneously. Table 13.5 furnishes the calculations and results for the exercise.

Table 13.5 Calculation table for example problem

Month (i-th)	Total nos failed in i-th month	Cumulative total nos. failed upto i-th month	Cost of individual replacement	Cost of group replacement made at the end of i-th month	Total cost	Average cost per month
1	100	100	1000	3000	4000	4000
2	210	310	3100	3000	6100	3050
3	441	751	7510	3000	10510	3503
4	276	1027	10270	3000	13270	3317
5	365	1392	13920	3000	16920	3282
6	282	1674	16740	3000	19740	3228
7	309	1983	19830	3000	22830	3261

Table 13.5 has been constructed by the as following consideration.

Let us we pursue the replacement policy for individual bulbs for the first month after the initial installation of 1000 bulbs. Then we replace again all the bulbs at the end of the first month. Then we replace 100 bulbs during the first month costing Rs 1000 plus group replacement of Rs 3000 at the end of the month totalling a cost of Rs 4000 per month. Now, let us continue with individual replacement up to the second month and then replace all the bulbs. Total number of bulbs that failed and thus replaced up to two months is 310 and the cost of replacement is Rs 3100. After this the total of 1000 bulbs are group replaced at the end of the second month costing Rs 3000, again totalling a cost of Rs 6100 for two months, averaging to Rs 3050 per month for this scheme, and so on. We observe that if there is an option of combining individual replacement up to second month and then a group replacement, the average monthly cost is minimum, which is Rs 3050. This is also less than the average monthly cost of Rs 3180 corresponding to a fully individual replacement, i.e., replacement only when a particular bulb fails. Hence, the most economic policy of replacement is a combination of individual replacement followed by a group replacement after every two months.

We can summarize the equipment replacement policy by means of a flow chart, as shown in Fig. 13.1.

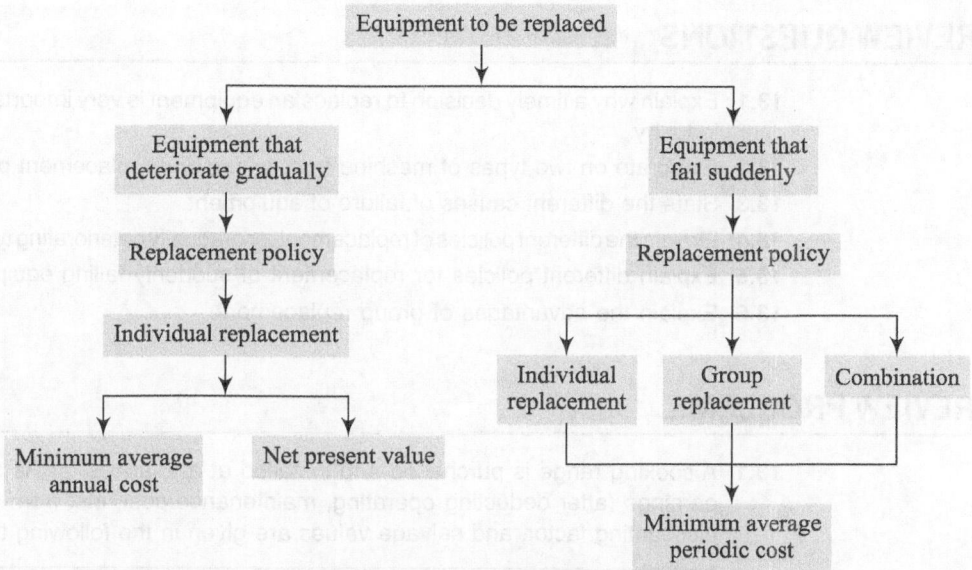

Fig. 13.1 Flow chart for equipment replacement policy

13.4 | CONCLUSION

Equipment replacement is, thus, a vital decision in any industry, usually taken by the engineering department in consultation with the finance department. It involves a lot of considerations such as technology suitability, operator's convenience, cost effectiveness, payback period, salvage value etc., making it really a complex decision to make. In case of replacement of a high value capital item, such as the refrigeration system or water-treatment plant, etc. for a large hotel, the stakeholders of the company may take the final decision in this regard.

KEY TERMS

Discounting	Discounting cash flow is a means of evaluating an investment by calculating the effective annual return on the investment (ROI) throughout its life.
Group replacement policy	The principle of replacing the entire group of an item, usually low cost, after a certain period of time, whether or not defective, is known as group replacement policy.
Net present value (NPV)	It is the present day equivalent of some return in future in monetary terms.

REVIEW QUESTIONS

13.1 Explain why a timely decision to replace an equipment is very important in the lodging industry.

13.2 Elaborate on two types of machine from failure and replacement point of view.

13.3 State the different causes of failure of equipment.

13.4 Explain the different policies of replacement for gradually deteriorating type of equipment.

13.5 Explain different policies for replacement of suddenly failing equipment.

13.6 Explain the advantages of group replacement.

REVIEW PROBLEMS

13.1 A cooking range is purchased and installed at a total cost of Rs 5000. Yearly net earnings (after deducting operating, maintenance cost, etc. from gross earnings), discounting factor, and salvage values are given in the following table.

Yearly	1	2	3	4	5	6
Net earnings	25,000	20,000	17,000	12,000	8000	3000
Salvage value at the end of the year	40,000	27,000	14000	10,000	5000	nil
Discounting factor	0.909	0.826	0.751	0.683	0.621	0.564

The initial cost of the range is 50,000. Calculate the economic period of replacement by net present value method.

[The piece of equipment should be replaced after the fifth year; NPV at the end of 5 years — Rs 18281]

13.2 The following mortality rates have been observed for a certain type of electric lamp, as used in lodging industry.

Month	1	2	3	4	5	6	7
Percent failing by the end of the week	5	15	28	45	70	90	100

There are 1000 such lamps in the hotel. It costs Rs 15 to replace a bulb. If all the bulbs are replaced at a time, the average cost per bulb comes to Rs 5 per bulb. Determine the following:

  (a) The average cost of replacement per month if replacement only when a bulb fails policy is pursued

  (b) The average minimum cost of replacement per month when a combination of group replacement policy and replacement only when a bulb fails policy is adopted and also the corresponding replacement cycle

The number of bulbs to be replaced in a particular month may be rounded off to a whole number for the purpose of finally calculating the cost of replacement.

[(a) Rs 3360/month, (b) Rs 3076/month; group replacement at the end of every fourth month]

14 Miscellaneous Utility Systems

Chapter

Learning Objectives

A hotel has numerous supporting utility systems for smooth, efficient, and safe functioning of the property.

The objectives of the present chapter are to:

- introduce a few utility items such as audio-visual equipment, computer systems, and sensors and detectors
- make the reader understand the importance and utility of various types of equipment discussed
- apprise a student of hospitality management about the general working and use of the equipment
- sensitize the reader about the needs and procedures for care and maintenance of the equipment

After reading this chapter, the reader should acquire a fairly good idea about the importance of audio-visual equipment, internal communication devices, computer systems, and sensors and detectors in the hotel industry, as well as their types, specific uses, and care.

14.1 | INTRODUCTION

Hotels have a large number and variety of audio-visual equipment and detectors and sensors. The actual varieties and number of such equipment, of course, depend upon the type of business being carried out by a particular property. For example, a convention hotel has a host of such equipment, while a primarily lodging hotel will have less of such equipment. With the advancement of technology in video and audio communication as also in sensor technology, the state-of-the-art equipment provides a high degree of flexibility in use and fidelity in the output. With the modern technology, a whole lot of such equipment can be connected and made to transmit signals in audio, video, and audio-visual equipment. For example, using computers with a Web connection and a Web camera, telephone, and projector screen, video conferencing can be done with the click of the mouse in a conference hall. Closed-circuit television (CCTV) can be connected to the public address system of the hotel and its output can trigger any control action desired. In recent times with the rapid progress in sensor technology, the hotel industry is increasingly using them for more efficient utilization of energy, improved service to customers, and better surveillance of property. As such, highly professional experts need to be employed to help identify the precise selection of equipment for the purpose in a hotel.

14.2 | AUDIO-VISUAL EQUIPMENT IN A HOTEL

Although hotels widely differ in the types and numbers of audio-visual equipment, the following are mostly included in standard hotels.
- Cassette player and recorder
- Video cassette player and recorder
- Follow spot light
- Karaoke system
- Loudspeaker
- Microphones (wireless, wired, hand-held, stand, table, and clip or lavalier microphone)
- Podium microphone
- Overhead projector
- 35-millimetre slide projector
- Portable screens
- Fixed projector screen
- Liquid crystal display (LCD) data/video projector
- Laptop computers
- Desktop computers or personal computers (PCs)
- Laser pointer and wireless computer mouse
- Closed circuit TV
- Plasma/LCD displays
- TV monitor (ordinary cathode ray tube or CRT, LCD, and plasma)
- CD/DVD player
- Sound and video mixers

- Conference phone: telephone only
- Speakerphone: telephone only
- Channel music system.

While small-sized hotels with primarily lodging facilities may not require many of the above-mentioned pieces of equipment, convention hotels require them in an elaborate manner.

Figures 14.1(a) and (b) show a typical overhead projector (OHP) and a slide projector. However, with the advent of computers and related software, LCD projectors have largely replaced the old slide projector system.

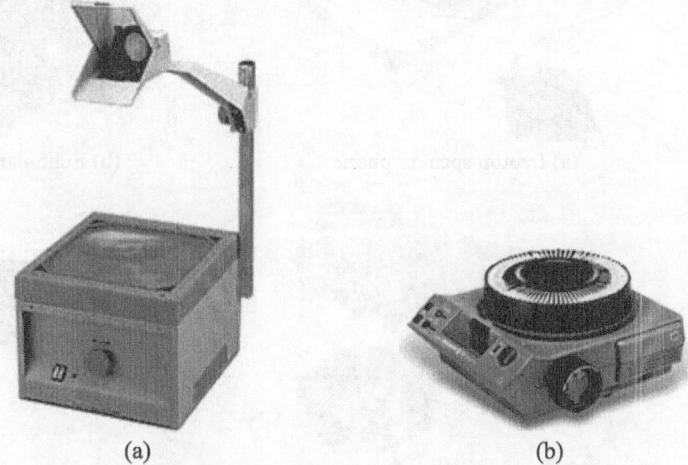

(a)   (b)

Fig. 14.1   (a) Overhead projector and (b) slide projector

(Source: www.indiamart.com and www.a1aaudio
visual.com)

Figure 14.2(a) shows a speakerphone that connects to PCs and laptops for rich, hands-free audio communication with Microsoft Office Communicator 2007. Figure 14.2(b) shows an echo-cancelling desktop microphone that connects directly to a set-top video codec to bring affordable, high-quality audio to videoconferences.

Figure 14.2(c) shows tabletop button-type microphones for unobtrusive placement on tables and are commonly used in applications in which microphones are permanently built into the table. These button microphones provide an excellent pick-up range. Figure 14.2(d) is also a tabletop microphone suited for teleconferencing with very little visibility on the table. Multi-screen multi-channel plasma TV screen is shown in Fig. 14.3(a), while a flat screen LCD data and graphic display monitor is shown in Fig. 14.3(b). Figure 14.4(a) and (b) show fixed-type and free-standing projector screens, respectively. Figure 14.5 is a BenQ LCD data and video projector. Through the power of six independent colour adjustments, one can precisely fine-tune saturation levels of even a single colour in any image.

Users can dictate complete colour customization by independently adjusting hues, and intensity and saturations levels of all colours without compromising any of them. If a scene contains an apple, a blueberry, a banana, and green grapes, users can intensify the apple's red hue without affecting the colours of the other fruits. In the economy mode, the lamp will have an average working life of 4000 hours.

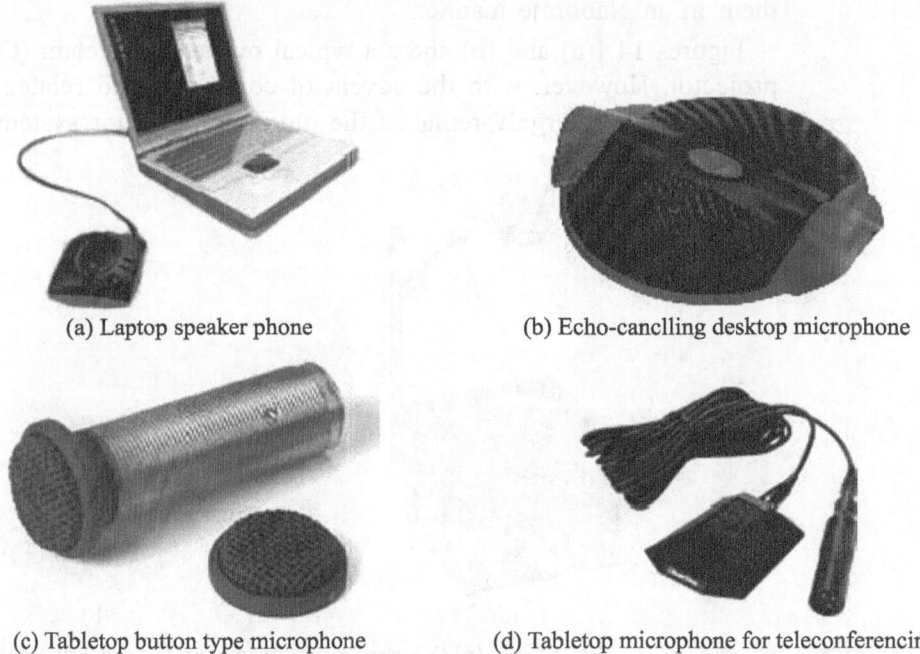

(a) Laptop speaker phone   (b) Echo-canclling desktop microphone

(c) Tabletop button type microphone   (d) Tabletop microphone for teleconferencing

Fig. 14.2 Microphones of the modern age

(*Sources:* blog.svconline.com, pc.pcconnection.com, clearone.com)

(a)   (b)

Fig. 14.3 (a) Multi-screen monitor
(b) A high-resolution graphic display monitor

(*Sources:* www.fahad.com and www.scan.co.uk)

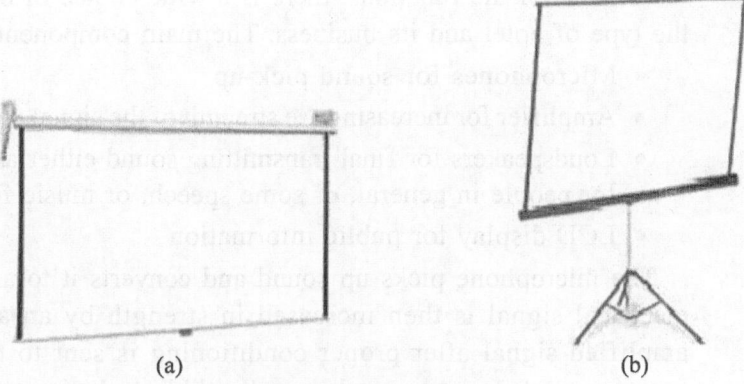

(a)   (b)

Fig. 14.4   (a) Fixed-type projector screen
               (b) Free-standing projector screen

Fig. 14.5   BENQ data and video projector

14.3 | PUBLIC ADDRESS SYSTEM AND INTERNAL COMMUNICATION

We have discussed, in general, audio-visual equipment used in the hospitality industry. Here, we discuss two specific areas of functioning where audio-visual equipment is employed. They are public address system and internal communication.

Public address (PA) system

A comprehensive PA system should have the following tasks to perform.

- Amplification for speeches and music in conference room, ballroom, and banquet hall
- A paging system in public places and rooms, which is a system of loudspeakers for making announcements to contact guests and other visitors, also in case of emergency. It is so named as it replaces the pageboys
- A babysitting arrangement, whereby a microphone activated in a lodging room could inform a staff about a child left unattended in the room when he/she cries
- An internal two-way communication system for the staff

For each of the functions, there is a wide choice of equipment depending on the type of hotel and its business. The main components of a PA system are

- Microphones for sound pick-up
- Amplifier for increasing the strength of the signal received from microphone
- Loudspeakers for final transmitting sound either as a piece of information for people in general, or some speech, or music for the general audience
- LCD display for public information

The microphone picks up sound and converts it to an electrical signal. This electrical signal is then increased in strength by an amplifier and then this amplified signal after proper conditioning is sent to the loudspeaker, which again converts it into sound waves audible to human beings.

Internal communication

Any commercial office and establishment must have an efficient system for internal communication, for effective management and coordination. While in majority of hotels, a simple internal telephone system is sufficient, large establishments may install one or more master units (manager's room or front office desk) consisting of a loudspeaker and microphones and a number of subordinate points (engineers' office, kitchen, stores, etc.), each having a telephone to interact. Selector switches on the master unit enable the master unit to communicate with two or more subordinate points simultaneously. Subordinate points can communicate with master units but, usually, not among themselves to avoid confusion and lack of coordination. There may even be tiers of such master units. For example, in a kitchen call system, there may be a central master unit in the control of the chef who communicates with outlying service areas of the kitchen and also with areas such as dining hall, banquet servery, and crockery room.

Nowadays, mobile cellular phones have revolutionized the way we communicate and has opened up many new possibilities of communication. Many properties also use wireless calling system in which the operator in the control room has a transmitter and the recipients of the call will have small pocket size receivers that they carry. Whenever someone is needed for passing instructions, the operator presses the particular key on the transmitter keyboard. The recipient will have a 'beep' sound in the receiver. He/she can either get the speech message through the receiver or respond to the call through a nearby telephone and get the message if the receiver set does not have the provision of receiving speech message. In both the cases, the called person can speak back to the caller. Messages remain confidential. Staff who normally carry receiver sets include managers and assistant managers of different service departments, banquet manager, head waiter, hall porter, night porter, head valet, and maintenance staff.

14.4 | CARE AND MAINTENANCE OF AUDIO-VISUAL EQUIPMENT

Audio-visual equipment are mostly sophisticated electronic equipment, and as such they must be handled with proper care and maintained regularly to ensure their proper functioning. Usually, the sound systems are maintained by outsourcing to contract maintenance. The video units, LCD monitors, computer systems, projectors, cameras, etc. usually have some kind of warranty and service contract. However, there must be some care in operating them and regular routine of cleaning them that greatly enhances the life and performance of the equipment. Let us discuss a few of them in brief.

14.4.1 LCD Data/Video Projectors

These projectors work with computers and have become extremely popular for making presentation on screen that were earlier done with overhead projectors (OHP) and slide projectors. Nowadays we can make a presentation in the computer through a Microsoft PowerPoint, save it in a pen drive, or post it to an Internet storage. We can then travel anywhere in the world, download the presentation from the Internet or from the pen drive to a computer at the venue anywhere in the world. Then with the help of the LCD projector, we can demonstrate the presentation with the click of the mouse.

Care and maintenance

Projectors are very sophisticated equipment and great care is needed in operating them. The start procedures are mostly controlled through various control buttons on the projector body and they need to be strictly followed in accordance with the instructions. The temperature inside may go up to 400–500°C and there are very efficient cooling fans inside. As such, strict protocol must be maintained while shutting them down. After use, they should be disconnected from the computer and properly packed inside the casing.

All projectors require some maintenance. Digital projectors, such as LCD, produce a lot of heat. To remove this heat, cooling fans are installed inside, which contain filters. These filters need to be cleaned or changed through regular maintenance. A Sony projector typically specifies that this should be done every 300 hours for normal conditions, and earlier for dusty ambience.

Filtration is important because any dust particles that end up in the wrong place in the projector can cause visual anomalies on the screen. These are most commonly caused by dust on the green LCD panel. Dust on the red or blue LCD panels are not so clearly visible. The symptom is green fuzzy patches visible in dark passages. These patches are called 'dust blobs'. If 'dust blobs' appear on the screen, the projector need to be opened and the filter and the crevices in the LCD and polarizer areas need to be cleaned with compressed air. Small cans of compressed air may be used for the purpose. However, this work must be done in a way as detailed in the users' guide and maintenance manual, provided with the equipment.

A few tips for care and maintenance of a projector are as follows:

- Projector lens should be cleaned with isopropyl alcohol or a photographic lens cleaning solution. Apply cleaning solution to cloth and wipe the projector lens in a spiral motion beginning in the centre and working outward. The cleaning solution should not be applied directly on to the projector lens.
- LCD or DLP projector must be used at temperatures between 5°C and 35°C.
- Dust should be blown off the projector lens by using a can of compressed air maintaining a safe distance of approximately 6 to 8 inches from the projector lens.
- Ceiling mounting of LCD projector is difficult and dangerous and has to be done by an audio-visual technician having sufficient skill and experience.

14.4.2 Overhead and Slide Projectors

With the advent of computers, digital projectors, and presentation software, such as Microsoft PowerPoint, the older devices for presentation, such as overhead and slide projectors, are becoming less commonplace. But some old presentation materials still require them and a brief mention of their care and maintenance will not be out of place to mention here.

Safety requirements

An OHP set is prone to cause accidents. It is extremely hot when in operation. Loose electrical connecting wires are also potential source of danger. While handling the equipment, some care must be exercised. Listed below are a few precautions to be taken while operating such a device.

- The appliance should never be left unattended, particularly when children are around.
- The operating temperature of OHP is very high and burns can occur from inadvertent touching of the hot parts.
- Care must be exercised not to trip over or stumble on the wires and cords.
- The appliance should be cooled before handling it for keeping in stores.
- The cords must not touch the hot surfaces.
- If an extension cord is necessary, it must be of compatible current rating to avoid overheating of the cord.
- If the projector lamp burns out, the projector should be allowed to cool for 4 to 5 minutes before replacing the old bulb with a new one as the bulbs become extremely hot while in operation.

Cleaning and maintenance of an OHP

Overhead projectors do not require much maintenance. The optical surfaces (glass surfaces) need to be cleaned with soft, lint-free cloth.

Cleaning and maintenance of slide projectors

Slide projectors are now rarely used for presentations. Images captured by digital camera and later shown through a computer-integrated projector presentation system have become much easier and flexible means of showing photographs in succession, as was previously done through photographic slides in a slide projector. Simple care and occasional cleaning of the optical surfaces with lint-free cloth and removal of dust by camel hairbrush are sufficient for good performance of the equipment.

14.5 | UNINTERRUPTIBLE POWER SUPPLY (UPS)

Modern age is the age of electrical power and almost all of the powered equipment are essentially electrically-powered. However, for effective and safe operation of many types of equipment, continuous supply of good quality power is essential. The following sections discuss in only a rudimentary fashion the essence of good quality power and means to achieve it.

14.5.1 Power Problems in Running Electrical Equipment

Good quality electrical power supply means a reasonably steady voltage under steady frequency (50 cycles/sec or 50 Hz in India) and continuous supply of power. Problems related to power supply include the following:

1. Power failure: momentary (dropout) and long-term (blackout)
2. Supply voltage drop or sag (brownout)
3. Overvoltage
4. Voltage surges and spikes
5. Supply frequency variation

We discuss the cause and effect of each of these, in brief, as follows.

Power failure

Power failures may be due one of the following causes:

- Dropout, which is a momentary stoppage in supply for a period ranging from milliseconds to a second or so due to some temporary fault in the line
- Blackout or complete power failure due to planned load shedding or power station failure

Voltage sag or brownout

Brownout or temporary drop in supply voltage is what we observe when lamps suddenly dim. This phenomenon is damaging to electric motors.

Overvoltage

It is the phenomenon when the supply voltage goes up. The effect of the higher voltage is to increase the flow of current through the appliance, resulting in higher temperature of the conductor and possible burning out of the equipment.

Voltage surge and spikes

These are momentary steep rises in the line-supply voltage. Spikes occur when lightning strikes a transformer. Surges occur when high-powered electrical motors are turned off, releasing extra voltage into the line. They are sources of major damage related to sophisticated electrical and electronic equipment.

Supply frequency variation

This is a problem related to faults in the operation of power stations. The supply frequency becomes more or less than the standard 50 Hz in India. Electric motors and other sophisticated electronic control components are severely damaged due to wide fluctuations in frequency.

14.5.2 Computers and Power Problems

An IBM study, referenced on the American Power Conversion (APC) website, reports that a typical computer is subject to 120 power problems per month. Most are not severe enough to damage a television. However, computer systems are more vulnerable to voltage variations than appliances such as washers, dryers, microwaves, or TVs. Spikes and surges can damage keyboard, monitor, hard drive, or processor. Blackouts and brownouts can ruin data saved on the hard drive and any work in progress and unsaved file is lost. This not only means loss of data but also loss of valuable man-hours. There are many telecommunication devices which require uninterrupted power supply. Electric motors and sophisticated electronic control systems are very sensitive to supply frequency.

14.5.3 The UPS

In order to protect computers and other important and sophisticated electrical and electronic equipment, it is necessary to devise and install a system or device which can address the problems of supplying conditioned voltage continuously. Here we discuss a very commonly used protection device called the uninterruptible power supply (UPS).

A UPS provides instant emergency power to connected equipment by supplying power from a separate source, in case of a failure of the main utility power supply. The normal main supply voltage can also be routed through them to condition it for unusual supply voltage and frequency conditions, as already explained.

The UPS internal battery takes over as soon as the main power supply goes off. The UPS provides instant protection even from a momentary (of the order of milliseconds to a second or so) power interruption in the main supply (termed dropout) and continues to supply power for some period so that the connected systems are not damaged due to abrupt power failure and no data is lost while working on the computer. Users get time to save all the open files during this period and can formally shut down all the equipment connected and the equipment continues normal operation. If an auxiliary power supply can be generated and supplied or main power is restored by this time, the battery mode is automatically changed over to normal mains supply mode of the UPS. If the auxiliary power is not turned on or main supply is not restored by this time, the users and operators may shut down all the connected equipment following proper shut down protocol, thus saving the equipment from damage and saving data and files in computer systems.

Conventional emergency power systems, such as standby generators, do not provide this important feature of instant power backup in case of power failure even for a moment. They take some time to generate power and the connected loads (i.e., connected equipment) get power from them only after this start-up period. A UPS, however, provides uninterrupted power to equipment, typically for 5–15 minutes.

A UPS is typically used to protect personal computers, main Internet and data processing main servers, telecommunication equipment, or other electrical equipment where an unexpected power disruption is not acceptable due to possibilities of physical injury, serious business disruption, or data loss.

Uninterruptible power supply units are specified by the power rating expressed in volt-amperes or VA, in short. 1 VA is less than but close to 1 watt. A UPS of 500 VA is sufficient for a PC with the load of a monitor, CPU, and printer, while for main servers, data centres, and telecommunication use, they may be of several hundred kilowatt size.

Functions of UPS

A UPS may be called upon to perform any one, or combinations, of the following functions:

1. Protection against power failure (both momentary and short term)
2. Protection against voltage drop in the main supply
3. Suppress voltage spike and surges

4. Correct undervoltage (brownout) and overvoltage

5. Reduce line noise in the main supply

While all electrical equipment that need conditioned power supply would need a UPS performing all the functions, for obvious reason of cost, manufacturers design and make UPS based on particular types of functions it has to perform and, correspondingly, equipment requiring similar kinds of protection is supplied with the particular category of UPS. For example, computers may not need costly protection against frequency variation, while costly motors would require frequency stabilizers to be incorporated in its protection device.

14.5.4 UPS Maintenance and Care

A UPS system essentially consists of an electric circuit with some electronic component for conditioning the incoming voltage and most importantly a battery, which is the heart of the system, as well as the most vulnerable. Most UPSs still use either sealed or flooded cell batteries. Flooded cell, or wet cell, batteries are typically found in larger, three-phase UPSs. These batteries are similar to car batteries, where the water level is routinely checked and water can be added if needed. Sealed batteries use a suspended gel electrolyte instead of water and do not require the fluid level to be replenished. Though sealed batteries are often called maintenance-free, they still require scheduled maintenance and service.

Small-capacity UPS as normally used with individual PCs are sealed units and almost maintenance free (except for keeping it dust-free and keeping the cooling air holes in the outer casing clean). Larger capacity units are used for comprehensive power backup for central data processing and network server, voice and data communications systems, and critical electrical-distribution equipment, including switchgear. Usually, lead acid battery is employed for UPS as the backup power source.

The battery is ultimately at the heart of the UPS in terms of reliability. The key to ensuring the reliable operation of the battery plant is a comprehensive predictive maintenance and testing program. A periodic maintenance program extends battery string life by preventing loose connections, removing corrosion, and identifying bad batteries before they can affect the rest of the string. Without regular maintenance and service checks, a UPS battery may experience: heat-generating resistance at the terminals, improper loading, reduced protection for its connected loads, and premature failure. Temperature, float level, cycling, and other factors all affect UPS battery life. Batteries can degrade to the point that they are unable to provide adequate run time for a UPS. The following steps can help establish a good maintenance program for the UPS battery.

1. Inspection: Periodic inspection should be carried out for of the battery installation.

2. Periodic measurements: Electrical measurements of the battery installation should be recorded periodically. Readings should include individual cell and bank voltages, electrolyte, and ambient temperatures and inter-cell resistance of all connections. Temperature-compensated charging process monitors the battery temperature and, through sophisticated algorithms, adjusts the rate of charging, compensating for heat. This process prolongs the life of the battery.

3. Load testing: The Institute of Electrical and Electronics Engineers (IEEE) defines the end of useful life for a battery as the point when it can no longer supply 80 per cent of its rated capacity in ampere-hours (Ah). When the battery reaches 80 per cent of its rated capacity, the aging process accelerates—and the battery should be replaced. The best method of performing a battery load test is to monitor each individual cell or unit during the discharge. Weak or bad cells can be easily identified and scheduled for replacement.

4. Record keeping: All data collected through the above-mentioned maintenance program should be documented and preserved for review and to track the performance of the battery over time. Based on these, a benchmark may be established that will greatly enhance the success in the maintenance of a reliable system.

Battery designers and manufacturers have, over the years, sought alternative measures to monitor batteries and extend battery life. Advances in monitoring and managing UPS batteries through software and other communications devices have continued to improve. Advanced battery management can double service life and optimize recharge time.

14.6 | COMPUTER SYSTEM AND DATA COMMUNICATION

Computer systems in an establishment like a hotel are used both as a part of the communication system and for accounting and other data processing jobs using commercial software and programs specifically developed for the particular property. Here we discuss desktop computers or PCs. Such systems normally comprise the following components:

- Monitor or video screen
- A tower cabinet that houses all the electronic components (motherboard for housing the central processing unit (CPU) and other data cards, computer processing job, hard disk, hard disk drive, CD/DVD drive, cooling fan,

USB ports for external communication, and a host of other components) necessary for functioning of the computer system

- Keyboard
- Multimedia components such as web camera, audio system, etc.
- Scanner
- Printer
- UPS
- Modems and data communication devices
- Operating system and application software

While all these components have a warranty period for their performance, within which any defective components will be replaced by the supplier, maintenance of all the hardware and software is usually put under annual maintenance contract with external agencies. However, a few dos and don'ts while handling many of them greatly enhance the trouble-free period and protect sensitive components from early malfunctioning. The following steps relate to the measures that can be taken in this respect for some relevant components.

14.6.1 Care and Maintenance of Computers and its Peripherals

Computers are other peripherals are quite robust and reliable. However, casual use of the different components of them may lead to serious problems ranging from faulty and unreliable operations to complete stoppage of operation. However, a few routine care and maintenance schedule ensure trouble-free operation for a reasonable period of time. But it should be borne in mind that service from the professionals must be hired in case of serious problems.

Monitor/screens—CRT and LCD in PCs and laptops

Computer monitors are either conventional cathode ray tube (CRT) or liquid crystal display (LCD) type. While the former is of rugged construction and does not need day-to-day care, LCD screens are possibly the most sensitive pieces of electronic component we normally handle. We discuss the care and precautions to be taken for LCD screens in some details.

Care and cleaning of LCD screens   Utmost care has to be exercised while working with devices having LCD monitors such as flat screen monitors in PCs or the display screen in laptop computers. The following are the do's and don'ts for LCD screens.

The dos

- To remove stains on the screen, properly clean the display surface by gently wiping with a lint-free cloth moistened with a small amount of non-ammonia, non-alcohol glass cleaning agent

- Avoid strong shocks and vibration
- Be sure to use and store LCD monitors within specified temperature and humidity range as high temperature and humidity can adversely affect the performance of LCD screen. The operating temperature range could be 5°C to 35°C with 85 per cent relative humidity, while the temperature range for storage could be –10°C to 60°C with a humidity of 85 per cent.
- The screens should be turned off or the brightness should be reduced to prevent 'burn-in', which is a permanent damage to certain pixels in the LCD due to overheating. Fixed pattern display for a long time can aggravate the situation

The don'ts

- Do not place any heavy object on the laptop
- Do not scratch, twist, or hit the surface of the LCD screen
- Do not spray cleaning liquid directly onto the screen
- Do not let any moisture to accumulate as any moisture is dangerous for the LCD screen. If it accumulates, it should be gently wiped off with a soft cloth before switching on the monitor
- Never disassemble the LCD monitor, as the internal backlight assembly can retain a charge of 1000 V and may prove fatal
- LCD screens should not be subjected to sudden temperature changes, particularly when there is chance of condensation

Multimedia components—Web camera and audio systems

These are rather robust units that require very little maintenance, except that they are to be switched off when idle and are to be kept in a dust-free environment.

Keyboard and mouse

The most common kind of mishandling a keyboard is to spill liquid such as tea, coffee, or water. In case of such incidents, the computer should be turned off immediately. The keyboard should be disconnected from the system and then turned upside down for draining the liquid. It should be then allowed to dry up for 24 hours at room temperature before being put to service again. Keyboards should be covered to keep them clean.

A ball-type mouse which was common earlier could not track the screen properly after a few months of use. In such a situation, the underside retaining cover of the mouse was opened, the roller balls removed, and cleaned thoroughly to remove dirt build-up with a moistened cloth. Then the balls were washed with soap water. The internal parts were also cleaned properly and then assembled for use. New designs with optical mouse have less problems, though after sometime even with these, the pointer tends to jump and wildly moves through the screen. The costs of mouse and keyboards, however, have come down a lot in recent years and, quite often it becomes wiser to replace the defective ones with new units.

Printers

Printers can be broadly categorized into the following types:

- Line printers: For heavy duty routine printing
- Dot matrix printer: For light routine printing
- Inkjet printer: For light work and good quality printing
- Laserjet printer: For light work and very good quality printing

The type of printer used in a particular facility will, thus, depend on the type of service it is called upon to do. While the running cost of the first two types is very low, the cost of ink cartridges for the last two is quite high. However, although the initial cost of laser jet printers is higher than (in recent times, the difference in cost has become much less) that of inkjet printers and the cost of laserjet ink cartridge is also much higher than that of inkjet cartridge, the life of a laserjet cartridge (in terms of the number of pages being printed) is much higher than inkjet cartridges and so laserjet printers are very fast replacing the inkjet printers.

Care and maintenance of printers   Printers will have warranty for a specified period and on expiry of this period, they are included in the annual maintenance service contract for computer and peripherals,. However, taking some care in handling them would help their smooth functioning and longer life.

Many of the troubles with printers can be avoided by regular use of the printers. These problems include poor print quality, streaking print, white lines through prints or no print. Most of these problems originate from ink drying up at the print head or nozzle and clogging them during printing operation. Running the printer at least once a week helps avoid this problem. The following tips help keep the printers in good running condition.

- Use the printer at least on a weekly basis with both black and white and colour text printing.
- Always use printer switch rather than the spike buster/UPS switch to shut the unit down. Most inkjet printers will have printer heads that are properly 'parked', triggered by the printer's own power switch so that cartridges are sealed properly and are not exposed to the outside dry environment.
- Although many of the inkjet print heads have auto head-cleaning feature, for others, it is a good practice to periodically open the cartridge and clean the head with soft cotton cloth.

- Refilling for colour cartridges must be done with proper ink in the proper chamber.
- It is always better to refill an inkjet cartridge before it is empty as the ink of dried up cartridges, if left for a long time, would clog the microholes at the outlet of the print head requiring rejection of the cartridge altogether.
- Problems with laserjet printers being much less, it is good enough to keep the printer in a dust-free environment and switching the printer off as soon as a print job is over. This enhances the cartridge life to a considerable extent.
- Line printers and dot matrix printers are quite rugged and require very little care except for keeping them in proper environment-free from dust and switching them off when not in use.

Scanners

Scanners are optical readers for scanning pictures and documents. They require only a few routine maintenance steps. These are (a) to keep them in a dust-free environment, (b) to switch them off when not in use, and (c) not to load the top cover with heavy objects. Scanners are usually served by manufacturers' warranty and after this period is over it will be, in general, under comprehensive computer maintenance program.

Modems and Internet

Modem (from modulator-demodulator) is a device that modulates an analog carrier signal to encode digital information and sends it to a distant host. The same device can also demodulate such a carrier signal sent to it from a distant source to decode the transmitted information for further processing. The objective of encoding is to produce a signal that can be transmitted easily and decoded to reproduce the original digital data. A modem is, therefore, transceiver, i.e., a transmitter and receiver moulded into one.

The most familiar example is a voice-band modem that turns the digital 1's and 0's (digits to represent a number in binary number system) of a personal computer into sounds that can be transmitted over the telephone lines of plain old telephone systems (POTSs), and once received on the other side, another modem converts those 1's and 0's back into a form used by a USB, Ethernet, serial, or network connection. Faster modems are used by Internet users every day, notably cable modems and asymmetric digital subscriber line (ADSL) modems (coming with BSNL broadband Internet connections). Asymmetric digital subscriber line modem or DSL modem is a device used to connect a single computer or router to a digital subscriber line (DSL) phone line, in order to use an ADSL service. Like other modems, it is also a type of transceiver. DSL is a very high-speed connection that uses the same wires as a regular telephone line.

The advantages of DSL are as follows:

- The Internet connection can be kept alive and still the phone line can be used for normal voice calls
- The speed is much higher than a regular modem
- It can use the existing phone line and hence does not require new wiring
- The Internet service provider (ISP) companies (such as BSNL) that offer DSL will usually provide the modem as part of the installation

The disadvantages of DSL are as follows:

- Asymmetric digital subscriber line connection works better when it is situated near the service provider's central office. With distance the signal becomes weaker
- The connection is faster for receiving data than it is for sending data over the Internet

Voice modem Voice modems are the modems that can record or play audio over the telephone line. They are used for telephony applications. In telecommunication, the word 'telephony' refers to the general use of equipment to provide voice communication over distances, specifically by connecting telephones to each other.

Modems (Fig. 14.6 and Fig. 14.7) are either inbuilt inside the computer motherboard or put as external devices. If connected externally, they need to be kept free of dust and handled with care so that they are not dropped.

Usually, computer systems along with the peripherals are put under comprehensive annual maintenance contract to a competent outside vendor. They also support software maintenance. However, a few routine measures will ensure smooth and reliable operation of a computer system. They are discussed in the following section.

 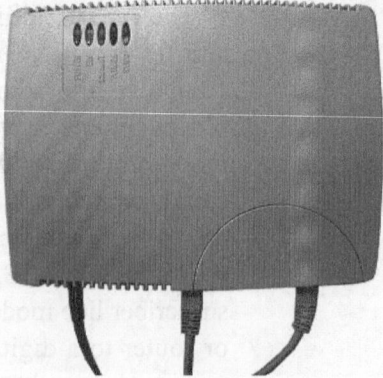

Fig. 14.6 28.8 kilobytes per second (kbps) serial port modem from Motorola   **Fig. 14.7** DSL Modem

14.6.2 A Few Good Maintenance Practices for Reliable Operation of Computer Systems

Computer systems are now a part and parcel of modern organizations. Smooth and efficient functioning of the organization nowadays depend a lot on continuous and reliable functioning of the systems. As such, it would prove highly beneficial to follow a few courses of care and maintenance of these systems and these are listed below.

Switch off computer only after the operating system (for example, Windows, etc.) has properly shut down the computer Never switch off your computer directly with only the main switch, except when the computer locks up or hangs. Such abrupt stoppage of machine can result in loss of data or system files.

Always use a UPS Use a UPS to avoid interruptions due to power cuts and voltage surges and spikes.

Keep the operating-system software and other important utility software in a secured place Quite often when a computer crashes, the maintenance people ask for system software for re-installation. So, these should be kept ready for such exigencies.

Run scandisk and defragment the file system once a month A hard disk drive saves files in sectors; the size of the file determines the number of sectors needed. When new, the hard drive can easily put all the pieces of a particular file in a contiguous block like the links of a chain, one after the other. But the more we revise and save files later, the more discontinuous or fragmented the file elements become. Thus, the hard drive must search for each element in order to re-combine them into a complete file. Constant searching increases drive wear and tear, shortening its lifespan and also there is delay in file processing. That is why it is strongly recommended to use a de-fragmenting utility regularly so that the fragmented parts are again brought closer together. All the standard operating systems, such as DOS (and, thus, Windows), Macintosh, and OS/2 operating systems come with in-built de-fragmentation disk utilities.

Disk Cleanup utility should also be run periodically to clear the temporary and unwanted files and folders from the computer. This is sure to increase the speed of the computer.

Take regular data backup The hard drive is one item that is very vulnerable and prone to crashing. The hard drive is the only integral component, which is an electro-mechanical device composed of moving parts. Processor, motherboard, and random access memory (RAM) chips are all static electronic components.

The advent of ever-larger hard drives means more data are at risk in case a hard drive crashes and the only protection is to have regular backup.

Keep at least 250 MB of C: drive free   The C: drive should be kept free for Windows to use. If the C: drive does not have enough free space, the operating system will choke and it will start dumping data to the hard drive and the system will become very slow. C: drive should contain only system files. All unneeded programs from C: drive should be deleted or moved to other drives.

Do not let many programs load up in the computer memory or RAM when the computer starts   These programs use valuable memory and Windows Resources (Windows internal workspace). Close them if you do not need them or configure them not to load when you boot up.

Avoid storing files on the desktop   It is a common practice to save a file initially (usually a downloaded file or a temporary file, which does not have a target folder) on the desktop. This practice severely affects the speed of the system. Move all such files to a destination folder to free the desktop.

Run system file checker twice a month

Remove all unnecessary files and folders from 'my documents' directory and empty 'recycle bin' once a month

Get all critical windows update

Scan computer with antivirus at least once a week   The computers should be scanned with computer at least once in a week with antivirus software such as Norton Antivirus, AVG, Kaspersky, Avira, Symantec, etc. available in the market.

Install firewall protection for Internet access   A firewall keeps those who want to hijack (or hack) computer from gaining access to the system. The latest firewall should be installed to avoid hackers for effective computer maintenance.

Clean computer and its accessories frequently   When cleaning a computer monitor, it must be disconnected from the power source. Dirt and fingerprints can be removed with ordinary household glass cleaner sprayed onto a lint-free cloth. The cleaning liquid must not be sprayed directly onto the computer screen. All the cooling fan outlet vents and holes in the units, such as monitor, CPU, UPS, etc., must be kept clear for ensuring a good cooling effect. Keyboards and mouse have to be cleaned with vacuum cleaners at least once a week so that they remain bacteria-free. One should avoid taking food or drink (water/tea etc.) while working on the computer.

If the above steps are undertaken regularly, it will keep most of the problems encountered in PC operation at bay. Even if the computers crash due to some unforeseeable reasons, vital data loss can be avoided.

14.7 | SENSORS AND DETECTORS

Though detectors and sensors seem to have almost similar connotations, they are, strictly speaking, two different entities. The term 'detector' is used to mean an item which detects physical phenomena, such as pollution, light, smoke, heat, flame, occupancy in a space, etc., with the help of sensors, which sense and in many cases measure physical quantities, such as intensity of light, temperature, degree of ionization of gas, speed of motion, humidity, amount of carbon monoxide, suspended particulate matter, noise level, etc., to help detection of the physical phenomena. Detectors are usually an assemblage of sensors and other components for triggering some control action either automatic and/or through human intervention. For example, a heat detector can trigger a fire-sprinkler unit in the particular zone and/or a fire alarm as part of its action programme.

The hotel industry, because of its very special nature of business, requires a host of sensors and detectors for reliable, efficient, and safe functioning. A number of them are listed as follows:

1. Smoke detector
2. Heat detector
3. Occupancy detector
4. CCTV
5. pH detector/pH meter
6. Photosensors
7. Gas leak detector
8. Gas detector
9. Flame detector
10. Flame monitoring system
11. Carbon monoxide level detector

We discuss each of the different types of detectors/sensors, in brief, as follows.

Smoke detector

Smoke detectors are used for fire safety. These are discussed in detail in Chapter 9.

Heat detector

A heat detector is a device that detects changes in ambient temperature in the space where it is installed. Typically, if the ambient temperature rises above a predetermined level, some control action is initiated through an alarm signal and/or activating a water sprinkler system to extinguish the fire. This type of detector is discussed in detail in Chapter 9.

Occupancy detector

Occupancy or motion detectors are devices that turn lights and other equipment, such as ventilation and air conditioners, on or off in response to the presence (or absence) of people in a specific area in the field of view of the sensor. A complete detector unit consists of a motion sensor, an electronic control unit, and a controllable switch/relay. Originally developed for use with security systems, occupancy sensors have been refined and enhanced to control lighting and HVAC in commercial and residential spaces. Figure 14.8 shows a typical occupancy sensor.

In a hotel, they are usually installed in guest rooms, conference rooms, and reading rooms. The positions of the sensors are very important for meaningful operation of the detectors. They may be wall mounted, or ceiling mounted, or both.

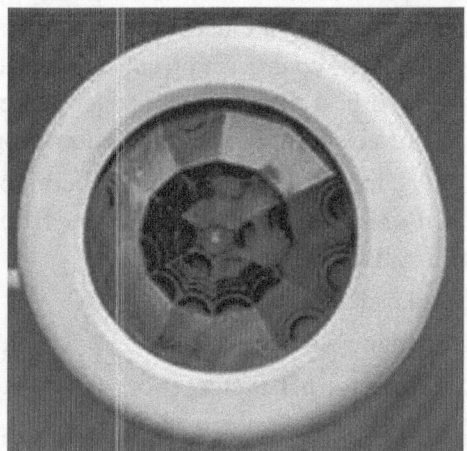

Fig. 14.8   View of an occupancy sensor
(Hager Limited, Ireland)

Types of occupancy sensors   Passive infrared (PIR) sensors are line-of-sight devices that cannot see around corners and crevices. Thus, they are ideal for meeting rooms, office spaces, and other support spaces.

Ultrasonic-occupant sensors use sound waves as a transmitter, which when reflected back from a moving object senses the object in motion. Sound waves are unhindered by the presence of intervening objects and can reach corners and crevices. They are, thus, ideally suited for guest rooms and public toilets.

The most important factors for the success of occupancy sensors are proper sensor specification, location in the space where it is fitted, and adjustments done to suit the actual conditions and, hence, deployment and installation of such devices calls for the expert service of professional agency.

Closed-circuit TV or CCTV

Closed-circuit television (CCTV) is an arrangement that employs the electronic eyes of video cameras to transmit a signal related to human activities in some specific place to some storage space and/or to a limited set of monitors. It is often used for surveillance in areas that may need monitoring such as hotels, banks, airports, military and police installations, consulates, and big departmental stores. Figure 14.9 shows surveillance cameras fitted on a rooftop of an establishment.

Fig. 14.9 Surveillance cameras fitted on the roof of a building

(*Source*: wikipedia.com)

The objectives of installing CCTV in a hotel property could be as follows:

Objectives of Installing CCTV in Hotels

- To protect customers
- To protect assets from pilferage and theft, such as in a liquor storage room, to prove that something was taken
- To prevent loss by way of legal compensation by identifying false claims by customers
- To help get insurance claims in case of property damage due to fire, etc.
- To gain control over activities of employees
- To monitor various activities in the hotel such as progress in a banquet hall or arrival of guests in the foyer
- To enable the audience to see more clearly in an overflowing meeting and conference
- To control access to selective places
- To identify offenders and help law-enforcing agencies by footage of the video and still images
- To act as deterrent to crime (using CCTV to monitor the front desk has helped reduce the number of armed robberies)

With improvement in technology and simultaneous decrease in cost, many properties are using digital video recorders, which improve their ability to store, retrieve, and transfer data. These video recorders offer a variety of quality and performance options and extra features (such as motion-detection and email alerts).

On an average, a hotel may install 10 to 30 cameras, depending on the property's size. Larger properties often have still more cameras.

Closed-circuit television cameras are being used more than ever in the hotel industry, but they are not a solution for all problems related to crime or loss prevention. And in some cases, they might do more harm than good. For example, if CCTV is used to protect customers, they should be properly monitored. Many CCTV cameras are for recording and storage rather than online monitoring and that means asking for trouble.

The system has to be able to respond to a the situation of a guest threat which is being recorded. The failure to respond might invite legal suits for huge compensation because judges cannot understand how a property can take the trouble of installing a CCTV to tape an act of crime, but not do something to help prevent it. This is worse than not installing it at all. Closed-circuit television to protect guests should only be installed where trained, competent staff members can monitor the cameras and respond if a guest is in danger. If not, the hotel at least should post signs that state premises are video recorded only and there is no monitoring. Potential liability is an issue that has caused some hotels to shy away from using CCTVs.

Given below is an excerpt from the January 2008 issue of the magazine *Hotels* of Reed Business Information, which is part of Elsevier Inc. This excerpt underlines the importance of a CCTV system in a hotel.

A few weeks before New Orleans was pummeled by Hurricane Katrina, the Holiday Inn French Quarter-Chateau LeMoyne was the victim of an early-morning armed robbery. Despite the presence of a security camera, the perpetrator never was identified. It was then that General Manager Kathleen Young decided it was time for an upgrade to digital closed-circuit television, or CCTV.

'We looked at our old security camera, and it was a 6-year-old system, one of those old technologies that took a snapshot every 15 seconds,' Young says. 'When we watched the snapshots, there was one where the night auditor was perfectly fine, then the next picture had the robber with a gun to his head, and the next shot he was gone. It was worthless to the police.'

Young called DefenderTech International Solutions, which installed its digital InSight CCTV system at the property—just before the hurricane hit. ...... Once Young and a team of maintenance and security personnel returned to the property, the 12-camera InSight system gave staff the peace of mind to work around the clock to repair the hotel and return it to operation, even as nighttime anarchy reigned in the surrounding neighborhoods. ......

(Contd)

(*Contd*)

In the subsequent two years, InSight has paid for itself several times over by providing visual evidence refuting false guest claims, Young says. Visitors to the French Quarter tend to imbibe more alcohol than they probably should, which can lead to an inordinate number of misunderstandings and incorrect memories. InSight's archival search allows hotel staff to quickly access video from a specific time and location.

'Sometimes, people will come down the next morning and say something like Spiderman must have scaled the wall and stolen their wallet from their room,' Young says. 'So we'll go to the tape and show them that they came back at 3 a.m. to the hotel lobby and asked for a new key with their new friend Spiderman, and they'll say, 'Oh, yeah, it's all coming back to me now...'

Young uses the InSight system to keep an eye on employees as well, making sure they stay on task and that their interactions with guests are appropriate. Because it is digital, real-time video can be viewed on computer monitors off-site, as well. New Orleans police even frequent the property more often now, because cameras trained on the hotel's exterior areas provide visual clues to incidents in the surrounding area. Guests and employees alike are keenly aware of Chateau LeMoyne's newfound reputation for superior security, which Young credits with helping to draw business and retain staff.

pH detector/pH meter

Although pH meter is strictly not a detector but a measuring device for the degree of acidity (or alkalinity) in water, we discuss it in brief to indicate its many areas of application in hotels and hence its importance.

We already know that pH value of water indicates the number of free or active hydroxyl (OH^+) or more precisely hydronium (H_3O^+) ions in a solution, which in turn indicates the degree of alkalinity. pH value usually ranges from 0 to 14; a value of 14 indicates fully alkaline solution, while pH 7 indicates a neutral solution.

pH can be measured using either pH indicators (such as phenolphthalein) in the form of solution or pH strips or using potentiometric method. Strips give accuracy to 0.2–0.5 pH units. If higher accuracy is needed, pH meter is the only solution. Further, pH indicators give only visual indications while potentiometric method lend itself to a feedback signal for auto-correcting a process that controls pH.

In potentiometric methods, the potential difference between known reference electrode and the measuring pH electrode is measured. The potential of the pH electrode depends on the activities of hydronium ions. pH metre is, thus, a precise voltmeter whose scale is graduated in such a way that it displays not the measured potential, but ready pH value.

Uses of pH sensors and pH meters   pH values are regularly monitored in the following areas in a hotel industry.

1. Water treatment plants: demineralization (DM)/reverse osmosis (RO) plants
2. Boiler: pH correction
3. Cooling tower recirculation water: pH correction
4. Swimming pool water: pH correction
5. Effluent treatment plant (ETP)/sewage treatment plant (STP): pH correction

Photosensors

These sensors are used to detect the amount of daylight available and accordingly vary the light output (lux) of illumination bulb and tubes and also switches on or off floodlight and light in the premises in a property. However, for street lights, for which timings are known, timer switch with electronic time clock may be used to control the hours of operation. A variant of such sensors are also used to automatically open or close entrance/exit doors usually in the reception lobby of a hotel.

Gas leak detector

Gas leak detection system employ detectors using infrared point sensors, IR open path sensors, and more recently ultrasonic gas leak detectors. In order to detect gas leaks with traditional methods, the gas itself must either be in close proximity to the detector or within a pre-defined area. To obviate this problem, ultrasonic gas leak detectors are employed. This detection method often uses a microphone technology, which detects outdoor leaks by sensing the distinct high-frequency ultrasound emitted by all high-pressure gas leaks. This means that the gas does not have to travel to the sensor—just the sound of the gas leak is enough for leak detection.

Basic detection can be accomplished using semiconductor gas detection. The semiconductor sensor method can be used for a variety of gases including combustible, toxic, and refrigerant gases. Semiconductor gas detectors are very economical and have a lifespan of five to eight years; however, they are prone to cross sensitivity.

Gas detector

A gas detector is a device that detects the presence of various gases within an area to warn about gases which might be harmful. Gas detectors can be used to detect combustible, toxic, and oxygen and CO_2 gases. Figure 14.10 shows a typical gas detector.

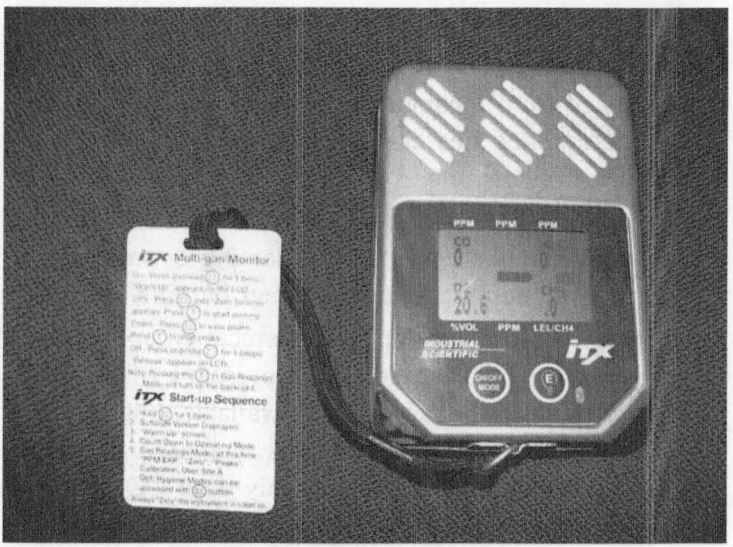

Fig. 14.10 A gas detector

Flame detector

A flame detector is a detector that uses optical sensors to detect flames. This optical sensor senses the wavelength of radiation emitted from a surface and identifies it as the surface of a flame when the wavelength pattern matches with the pattern given as reference input pattern for a flame.

Closed-circuit televisions or web cameras can be used for video detection (wavelength between 0.4 and 0.7 μm). But the camera lens can be obscured by dust and smoke like human eyes and loose its effectiveness.

Flame monitoring system

A flame monitoring system is used for monitoring the condition of a flame in a burner mostly employed in a kitchen. This usually comes as a built-in sensor and monitoring system with a gas cooking range. Flame ionization detectors (FID) are best for detecting hydrocarbons and other easily flammable components as in kitchen-burner flame. If a flame is extinguished unintentionally, the electronic flame monitoring system reignites it immediately. If reignition fails, the gas supply to all burners is automatically interrupted and shut off.

Carbon monoxide level detector

If the level of carbon monoxide increases beyond a certain level, it becomes very dangerous for human life and eventually may prove fatal. This is known as carbon monoxide poisoning. A carbon monoxide (CO) detector detects the presence of the carbon monoxide and sets in some kind of alarm so that the situation can be controlled by measures such as improved ventilation of the area and/or evacuation of occupants from the affected area. Carbon monoxide is produced from incomplete combustion and may originate from open flames in kitchen, space heaters, water heaters, blocked chimneys for exhaust from furnaces, ovens, etc.

Dual smoke/CO detectors are also available in the market. Smoke detectors detect the smoke generated by flaming or smoldering fires, while CO detectors can alarm people about faulty fuel burning devices in a kitchen.

Early CO sensors were of chemical type where a white pad would fade to a brownish or blackish colour in presence of carbon monoxide. Such chemical detectors are cheap and widely available, but only provide a visual warning. As carbon monoxide-related deaths increased during the 1990s, audible alarms became a standard. The commonly used CO concentration sensor technologies may be one of the following types:

Chemical type   The detector consists of a pad of white chemical that blackens by reaction with carbon monoxide. It indicates qualitative presence of the gas and provides only visual warning to the problem.

Biomimetic type   A biomimetic sensor works with a form of synthetic haemoglobin, which darkens in the presence of CO (just like human haemoglobin behaves), and lightens without it. This can either be interpreted through visual inspection or connected to a light sensor triggering an alarm.

Electrochemical type   A type of fuel cell is designed to produce a current that is related to the amount of the target gas (in this case carbon monoxide) in the atmosphere to a very high degree of sensitivity. The magnitude of the current is a measure of the concentration of carbon monoxide in the ambience. The current can very easily be transformed to some electrical signal for triggering control action.

Semiconductor type   Thin wires of tin dioxide, which is a semiconductor material, on an insulating ceramic base provide a sensor based on the measurement of its electrical resistance. Oxygen increases resistance, but CO reduces resistance and it allows a greater current. If the magnitude of this current exceeds a set limit based on the maximum allowable concentration in the area, it would lead to triggering an alarm.

14.8 | CONCLUSION

The degree of sophistication of modern sensor and detector technology is increasing in gallops and newer areas of applications are emerging. In the face of new challenges of safety, security, enhanced performance, and energy conservation, hotels are also installing modern sophisticated sensors and detectors in greater numbers and varieties for various functions as outlined.

KEY TERMS

Audio-visual equipment	It is a device for communication employing both audio and visual means.
CCTV	Closed-circuit television (CCTV) is an arrangement that employs the electronic eyes of video cameras to transmit a signal related human activities in some specific place to some storage space and/or to a limited set of monitors.
CPU	CPU or central processing unit is the heart of a computer which is a hardware that does the vital functions of arithmetic and logical operations, control and memory.
Detector	Detector is a comprehensive system that detects the presence of some physical conditions using sensors and some control circuits to trigger control action.
LCD projector	Liquid crystal display (LCD) projector is an electronic projector that projects video, images either directly from computer data or from separate storage device input onto a screen for viewing.
Loudspeaker	It is the reverse of microphone and converts electrical signal into sound.
Microphone	It is an electro-acoustic sensor that catches soundwave signal and converts it to electrical signal for subsequent processing.
Modem	It is a piece of electronic equipment for modulating/demodulating analog carrier signal to encode digital information and send to a distant host.
Motherboard	It is the central circuit board of a desktop computer into which all other components and peripherals are plugged, including the BIOS (basic input/output system), the memory, CPU, graphics card, sound card, hard drive, disk drives, along with various external ports and peripherals.
Scanners	Scanners_are optical readers for scanning pictures and documents.
Sensors	These are devices that sense and in many cases measure physical quantities such as intensity of light, temperature, speed of motion, presence of some objects, etc.
UPS	Uninterruptible power supply (UPS) is an electrical device that supplies conditioned power to vital electrical equipment without any interruption.

USB Port   Universal serial bus or USB port is a data communication port of the computer through which external devices, such as printer/plotter/pen drive/ digital camera, etc., can be linked to the computer.

REVIEW QUESTIONS

14.1   Name some audio-visual equipment used in the hotel industry.

14.2   Citing some representative examples, discuss about the utility of audio-visual system in the hotel industry.

14.3   Discuss the system of internal communication in the hotel industry.

14.4   Discuss the system of the care and maintenance of an LCD data/video projector.

14.5   List the safety requirements in using an overhead projector (OHP).

14.6   Discuss the system of care and maintenance of LCD screens, printer, and keyboard in a computer system.

14.7   Discuss the various types of power problems and the role of UPS in continuous supply of good quality power.

14.8   Discuss the system of care and maintenance of a UPS.

14.9   Sensors and detectors are important for efficient and safe working conditions in the hotel industry. Elaborate on the statement.

14.10   List the sensors and detectors commonly used in hotel industry, with reference to their applications in the hotel industry.

14.11   Discuss the various types of occupancy detectors, citing their individual areas of application.

14.12   Discuss smoke detectors, with particular reference to kitchen application.

14.13   Write down the objectives for installation of CCTV in a hotel indicating advantages and also disadvantages, if any.

14.14   State the uses of pH meters/pH measurements in a hotel.

15 Chapter

Safety and Security

15.1 | INTRODUCTION

Safety and security are factors that are of paramount importance in the hospitality industry, as guests put a high priority to a safe and secure environment where they stay. Safety involves avoiding those causes of injury and damage that may be called accidental. Safety issues are very important in a lodging industry due to their impact on both guests and employees. Studies have shown that guests put security very high in their ranking for a hotel. Safety becomes more so important in the context of recent rise in legal cases, i.e., the compensation the management has to pay the guests and employees in case of accidents due to lack of safe environment. In addition, such accidents entail a consequent loss of reputation. Security in hotels is also a very critical issue. This has assumed newer dimension in view of the recent attacks on big hotels like Hotel Marriott in Islamabad, Pakistan and The Taj Mahal Palace & Tower in Mumbai, India in 2008, the JW Marriott and Ritz-Carlton Hotels in Jakarta, Indonesia in 2009, etc. The notion of secured environment has changed a lot from the stout walls with barbed fencing and entrance gates with barred design to the modern electronic lock, water sprinklers and closed-circuit television (CCTV), burglar alarm, etc. As such, the management must carry out the task of assessing the safety and security needs and accordingly design and manage the facilities by continuously monitoring the safety conditions and standards. In this chapter we will discuss the safety aspects, as well as the security aspects in the hospitality industry.

15.2 | SAFETY IN HOTELS AND ITS MANAGEMENT

Safety essentially means creating a working and living environment that eliminates causes of physical injury and damage that can occur accidentally. They relate to both employees of the property, as well as guests. These accidents could include slips, falls, burns, cuts, and other personal injuries. All these contribute to compensation, workforce unrest, loss of goodwill, and loss of working man-hour. As such, safety aspects should attract a high degree of priority from management. Statistics indicate that there has been a high incidence of accidents in hospitality industry. The areas of human activities where such accidents might occur are shown in Table 15.1.

Table 15.1 Areas of activities in the hospitality industry

Areas mostly occupied by guests and boarders	Areas in support service, mostly manned by in-house and also by contractor employees
Guest rooms	Engineering plant room (boiler, calorifier, engine, pump, etc.)
Restaurants	Low-tension electrical panel room
Cafeteria	High voltage area such as near transformer, etc.
Bar	Air-conditioning plant
Banquet hall	Chemical storage areas
Health clubs, if any	Water treatment plant
Shopping arcade, if any	Sewage treatment plant (STP)
Reception and lobby	Chemical testing laboratory
Lift	LPG Bank/HSD (fuel oil) storage areas
Swimming pool	Kitchen including walk-in freezers
	Laundry
	Managerial offices
	Computer room and accounts office
	Stores
	Rooftop

There are possibilities of accidents in all the areas mentioned. The typical major activities undertaken in-house in the hospitality industry acting as potential sources of accidents can be listed as follows.

15.2.1 Typical Activities and Possible Accidents in Hospitality Industry

Typical activities in a hotel subject to risk of accidents

Some of the typical jobs in the hotel industry exposed to relatively high risk of accidents are

- Painting of windows, walls, etc., particularly at heights
- Loading/unloading of materials from vehicles, particularly service vehicles
- Storage and stacking of goods
- Opening glass windows
- Baggage shifting
- Changing of electrical fittings
- Cleaning sewerage system
- Fire-service operation during fire
- Vegetable and meat cutting and chopping
- Slicing operation
- Making chapattis and roti in *tandoor* oven
- Dish washing
- Laundry (pressing)
- Laundry (washing machine)
- Laundry (tumbler)
- Laundry (hydro)
- Chlorination in swimming pool
- Serving at the table
- Refrigeration plant operation and servicing
- Boiler and engine room operation and maintenance
- Regeneration and backwash in water treatment plant
- Air handling unit (AHU) in air-conditioning plant servicing
- General engineering maintenance activities

Major types of accident in a hotel

Accidents that occur in the hospitality industry may be categorized as follows:

- Slips and falls
- Striking against objects

- Burns
- Electrocution
- Cuts
- Car accident

We now discuss the sources and preventive measures related to each of them as follows.

Sources of accidents and their prevention

Slips and falls Slips and falls can occur due to a number of reasons. Most of them occur due to buildings, and facilities therein, being designed with inadequate safety considerations. Slips and falls can occur due to slippery floors (more so with modern glazed floor tiles used to reduce dirt accumulation) in walkways and bathrooms. Faulty design of stairs can also cause severe tumbling. In boiler room, engine room, refrigeration plant, water treatment plant, kitchen room, etc. floors can be very dangerously slippery due to spillage of oil, grease, and water. Accidents in such places could be very serious in nature.

Preventive measures for slips and falls start from an adequate consideration of safety while designing a building and the facilities therein. Stairs should be of such height and design that chances of stumbling while climbing up and slipping while coming down are minimized. Anti-skid arrangements are often provided at the end of the flights. Floor materials, particularly in the toilets and machine room should be anti-skid type. Toilet-plumbing arrangement should be such that there is very little water coming on the floor. Any oil and grease in machine room floors should be always wiped off by floor cleaning personnel.

Striking against objects Quite a number of accidents have occurred due to people striking against a transparent glass partition wall between rooms. Glasses installed in sliding doors, other patio doors and exit/entrance doors are potential sources of accident due to impact. People, particularly the taller ones, also get head injury if the heights of doors and head-rooms in basement stairs, etc. are not adequate.

Preventive measures are to use tempered glass, which does not break into sharp glass pieces but into round pieces when shattered. When a glass doorway can be mistaken for open doorway, it should have labels/signs or a bar of some type affixed on it to warn people of its presence and consequent proper handling. Door heights and head room provision should be enough to avoid head injury.

Burns Burns can be caused by handling of acids, contact with very hot surfaces such as bare hot steam pipes, cooking utensils, hot ovens, and inadvertent contact with hot water while bathing.

Personal alertness is the best way to prevent burns. Acid should always be handled wearing proper hand gloves. Bare hot pipes should be as less as possible and be clearly painted with proper colour to indicate that it is hot. Proper colour indications must be maintained to indicate hot-water supply line and fittings in the bathroom.

Cuts Cuts in limbs are often caused by sharp objects such as scissors, knives, choppers and kitchen personnel often have such accidents while processing raw food materials. Electricians and maintenance people also do have such accidents during the course of their work with chisels, screw drivers, knives, etc.

Prevention against such cuts can be best ensured through conscious personal alertness.

Electrocution This accident occurs when human body comes in contact with live electrical wire. Potential sources of such accidents are equipment which not properly earthed and electrical wires having a deteriorated insulation. Operators who handle electrical equipment such as mixer, grinder, electric cooking range, etc. have a high risk of getting electric shock. Electricians who work in electric lines while electrical power line is still active may get shock if proper safety rules are not followed. Guests may also get shocks if electrical fittings such as electric switches and electrical gadgets in the room are not properly maintained and earthed.

Preventive measures for electrocution include checking for proper insulation in the electric lines periodically, maintaining proper safety regulation by electricians while handling electrical equipment, obtaining a power off token from appropriate authority before handling electrical machines.

Car accidents In big hotels, there is a continuous movement of cars and goods trucks. There are chances of accidents if drivers and pedestrians are not alert. Poor driveway designs with inadequate space for traffic movement can also cause accidents.

To prevent such incidents, the traffic way for cars, goods vehicle and guest and employees walkway should be separated as far as practicable. Speed breakers can also be put on the roads meant for car and other vehicle movement. There should be sufficient space for manoeuvring vehicles inside the premises.

Since the stakes arising out of inadequate safety practices are very high in hospitality industry, it is always advisable to employ professional service to have risk assessment in the property and take appropriate measures as safety cannot be compromised under any circumstances.

15.3 | FIRE SAFETY

The lodging and food services industries face several challenges when dealing with fire safety.

The guests and boarders stay only for a short time and are not familiar with the layout of the building, the knowledge of which is so vital in case of a fire. They are also ignorant about the standard fire-safety practices and do not know how to use them in case of an emergency. It is also very difficult to train them during their short period of stay. Many of them spend most of the time outside and remain asleep. However, it is a good practice to have a small guide booklet regarding all the safety practices including fire safety and hand it over to them in the reception when they check-in.

A fire safety program involves the following components:

1. Fire prevention
2. Fire detection
3. Fire suppression
4. Fire control
5. Fire notification.

We discuss them in some detail as follows:

Fire prevention

Fire prevention is a culture and no amount of fire control and suppression device will be sufficient if the aspect of fire is not always present in the minds of the employees and guests. Fire prevention and maintenance are closely related. The following measures do well to prevent occurrence of fire in a hotel.

- On-site laundry facilities need to have regular cleaning of the dryer duct and removal of lint from washing-machine filters, as these could pose threat of fire. If a linen chute is used, it should be cleaned periodically. All linen chutes should be locked.
- Regular inspection (including the state-of-the art thermal mapping of them for any abnormal heating) of electrical wiring and proper earthing should be regularly carried out.
- Regular maintenance of kitchen ducting and exhaust system is a key defence against fire in this fire-prone area. Because of its importance, kitchen safety has been dealt with in some more detail separately in a subsequent article.
- Trash storage and disposal should be designed and implemented keeping fire in consideration. Trash rooms and other storage areas should be properly secured.
- During any renovation, items and objects should be replaced/changed keeping the old safety aspects as well as new ones specific to the changes in place.

Fire detection

Fire control and suppression, to a large extent, depends upon efficient functioning of a fire detection system. Heat detectors and smoke detectors are the two types of sensors that sense the occurrence of fire and trigger suppression and control actions. Regular cleaning of the surfaces of such sensors is very important for their proper functioning

Fire suppression

In spite of the best efforts to prevent fire, there will always be chances of fire and proper suppression and control measures must be taken and proper equipment must be installed to tackle fire in case it breaks out. Various fire suppression measures include installation and maintenance of items such as sprinklers, standpipes and water-hose system, portable fire extinguishers, fire pumps, emergency generators, and related equipment.

Fire control

Fire-control systems relate to measures so that new areas are not exposed to fire and further spread of fire is prevented. It also involves design and construction of buildings such as making a firewall, etc. so that the fire does not spread from one part to the other part of the building. The various control measures include the following:

- Installing fire and smoke actuated dampers in the air handling system that penetrate through the walls so that fire cannot propagate through the ducting.
- Automatic guest-room closers.
- Installing stairwell pressurization system. This means keeping the stairwell area slightly pressurized so that the space can remain relatively smoke-free.
- Fire- and smoke-door closers that are triggered by fire alarms.

The foregoing discussion relates only to the basics of fire safety in a hotel. Most of these establishments have elaborate arrangements for fire safety according to building codes and practices with detailed instructions and training program being practised.

Fire notification

This is possibly the most important part in the fire-safety measures as far as guests and employees are concerned. It refers to all means to inform them to deal with a fire that has already broken out, as well as the correct procedures to follow in case a fire breaks out. The notifications include

- Emergency instructions and plans
- Horns and alarm bells
- Visual alarms such as emergency lamps
- Light signs indicating exit ways both normal and emergency
- Voice alarms for guests

In recent times, properties are increasingly using voice alarms, whereby recorded messages are relayed to the guest rooms explaining the procedures to follow in case of a fire. For the hearing impaired, visual signaling may accompany the voice in the speaker inside the guest room and also on all sign signals including the exit ways such as a flashing fire signal, etc. Doors that are dead ends should be appropriately marked.

15.4 | A FEW SAFETY ISSUES IN HOTELS

In a hotel, many accidents have occurred in guest bathroom, kitchen, and machine rooms areas. These areas are very accident prone because of the working environment and the condition of the people (weak, aged, physically challenged, etc.) involved therein. We, therefore, discuss the safety aspects for these areas in some detail.

15.4.1 Guest Bathroom

This is an area of particular concern for guest safety in a hotel. Some of the accidents occurring in guest bathrooms are as follows:

- Hot water scalding (burning skin, etc. with hot liquid or vapour)
- Slip and fall
- Electrical shock

In the past, scalding of guests had occurred due to faulty system design and operation of hot-water system. This has resulted in injury and in some cases even death. To prevent scalding the following measures are recommended.

- The temperature of hot water for the use of guests should be maintained at not more than 43°C at the taps.
- A separate hot-water supply line should be installed for guest bathrooms for the proper control of temperature.
- Pressure- and temperature-compensating bath and shower valves should be installed for maintaining a proper mix of hot and cold water automatically even when there is a change in the system conditions such as variations in the supply pressure and temperature.

Slip and fall are very common occurrences in guest bathtubs and shower chambers while stepping in. Initially, the surfaces of bathtub and shower chambers may be made of non-slippery materials. But over the passage of time and due to use of cleaning chemicals, these surfaces may become slippery. As such, periodic resurfacing of these systems with proper material as per recommendations of the manufacturer should be done.

Grab bars are fitted in bath and shower chambers to facilitate getting in and out of them. The location and installation of these grab bars are specified in relevant codes and should be followed while installing them. These should be very securely anchored on the wall.

Any glass used in baths, including mirrors, should be glazed. Shower entry glass doors should be made of toughened glass material so that there is less possibility of cuts in case of any breakage.

Bathroom floorings should be made of anti-skid materials and the manufacturer of such floorings should also specify cleaning materials which will not increase slipperiness.

Proper ground fault protection must be ensured for all electrical gadgets and electrical outlets. The fan and coil of a hair dryer, if provided in the bath, should be mounted at some height on the wall.

15.4.2   Kitchen Safety

Kitchen areas are very prone to accidents. The floors are often wet and slippery due to spillage of oil. There should be a janitor to clean and dry up the kitchen floor at regular intervals of time. The kitchen employees should also bring any such conditions of the floor, when they occur, immediately to the notice of the janitor. As already discussed, operations such as slicing, chopping, etc. are activities susceptible to injury due to cut. The persons performing these operations must be very alert and not engage in talks with others while on the job.

The greatest risk in kitchen areas, however, relates to fire. Kitchen areas deal with cooking oil, fat, gas, gas flame, and trapped oil and grease in the ventilation duct. These components combine together to make kitchen a favourite playground of fire. Constant vigil and regular maintenance of gas line, burners, ventilation ducts, and grease filters are the means to combat this menace. We discuss safety related to ventilation system and gas system as follows.

Ventilation system maintenance

Ventilation system is critical in maintaining proper hygiene and safety in a kitchen. Gas appliances should not be run if the ventilation system is not functioning properly. The following steps should be taken to minimize the risk of fire and reduce their severity in case it occurs:

- Kitchen ventilation hood system must be periodically cleaned.
- Old hood systems, if changed, should be replaced by a system with new improved designs which incorporate automatic wash cycles, minimizes pockets and crevices in the ductwork where dust and grease could accumulate.

- Entire ducting and the exhaust fan for the ventilation system should be periodically cleaned.
- Grease 'build up' on the walls of the hood ducting/fan outlet exit should be regularly cleaned.

Factors that will reduce risk include

- Well-maintained ventilation system
- Good natural ventilation
- Satisfactory fume removal by natural draught alone
- Good awareness of risks among kitchen staff
- Minimal use of gas-fired appliances
- Large room size
- Proper training of staff about safety hazards and their sources
- Displaying clear notices explaining the dos and don'ts in a kitchen

Safety in gas work and gas appliances

Everyone, whether contractor worker or in-house staff, who is engaged in maintaining gas lines and equipment must be competent to do so and must have a valid certificate of competence relevant to the particular type of gas work involved.

Gas appliances, flues, pipework, and safety devices are to be maintained in proper safe operating condition. They should be periodically inspected by competent persons in accordance with current industry norms and they should also follow manufacturer's recommendations in this regard.

All catering and hospitality staff who use gas equipment should be trained for their proper use and be able to identify common faults, which include such things as damaged pipework and connections, inoperative flame supervision devices (these devices stop the gas supply automatically if the flame goes off), missing safety devices on equipment, malfunctioning of gas control valves (for example, gas leaking through valves even when they are closed).

Some kitchens have carbon monoxide alarms installed. These detectors shall give an audible alarm and be linked with an automatic gas shut-off system.

15.4.3   Accidents Common in Machine/Boiler Room

Quite a few accidents have occurred in the machine room and boiler room. Machine- and boiler-room conditions and the type of activities therein are very much favourable for accidents of very serious nature. Table 15.2 furnishes the types, causes, and preventive measures related to some of such common accidents in these rooms.

Table 15.2 Types, causes, and preventive measures related to accidents in machine/boiler room

Types of accident	Causes/source	Preventive measures
If a machine for maintenance starts suddenly, the limbs of the worker may be severely damaged and in some cases, the accident may be fatal.	Operator of the machine unknowingly starts the machine or electrical department may restore power to the machine	Proper shut down must be obtained from the operations department as well as the electrical department and there should not be any kind of communication gap in this regard. Machine should be started only after getting green signal from maintenance department.
A slip in the machine room can be very dangerous as the person may fall on a running machine, hot pipe or strike against a machine/equipment.	Oil (fuel oil, lubrication oil, etc.) may be spilling on the floor	The machine/boiler room floor must be inspected for any oil spillage after every work with oil and washed with detergent if necessary and thoroughly dried.
Inhaling toxic ammonia gas used in refrigeration plant	Leakage in the pipelines of ammonia gas in a refrigeration plant	All the pipes and joints of ammonia line should be properly maintained to prevent such leakage.
Cold burn due to contact with cold ammonia liquid	Usually occurs during recharging of ammonia in the refrigeration plant	The operator doing the recharging must wear safety gloves and be careful not allowing the very cold liquid ammonia to come in contact with skin.
While walking through the machine room, inadvertent collision with machine parts/pipelines	Lack of alertness on the part of the person. Lack of visible danger sign on areas with protruding parts prone to such accidents	One should be very alert when walking through the machine/boiler room. Visible danger signs should be displayed in zones with high probability of such accidents.
Head striking against pipelines running through or the underside of staircase	Not enough head room for such installations. Lack of alertness. Visible signs absent	While laying out pipe lines and constructing stair cases, enough head room should be provided so that people can walk past them unobstructed. One should remain very careful when such headroom is impossible to be provided Visible danger sign should be displayed.
Cuts and injury	Sharp edges of machinery and equipment	Free sharp edges should be properly guarded and avoided as far as practicable.
Serious injury to limbs	Entangling with rotating parts of machines, e.g., flywheel of engine, driving belt and pulleys in belt pulley drive, exposed gearing arrangement	All exposed rotating parts must be properly guarded with protective shields to avoid such accidents.
Fire breaking out during welding/gas cutting	Welding/gas cutting being done in a fire prone area without proper precaution	Proper fire fighting arrangements must be provided when welding/gas cutting job is carried out in zones prone to fire like near a liquid/gas fuel storage area or an area containing combustibles like paper, jute, etc.

(Contd)

Table 15.2 (*Contd*)

Types of accident	Causes/source	Preventive measures
Boiler may burst	Very high steam pressure	Safety valves and fusible plugs must be inspected periodically for effective venting out of steam in case the pressure exceeds the set safe pressure.
Fall from a height while doing maintenance work at height	Lack of alertness and failure to abide by safety regulations like not wearing safety belt or not using proper scaffolding. Maintenance crew is sometimes under pressure to complete a job quickly and tends to ignore safe working practices	Maintenance people must be very alert and abide by the proper safety norms. Trying to do something in haste is inviting danger.

15.5 | SECURITY IN HOTELS AND ITS MANAGEMENT

The hotels are responsible for the protection of property and personal security of their guests. A hotel with the best service but poor security would expose guests and employees to crime risks. There may be a costly loss of reputation and goodwill. In recent times, big hotels of repute have become a very soft target for attack. India has also witnessed such attacks in recent times. These incidents have showed the vulnerability of security in hotels under such situations. However, the present form of security system is, of course, mostly not designed to tackle such incidents. In this context, security of hotels has now become a complex issue with newer dimensions.

Security is, thus, an essential investment and must not be viewed as a secondary service. The steps that can be taken for proper security arrangements can be listed as follows.

15.5.1 Security Measures at Different Areas of the Property

Security measures to be adopted at different areas of the property are discussed as follows:

Security in public places
The following security measures should be adopted in public places:
- Installing devices such as surveillance cameras and adequate lighting should be implemented in all places of public circulation including public corridors.
- Patrolling guards should be employed in public places.
- Multi-tier entrance system should be implemented both for cars and visitors.
- Intelligent video analytics with trip wire feature makes it possible for the security to be alarmed in case someone jumps over the wall.

- Delivery vans should preferably have separate entrances and should also be checked thoroughly.
- Suppliers and service providers should be checked and granted access through separate entrances under proper visitor management systems such as identity cards, etc.
- Unauthorized vehicles should not be allowed in the premises.
- The use of baggage x-ray machines, metal detectors, surveillance cameras, access control systems, and public address systems are recommended, particularly in view of the recent incidents of attacks in big hotels.
- Separate lifts should be earmarked for hotel guests with proper access control systems.
- Fire alarm systems and fire fighting arrangements are mandatory by law and are to be kept in top working condition.

Security in non-public areas such as guest rooms and corridors

Measures to be adopted in non-public areas are as follows:

- The doors should also be installed with camera and alarm systems to monitor abuse.
- Lift doors exiting into non-public areas and corridors in the guest-room areas should be installed with closed-circuit cameras to monitor any unauthorized /suspicious visitors.
- Every hotel room should be equipped with a video door phone so that the guest can see the visitor and talk before opening the door.
- Staircase exits should be installed with panic-bar bolt doors for one way exit only.

Surveillance of guests

In recent times, the attackers entered the hotel as guests and operated from inside the guest rooms. So, it has become very important to keep vigil on activities of guests and their visitors. However, this is a very sensitive issue for the guests as it may amount to encroaching on their private life and may entail many issues, including legal compensation and in the worst a potential loss of business. Therefore, proper methods should be designed so that the security for the hotel and its guests and employees as well as the privacy of the guests is ensured. Any surveillance inside the room will be quite objectionable. Guests may be requested to establish their identity at the reception itself before check-in. Any visitor to the guests may also be passed through such identity check. Any luggage of the guests/visitors should be thoroughly screened. Guests may be monitored for their activities and movements outside their rooms, in public places and any suspicious activity should be brought to the notice of the law-enforcing agency.

While for small hotels, implementing all the measures as outlined may be cost prohibitive and may not be essential, big hotels under the threat of changing crime trends and risks need to be fully guarded with the most modern and effective security system and this will instill a sense of confidence and well-being in the minds of both guests and employees.

15.5.2 General Administration of Security

All security guards and hotel staffs are to be appropriately trained about the security systems in place and how to implement them in a routine manner, and also in case of emergency. While technology can help in the detection of crime response to detection signals comes from human beings. Without properly trained security personnel and general staff all the security arrangements may get less effective. Newly-arrived guests are unfamiliar with the surroundings, the staff, and hotel routines. A short briefing on security tips should be given to them when they check-in. A full-time security manager should be employed by the management for effective administration and control of the security system.

All security systems must be reviewed regularly by professionals and upgraded as well. Thus, a mix of proper facility design and good managerial practices is needed for maintaining a high level of security in a hotel. Every employee should keep security in mind every time a decision is made. Training the hotel employees to recognize and report suspicious individuals and unsafe conditions should be conducted by professional agencies.

15.6 | CONCLUSION

Safety and security is, thus, seen to be a critical issue for guest comfort and secured stay. Guests must be able to have a secured sleep at night and the employees should have a tension-free working environment. Hoteliers must address the safety and security issues with supreme importance and priority. Many a hotel has left the job of security in the hands of professional security agencies and has installed safety and security equipment and practices according to expert guidance and supervision.

KEY TERMS

Closed-circuit television (CCTV)	It is usually used for surveillance in hotels and commercial establishments, offices, etc.

| Safety | It essentially means creating a working and living environment that eliminates causes of physical injury and damage that can occur accidentally. |
| Security | Secure conditions or feeling; safety against espionage, theft, etc. |

REVIEW QUESTIONS

15.1 Discuss the necessity of safe environment in the hotel industry.

15.2 List some important functional areas in hotel industry related to safety issues.

15.3 Narrate different types of accidents and their prevention with respect to the hospitality industry.

15.4 List the components of fire safety management program usually adopted in a typical hotel.

15.5 Discuss the different measures taken in a fire-prevention program.

15.6 Explain the importance of guest bathroom safety and various measures taken to ensure it in a lodging industry.

15.7 List the steps to be taken for maintenance of a kitchen ventilation system

15.8 Discuss kitchen safety practices.

15.9 Discuss the importance of security in a hotel citing how building and facility designs are made keeping security in view.

15.10 Discuss what security measures can be taken in public areas in a hotel.

15.11 Discuss guest-room security in a hotel.

15.12 Discuss, in brief, general administration for security system in a typical hotel.

Care, Maintenance, and Troubleshooting of Select Engineering Equipment

Learning Objectives

This chapter deals with care, maintenance, and troubleshooting of some important engineering utility systems usually handled and operated by not-so-technical people.

The objectives of this chapter are to:

- sensitize students about the necessity of routine care and maintenance
- identify common problems occurring during operation of engineering equipment used in the hotel industry and to provide a list of possible causes and remedies for them

16.1 | INTRODUCTION

As outlined in the preceding chapters, there is a large number of engineering equipment used in the hospitality industry. The degree of complexities and sophistication, of course, vary greatly. While many of them are sealed units such as compressor units in small refrigerators, electric lamp, some detectors and sensors, many of them are subject to regular maintenance. However, for all of them, a little care in handling and following proper operational principles will help smooth and efficient operation and enhance the working life. In spite of best efforts, equipment exhibit troubles in operation. The maintenance department is usually responsible for dealing with such problems either in-house or through external service people or by the manufacturer during warranty period, if any. But for the operators, it is always helpful to know the possible causes and means to address the problem, if the nature of the problem is not serious. In what follows, we identify some typical equipment mostly handled by hospitality operational staff and discuss the measures to be adopted for the care, maintenance, and troubleshooting of those pieces of equipment. In this connection, we note that the care and maintenance of swimming pools, audio-visual equipment, public address system, computers systems, and sensors and detectors have already been discussed in the concerned chapters. Here we include mostly thermal, mechanical, and electrical systems as follows:

1. Refrigeration system
2. Air-conditioning and ventilation system
3. Gas and electric ovens/ranges   4. Washing machines
5. Clothes dryer   6. Dishwasher
7. Pumps

This chapter provides a general guide to the care, maintenance, and troubleshooting for the selected list of equipment. This guide may be incomplete as manufacturers use different parts on different models. The author does not take any responsibility for any action taken on the users' part as a result of using the guide. The manufacturer's manual must be consulted and maintenance people and service contract people must be involved in most of the situations of operating problems.

16.2 | REFRIGERATION SYSTEM

We have already learnt the meaning of refrigeration, refrigeration systems, and their utilities in modern day, with particular reference to the hotel industry. They are essentially engineering systems which create a cool environment (including sub-zero temperature conditions) for storage and preservation of food items and also for producing cold air used in air-conditioning systems. They are, therefore, very important for the effective functioning of the hospitality industry and have to be maintained in their topmost operating conditions. A few measures towards their care and maintenance would provide reliable trouble-free operation for a long period of time and avoid major failure of the systems.

16.2.1 Care and Maintenance

A few basic activities are to be undertaken for proper care of refrigerators of all kinds. The activities are aimed at ensuring uninterrupted reliable service of the refrigeration system. Some of the basic minimum activities could be listed as

1. Closing of door   2. Avoiding overloading
3. Proper defrosting   4. Daily routine care
5. Periodic inspection

Closing of door

Quite often, people forget to close the door after opening it for taking out or putting in articles. On other occasions, the door is not closed tight. The result is that the cold air inside being heavy comes down and out of the freeze and warm light air from outside enters and fills its place. This increases the internal temperature. The fall out leads to more compressor work and possible harm to the condition of the goods being stored due to temperature fluctuation.

Overloading

Each refrigerator compressor and the refrigerant it uses have some capacity to remove heat (called 'heat load') efficiently. Overloading occurs when the system is called upon to handle more load than it is designed for. Such a situation could arise if (a) there is too much foodstuff inside the unit, or (b) food is put in too high a temperature, or (c) the doors are kept open inadvertently for too long a period. Certain categories of hot food need to be cooled very fast to avoid deterioration in quality due to multiplication of germs. However, an introduction of such hot items immediately increases the local temperature inside possibly affecting other foodstuff there. Air-lock room or a completely separate chamber may be used for such situations.

Defrosting

The evaporator coil is the coldest space in a refrigerator system. The chamber air comes in contact with the cold evaporator tube through which the cold refrigerant is flowing. There is always some moisture in air and the refrigerated foodstuff also emits a lot of moisture. The cold air cannot hold much moisture which tends to condense and come out of air as water droplets. Thus, when the moisture in the cold air comes in contact with the cold evaporator tubes, it condenses and then freezes (temperature of the evaporator coil surface is lower than 0°C) to form layers of ice on the tubes. Ice is a bad conductor of heat, and hence these layers of ice make the refrigerator sluggish in its cooling action and the compressor/generator also consumes more power. As such, the thick layer of ice should be periodically removed for efficient operation. This process of removal of ice from the evaporator coils is called defrosting. There are two ways of defrosting. They are (a) manual and (b) automatic defrosting.

Manual defrosting It consists switching off the compressor motor, opening the door, and allowing the inside temperature to rise and wait till all the ice has melted away.

In some designs of walk-in refrigerators evaporator-pipe coils are mounted on the ceiling of some designs of the walk-in refrigerators and the cleaning can be done by brushing. Manual defrosting requires 2–4 hours and about 1 man-hour of labour. The frequency of defrosting depends upon usage pattern, number of door openings, outside ambient temperature and humidity, and the amount of foodstuff inside. Normally, as ice-layer thickness builds up to about 125 mm, the unit should be defrosted. Although manual defrosting is the cheapest of all defrosting methods, the principal disadvantage of the manual method is that while defrosting, the temperature of the refrigerated space increases to several degrees higher than the melting point of ice. This condition may spoil some food items. People may also forget to do the defrosting operation at regular intervals.

Automatic defrosting It overcomes the problems associated with manual defrosting and accomplishes defrosting with practically no rise in temperature. There are several ways by which automatic defrosting is accomplished. They are as follows:

Ambient-air defrosting A fan draws relatively warm outside air through a flap and across the evaporator coil.

Hot-gas defrosting In this the refrigeration cycle is reversed. This method has a low energy requirement but can cause a large temperature variation inside, causing harm to sensitive refrigerated foodstuff.

Electrical heating defrosting The most common defrost system is electrical heating type where electric current is passed through a wire with high resistance (called heater) so that its temperature goes up sufficiently high to melt the ice. This wire is located in the vicinity of the evaporator coil. Usually a timer clock will trigger current to pass through it, usually, every 12 hours. This time interval can easily be set at any desired value to suit individual conditions.

Daily routine care

Routine care includes checking for the proper temperature being attained in the space especially for large walk-ins. Thermometers may be put inside the space for the purpose. Condensed water should be removed daily and the walls and other surfaces such as tray, etc. should be cleaned thoroughly for better heat transfer.

Periodic inspection

Periodic inspection of the refrigerating units is very important as a part of the preventive maintenance programme. Some of the components and activities relating to periodic maintenance are

- Proper lubrication of the compressor and electric motor
- Checking for the condition of bearings of compressor and motor
- Cleaning oil filter for the compressor
- Checking for heat leakage from cabinet
- Checking for any possible leakage of refrigerant from pipelines, valves and other fittings, and compressor
- Cleaning the exterior surface and the internal tubes of the condenser
- Checking for correct working of all the controls and instruments such as pressure gauge, thermostats, etc.

All these activities related to periodic inspection as listed require specialized knowledge. Normally, this job will be contracted out and the contractor personnel will regularly, usually thrice a year, carry out the inspection work.

Temperature control in freezers

Food quality, storage life of food (the time for which the food can be stored maintaining the desired quality of food) and dehydration (loss of moisture) are the factors that determine the freezer quality. A range of temperature of –23°C to –18°C seems to achieve good freezer-quality. However, lower the temperature lower will be the efficiency and operating cost. Table 16.1 furnishes data on the relative efficiencies of the units at different freezer-operating temperature.

Table 16.1 Relative operating efficiencies of freezer system at different evaporator temperatures

Evaporator temperature (°C)	Relative system efficiency	Percent increase in operating cost
−17.8	0.8718	Base condition
−23.3	0.7424	14.8
−28.9	0.6557	24.8
−34.4	0.5797	33.5

(*Source:* Frank D. Borsenik, *The Management of Maintenance and Engineering Systems in Hospitality Industries*, John Wiley & Sons, 1979)

16.2.2 Troubleshooting in Refrigeration System

Any engineering equipment is subject to malfunctioning during operation in spite of best efforts in care and maintenance. Table 16.2 furnishes information on some common and frequently problems encountered during operation and their possible causes and remedies.

Table 16.2 Troubleshooting in refrigeration system

Problem/symptom	Possible causes	Possible remedies
Unit does not run	No power	Power cord, plug base and socket, fuses or tripped miniature circuit breaker (mcb) to be checked.
	Controls settings not proper	Set controls properly. If there is no result, controls are to be tested and replaced; if found faulty service people need to be called
	Compressor fan faulty	Service people need to be called
	Timer faulty	Service people need to be called
	Compressor faulty	Service people need to be called
Unit does not cool	Very hot weather	Set thermostat several degrees lower
	Door gasket faulty; cold air inside escaping out	Check gaskets for leak and if faulty replace
	Unit needs defrosting	Defrost, reset, and test run
	Door does not close properly	Level the unit such that the door easily closes on giving a slight push inwards. Check door alignment also—hinges may have to be replaced if found faulty
	Internal light, particularly in case of walk-in refrigerators, remain lit giving off heat, even when the door is closed	Replace switch, if defective
	Overloaded unit	Reduce the amount of stored food
	Unit placed near hot environment, very close to windows, wall	Relocate the unit; move units at least 50 mm away from the wall
	Condenser fan clogged	Clean fan assembly. If there is no result either replace fan or call professional
	Timer faulty	Test and replace timer. If it is a complex timer, disconnect and take it to a professional
	Coolant leak	Call service people
	Defroster heater faulty	Call service people
	Frost accumulated on evaporator coil	Defrost and then defrost frequently
Frost forms quickly or unit does not defrost	Incorrect controls setting	Reset thermostat control to higher temperature
	Defrost heater faulty	Test heater, replace if faulty
	Defrost limit switch faulty	Call service people
	Door opened too frequently	Open door less often
	Door gasket faulty	Check gasket for damage-replace if faulty
	Door sagging	Level unit so that door closes by itself. Check door alignment; if necessary door hinges are to be properly reset or replaced
	Drain may be clogged in frost-free units	Defrost freezer; clean drain port
Noisy operation	Unit not level	Level unit from front to back and side to side.
	Drain pan vibrating	Re-position pan; replace if damaged or warped

(Contd)

Table 16.2 (*Contd*)

Problem/symptom	Possible causes	Possible remedies
Hissing and popping is normal in frost-free refrigerator	The condenser fan and/or evaporator fan blades may be fouling	Check the condenser fan Check the evaporator fan
	Compressor mounts may be faulty	Check for alignment and anti-vibration pads in compressor base mountings.
Continual problem with ice formation at the bottom of freezer	The drain under the evaporator is frozen/plugged	Take off the back panel inside the freezer so that ice is exposed bare then and run hot water down it to melt it out completely
Condensation	Controls set incorrectly	Reset thermostat control to a higher temperature
	Door opened too often	Open door less often
	Door gasket faulty	Check gasket for leak; replace if faulty
Water leaks	Drains clogged	Defrost and clean drain ports
	Drain hose clogged or split	Replace drain hose
	Drain pan cracked	Replace drain pan
Unit runs continuously	Door gasket faulty	Check gasket for leak, replace if faulty
	Controls set incorrectly	Reset thermostat control to higher temperature
	Condenser oil dirty	Pull unit away from wall and vacuum condenser coil or remove bottom access panel and clean coil
	Unit in bad location	Relocate the unit away from windows or heat sources; move units at least 50 mm away from wall
	Door opened too often	Open door less often
	Coolant leak	Call service people
ON-OFF cycles of running for the compressor too frequent	Compressor faulty	Call service people
	Thermostat temperature margin for on-off control too small	Call service people
	Condenser coil may be unclean	Clean the surface of the condenser coil for higher heat transfer
	Condenser fan may not work	Test the compressor fan
Freezer does not defrost automatically	Fault with defrost timer	Check the timer setting and reset. If the timer is defective, replace it.
	Defrost heater may not work	Check the heater; replace if necessary
	Defrost thermostat may not work	Check and repair/replace the unit, if necessary
Refrigerator has an unpleasant odour		• Remove any spoiled food • Clean refrigerator interior with a solution of hot water and baking soda • Clean the door gaskets • Remove breaker strips and check for wet insulation
Bulbs do not light up	Bulbs burn out	Replace bulb
	Sensor sensing the opening of the door may be faulty	Call service people

(*Contd*)

Table 16.2 (*Contd*)

Problem/symptom	Possible causes	Possible remedies
Fridge has not stopped running for a few days; has an odour while it is running and the outside surface on the freezer side gets very hot	The condenser fan motor is bad.	Take off the back cover and put a fan blowing on the compressor; this will help until the fan is replaced
Refrigerator runs continuously— If temperature control is newly adjusted, the refrigerator is loaded or is located in a humid location, a refrigerator may run for 24 hours or more before getting cool.	Defrosting may not be proper	Defrost the freezer and check defrost setting
	The condenser may not function properly	Clean the condenser coil for better heat transfer
	There may be excessive heat loss through the doors	Check the door gasket for any fault; replace, if necessary
	The internal light may remain 'ON' always	Test the door switch
Refrigerator side getting warm but freezer still cold	Evaporator fan motor not running	Check motor for any fault
	Defective defrost timer	Check with the help of service people
	Defrost heater defective	Check heater; replace if faulty
	Defrost thermostat defective	Check defrost thermostat-replace, if faulty
	Airflow blocked from freezer compartment	Clean/unblock air flow passages from freezer to refrigerator space
	Freezer control turned to 'coldest' setting blocking air flow to refrigerator side	On most refrigerator/freezers, the freezer control closes a baffle when turned to coldest setting, and this blocks off the air flow to the fresh food compartment.
	Fan blade broken or mechanically damaged	Replace fan/repair blades
	Light staying 'ON' even with door closed thus warming the inside space of the refrigerator	Check door switch system; replace/repair
Refrigerator and freezer compartments both getting warm	No power connection	Make sure refrigerator is plugged in and has power
	Condenser fan motor under the refrigerator not running.	Check and take necessary action, if found faulty
	Condenser coil clogged	Clean the coil if clogged
	Defective defrost timer	Check and mend with the help of service people
	Defective defrost heater	Check heater; replace if faulty
	Defective defrost thermostat	Check and replace, if found faulty
	Evaporator fan motor not running	Check for power connection. Repair/replace if found faulty
	Cold control defective	Check for the control

(*Contd*)

Table 16.2 (*Contd*)

Problem/symptom	Possible causes	Possible remedies
	Compressor relay defective	Check and replace the relay, if defective
	Light remaining 'ON' with door closed	Check door switch system-replace/repair
	Compressor defective	Check and replace if defective
	Low on refrigerant	Call service people
Freezer compartment getting hot	Defrost timer stuck in defrost cycle	Check and replace the timer
	Ice maker stuck in harvest cycle	The ice cubes themselves might be causing the jamming. Make sure the ice maker is level, and the cubes are not too small or large.
	Light remaining 'ON' even when door is closed.	Check door switch system; replace/repair
Ice maker not making ice	Defective inlet water valve	Check and replace/repair defective valve
	Freezer temperature not cold enough	See problem under 'Freezer compartment getting hot'
	Defective thermostat in ice maker	Check thermostat- replace, if necessary
	Defective drive motor	Check and repair/replace, if necessary
	Water inlet tube may be clogged with ice	Check and melt/clean ice
Refrigerator sweating around door edges	Door gaskets leaking air	Check and replace gasket
	Defective case heaters	Check and change heater
	Energy saver switch not set to reduce exterior moisture	Check and reset
Water on floor outside the refrigerator	Drain pan may be overflowing	Check for any blockage in the drain pan
	Water inlet and outlet lines for the ice-maker may be leaking	Check for the water lines for any leak-repair/replace if there is any leakage
Water accumulating inside refrigerator floor	Drain tube may be clogged	Check drain tube and clean
	Refrigerator and ice-maker level may not be proper	Adjust to proper level

Some information on problems related to defrosting timer/thermostat/heater and checking the units for faults are given below. (Some newer refrigerators use electronic devices to control the defrost system. If the refrigerator does not have a defrost timer, the following is not applicable.)

- Turn the timer to the defrost mode after the refrigerator is stopped.
- Wait about 5 minutes, then see if the defrost heater is heating. If it is heating, the defrost heater and defrost thermostat are good. The timer should be replaced.
- If the heater is not heating, then the timer is probably good.
- Mechanically or using a hair dryer, defrost the evaporator (or cooling) coil.
- Inspect the defrost heater. Check for a broken or burned area on the heater. Replace the heater, if found.

- If no breakage is observed, the resistance of the heater coil is measured by a multimeter. If there is no resistance reading, the heater is to be replaced.
- If a resistance reading is observed, the heater is good and replace the defrost thermostat.

The above-stated procedures for checking are equally applicable for dishwasher application as well.

16.3 | AIR-CONDITIONING AND VENTILATION SYSTEM

Air conditioning and ventilation system in the hotel industry is a complex and elaborate engineering arrangement. While unit air conditioners are relatively simple, central air conditioning with full control is complex. Keeping these systems in good working condition is vital for guest comfort and hygienic environment. Maintenance of these units is done by skilled personnel, either in-house or under maintenance contract. However, a few tips on routine care and maintenance will help in smooth operation and reduce downtime.

16.3.1 Central Air Conditioner

The central air-conditioning system is a complex array of subsystems. Here, only a few simple care and maintenance measures as well as some tips for troubleshooting are discussed. It is iterated that troubleshooting must be done by experienced and skilled personnel.

Care and maintenance

A central air conditioner is a full-fledged complex engineering system and not a household appliance. It requires professional maintenance and repair. That is why attempts for repair/maintenance during the warranty period may make the warranty void for the remaining period. Other than performing simple maintenance, one should not attempt to make any adjustments to an air-conditioning system. Following are a few tips for routine care that can be easily undertaken for general upkeep of the system.

Clean or replace filter monthly; clean or replace filter twice a month during seasons when the unit runs more often When the air conditioner circulates and filters the air dust, dirt particles build up on the filter. Excessive accumulation can block the air flow forcing the unit to work more to maintain the desired temperatures consuming more power.

Keep outdoor unit clear of debris, leaves and shrubs/creepers Efficient operation of air conditioner depends on the free flow of air over the coil. Anything that blocks the air flow causes the compressor to work more to move the warm air out of the conditioned space, consuming more power and shortening life.

Keep the unit cool Shading the central air-conditioning unit is a good idea so that it does not have to work as hard to cool the coils inside it.

Never stop the system by shutting off the main power

Troubleshooting

Air-conditioning systems like any other engineering systems are subject to operational problems even after proper care during operation and maintenance activities are undertaken. Table 16.3 furnishes a few troubleshooting tips for reliable and sound operation.

Table 16.3 Troubleshooting chart for central air-conditioning units

Problem	Possible cause	Remedies
Air conditioner runs but does not cool	A central air conditioner that runs but does not cool may just need to be cleaned	A central air conditioner that runs but does not cool may just need to be cleaned. Plan to do this on a relatively warm day. Brush away leaves and debris from the outdoor condenser. Remove any protective grille or cover from the condenser's fins and clean the fins.
	Low on refrigerant	The refrigerant may need to be recharged-with the help of an air-conditioning technician
Blower does not operate	Blower door open or kept ajar	Close door securely to restore power to blower
Air conditioner does not start	Thermostat setting may be wrong	If central air conditioner does not start automatically be sure the thermostat is set to 'cool' and set below the current temperature
	There may be no supply power	Usually a central air conditioner is powered by a dedicated 240-volt circuit. Check the main electrical panel and any secondary circuit panels for a tripped breaker or blown fuse. Reset the breaker or replace the fuse if necessary. Make sure that the outdoor condenser's power switch mounted on the outdoor unit, has not been shut off.
Major room temperature swings	When room temperatures swing more than about 3 degrees between when the air conditioner goes off and on again	Thermostat heat anticipator has to be adjusted by service people
Room temperature drops too low below the set temperature	The thermostat is improperly calibrated or installed	It does not sense a proper sampling of room air; call service people for rectification of the problem

(Contd)

Table 16.3 (*Contd*)

Problem	Possible cause	Remedies
Water pool forms next to air conditioner	Air conditioners create considerable amount of condensation, which goes out through a plastic drain tube. The floor drain may be blocked	Check if one of the tubes is leaking. If it is, replace it. If it appears that a condensation drain tube is clogged with algae, remove it if possible. To kill the algae, pour a dilute solution of bleach (1 part bleach to 16 parts water) through the pipe. Ice may be blocking the tube. If this is the case, be sure the filter is not dirty. If the filter appears to be fine, the air conditioner's refrigerant supply is probably low. Call an air-conditioning technician.
	Condensate pump, if used to drain water, may not work properly	Test the condensate pump by pouring water into its pan. If the pump does not start, either it is not receiving power or it is broken. Check for power input to the pump. If OK, pump may be broken-either get it repaired or replace it by service people. If the pump runs but does not empty the pan, the ball check valve just before the discharge tube is probably stuck. Unscrew the check valve, loosen the ball inside, and look for an obstruction.
Air ducts are noisy	Many heating/cooling ducts are fabricated out of thin sheet metal so they conduct noise quite readily from the air-handling unit into the rooms	To break the transmission of sound, insert flexible insulation ductwork between the heating/cooling system and the ductwork lengths.
	A pinging or popping sound coming from the ductwork may be due to thermal expansion or by air blowing past a loose flap of metal.	Track along the duct runs, listening for the sound. If located, make a small dent in the sheet metal to provide a more rigid surface that's less likely to move as it heats and cools.

16.3.1 Unit or Package Conditioner

Care and maintenance

In order to maintain proper performance of packaged terminal air conditioner or heat pump, it is very important that the fan and outdoor coils, the blower wheel, blower scroll, electric heater, and all drain passages are thoroughly cleaned at least once every year. The company manual usually recommends that cleaning should be conducted prior to the start of each heating season.

Depending on local conditions, more frequent cleaning of the unit may be required to ensure optimum performance and long operating life. Examples of these special conditions include areas where construction dust or heavy airborne dirt is found, or environments that promote the growth of fungus. The following routine maintenance should be undertaken for smooth efficient operation of the units for a long time.

Air inlet filter It should be cleaned once every month by vacuum cleaner or by a brush.

Outdoor vent filter It should be cleaned in the same manner as the air inlet filter, but only once during a cooling or heating season.

Troubleshooting of room/window air conditioner

Many of the troubles in operation of these air-conditioning units can be addressed in-house without calling the service personnel, thus saving time and money. Table 16.4 provides information about the possible causes and remedies of such common problems.

Table 16.4 Troubleshooting chart for room/window air-conditioning units

Problem	Possible cause	Remedies
Room air conditioner does not cool	A room air conditioner that runs but does not cool may just need to be cleaned	Remove the grille and filter. Wash the filter or replace it with a new, inexpensive filter. Clean the inside coil's fins a vacuum cleaner with soft brush attachments Then, from the fan side, spray water back through the fins (protect the wiring and motor with appropriate cover). Clean the unit with a piece of cloth, making sure all drains that allow condensed water to drip away from the unit are open. Allow it to dry thoroughly. Lubricate the motor bearing Clean the evaporator and condenser coils
	Low on refrigerant	The refrigerant may need to be recharged—call an air-conditioning technician
Air conditioner does not start	There may be problem related to power supply. Room air conditioners draw a lot of electrical current, which can lead to complete failure due to overload.	Be sure the unit is plugged in and turned on. If the light does not glow, the circuit has probably overloaded-check the electric panel or fuse box and reset the breaker or replace the fuse, as the case may be.
	If the light works, it is likely that the air conditioner's switch is faulty or the thermostat needs adjustment or repair.	Be sure the thermostat is set to 'cool' and below room temperature
	Other reasons	Call air conditioner service people

16.4 | COOKING STOVES AND RANGES

Cooking stoves and ranges are among the most vital equipment in the hotel industry. Efficient operation of these sets of equipment is critical for economy of operation and reliability of service. These systems are directly handled by kitchen personnel and they should have a fairly good knowledge of the working principle and care and maintenance practices.

Care and maintenance

The single-most important step to keep a kitchen range operating properly is to clean it regularly. Beyond that basic maintenance, only a couple of repair jobs are within the skills of most users.

With the help of an owner's manual, one can replace door gaskets, adjust the feet beneath a range to correct for uneven cooking, and re-light a pilot. Of course, one can also re-position racks if the food is cooking unevenly, and can check cords, circuit breakers, and fuses.

Before doing any work on the stovetop, always unplug the range or turn off the power at the main electrical panel and shut off gas supply as the case may be.

The following are tips for good operational health of cooking ovens and ranges:

- Clean up spills soon after they occur. Some acidic foods, such as tomato products, can damage the surface of the cooktop or interior of the oven.
- Loosen cooked-on creamy substance with water or baking soda and water.
- Keep the reflector bowls (also called drip pans) of gas burners clean; the shiny surface not only gives a nice look, but also makes cooking more efficient by evenly reflecting heat back up to the pot or pan aluminum foil does not serve this purpose as well and may cause overheating.
- Wash burner controls and other operating knobs with warm water and detergent. Some knobs have markings that may be damaged in the process, so take care not to soak or scrub them too much.
- Clean gas burners and grates thoroughly.
- Use commercial oven cleaners very cautiously because they are highly toxic.
- Faulty seals cause uneven cooking, waste energy, and make the kitchen hot. With an electric oven, badly leaking seals can shorten the life of the heating elements. So replace oven door seals when they become stiff with age.

Troubleshooting in gas ranges/ovens

As more and more features of ranges become computerized, the list of things one can attempt to repair shrinks more and more. If components of the range do not work, check the circuit breaker or fuse that governs the circuit first. Table 16.5 shows a list of problems commonly encountered in gas ranges/ovens and their possible remedies.

Table 16.5 Troubleshooting chart for gas ranges/ovens

Problem	Possible cause	Remedies
The set temperature does not match the temperature inside the oven	Heat may escape through oven doors	Make sure heat is not allowed to escape by repeatedly opening the oven door.
	The problem may also be caused by an improperly calibrated or defective thermostat	Check and replace/recalibrate the thermostat, if necessary
Oven temperature drops below set level; food ends up soggy or frozen at centre	The problem may be caused by an improperly calibrated or defective thermostat	Thermostat needs recalibration or replacement
Food burns on the outside; oven temperature exceeds set level; oven produces condensation on the glass window	The exhaust vent could be blocked	Check to see if the exhaust vent to your range is blocked and do the needful
	There may be heat leakage through oven door	Inspect the oven door gasket; replace gasket, if worn out or defective
	The problem may be caused by an improperly calibrated or defective thermostat	Thermostat needs recalibration or replacement
• Food bakes on one side faster than the other • Bakes unevenly or burns food	Heat is not being distributed evenly throughout the oven	Remove any aluminium foil from racks or the bottom of the oven
	Heat is escaping out the door	If food is colder toward the front, check the gasket for wear. If the seal is worn out, replace it.
	The controls have lost their adjustments	Call service people
	Baking pan material at the base may absorb heat very quickly	Check the baking pans. If they are dark, they may be burning food. Reduce the temperature by 15° Celsius when baking in dark metal or glass pans.
	The vent may not be clear	Be sure nothing is obstructing the vent
	Oven racks may be very near hot zone	Reposition the oven racks so that baking pans are not very near the heat sources.
	Flame ports may be blocked	Check that the flame ports for blockage and ream with a piece of wire or needle, if needed
	Range may not be level	Make sure that the range is set level. Adjust the feet until the range is level
Oven or range seems to be receiving power (the light works, for example) but does not get hot	The heating circuit may have separate fuse or miniature circuit breaker (mcb) that might have blown off or tripped.	Reset mcb or replace fuse

Troubleshooting of Electric Ranges/Ovens

Electric ranges may seem complicated but they really are not. There are many day-to-day problems with them which can be easily set right. In case of major problems, however, the service people must be called. There are minor differences between brands and models, but they work on the same basic principles and comprise elements which are very similar in nature, thus troubleshooting is rather generic in nature. Table 16.6 furnishes guidelines for troubleshooting some common problems encountered with electric ranges/ovens.

Table 16.6 Troubleshooting chart for electric ranges/ovens

Problem	Possible cause	Possible remedies
Oven temperature drops below set level; food ends up soggy or frozen at center	Thermostat may not work properly	Undertake recalibration or replacement of thermostat
	Oven door seal may give way	Check to see the condition of the gasket (the covered flexible seal inside the oven door) - if it is worn out/damaged replace or call service people to do the job.
	Thermostat may not work properly	Undertake recalibration or replacement of thermostat
Food bakes on one side faster than the other, Bakes unevenly or burns food	Heat is being distributed evenly throughout the oven	Remove any aluminium foil from racks or the bottom of the oven
	Heat is escaping out the door	If food is colder toward the front, check the gasket for wear. If the seal is worn out, replace it
	The controls have lost their adjustments	Call service people for rectification
	Baking pan material at the base may absorb heat very quickly	Check the baking pans. If they are dark, they may be burning food. Reduce the temperature by 15 degrees Celsius when baking in dark metal or glass pans
	Oven racks may be very near hot zone	Reposition the oven racks so that baking pans aren't very near the heat sources.
	Range may not be level	Make sure that the range is set level. Adjust the feet until the range is level
Range surface elements do not heat at all	Problems should lie with electrical circuit and elements	Check for a blown fuse of tripped circuit breaker as the case may be Element terminals may have loose connections. Reposition the elements Burner switch may be bad. Replace, if necessary. Make sure that the power cord is plugged in properly
A single burner does not heat properly or works intermittently	If the burner is plug-in type, The burner receptacle or the burner may be faulty	Test the seemingly faulty burner in another receptacle. If it works, replace the original receptacle. If it doesn't work, replace the burner.
	If the burner a flip-up type burner, the burner may be faulty	If the burner a flip-up type burner, disassemble and replace the faulty burner and any other faulty parts.
	Other causes	Call service people
A single burner works only on high heat setting	Problem with burner control switch	Test the burner control switch
	Problem with wiring	Check the wiring for short circuit

(Contd)

Table 16.6 (*Contd*)

Problem	Possible cause	Possible remedies
An indicator light does not work	Bulb may be faulty	Check and replace the bulb
	Connection may be faulty	Check electrical connection to the bulb
Oven exterior is very hot	There may be heat leakage from inside the oven	Inspect the door seal gasket and replace, if faulty
Convection features do not work properly	There may be problem with circulating fan	Check for any obstruction to the fan Check the fan motor
	There may be problem with oven selector switch	Check the oven selector control
Running the oven produces excessive condensation	Oven vent may be clogged	Check and clean the vent and the duct

These diagnostic tests should cover most of the electric stove troubleshooting situations. For a more detailed explanation, always look to the model-specific literature.

16.5 | MICROWAVE OVEN

Microwave oven is a very sophisticated piece of equipment and except for a few dos and don'ts, the operator really does not have much to do.

Care and maintenance

Given below are a few tips on routine care and maintenance of microwave ovens.

- Always use only those trays or containers that are made of materials specifically suited for microwaves.
- If you are unsure whether a glass container is safe for use, put the empty container inside and run the microwave for one minute. If it is warm, it is unsafe for the microwave. If it is lukewarm, it is safe for reheating. If it is cool, it is safe for cooking.
- Always keep the door seal and the surface to which it abuts clean.
- Do not mount microwaves too high as there is risk of spilling hot liquids on your body.

16.6 | WASHING MACHINES

Washing machines do wonders in the hotel industry where guests need their clothes cleaned and laundered within a very short time. It is, therefore, very important to keep them in good operating condition. A washing machine is basically a big tub that periodically fills and drains with water, spins to wring clothes dry, and has a device for stirring things up. The device could be either an agitator installed in the middle of a top-load machine or a rolling drum in the case of a front-loader type machine. The four cycles that every washer performs are fill, wash, drain, and spin. A few simple steps in handling them with proper care would maintain their reliability and efficiency very high. Given below is a list of some procedures to follow while operating and maintaining them.

Care and maintenance

A few routine care and maintenance activities help washing machines run trouble-free for a long period of time. Some tips regarding the operation of the machines are given below.

- Regularly clean the top and door of the washer to prevent the build-up of dirt and detergent.
- When very linty materials are washed, remove lint from the tub after removing the laundry.
- Soap deposits may be removed by filling the tub with water and adding 500 gm of water softener or 3 litre of white vinegar; and then running the machine through the complete wash cycle
- For best results, put detergent before the washer is loaded.
- For majority of items, lowering the water temperature to warm wash and cold rinse will get clothes clean with reduced energy requirement.
- Small loads take up almost as much energy as large loads. Therefore, whenever situation permits, wait until you have a full load before running the washing machine.
- Most washer drain hoses hook over the side of a sink or into a pipe especially for that purpose. To prevent floods, make sure the hose is secure and cannot go out of position.
- Rags, socks, or similar items should never be kept on shelves over the laundry sink. They can clog the drain and cause flooding the floor, if they fall into the sink.
- If the washer is not in the basement, installing a washer pan and drain can help prevent flood damage.
- Check for condition of water hoses—if they are old reinforced rubber hoses, they might burst causing flooding of laundry floor. Consider replacing them.
- While replacing old reinforced rubber hoses, use tougher metal hoses.
- If the machine makes a rattling sound during operation, it may need leveling— adjust the legs until the machine is level with all four legs touching the ground.

Troubleshooting

Washing machines comprise rotating mechanical elements with sophisticated electronic control, and problems with them are generally addressed by trained service personnel from the manufacturer. However, a few commonly encountered problems can be tackled with a little knowledge of the system as furnished in Table 16.7.

Table 16.7 Troubleshooting chart for washing machine

Problem	Possible cause	Possible remedies
Washer fills with water but does not agitate	A faulty lid switch	Check the lid switch and the tab on the lid that strikes it. Press and release the switch; if it does not click each time you do this, it is probably broken
	A snapped drive belt *Note:* Direct-drive washing machines do not have belts	Check the spin cycle. If this works, the motor is operating and the belt is not broken. If it does not, the belt may be loose or broken.
	A problem with the motor	Check for motor fuse, etc.
	There may be a problem with the controls, or the agitator solenoid may be broken.	Call service people.
	Something is wrapped around or stuck beneath the agitator	Lift the agitator a little and remove the material.
Tub does not fill	Water supply hose may be blocked or kinked	Clean hose for any clogging and straighten hose for kinks—replace if necessary
	Water level switch is faulty. This switch with a clear tube attached to it in the control panel	Remove the switch and get it repaired/replaced by service people
	A faulty lid switch or timer could also prevent the machine from filling	Inspect the controls; call service people
	Filter screens may be clogged	Check, clean the filter; replace if necessary
Washing machine fills and agitates but does not drain	Drain hose may be faulty	Check the drain hose for any blockage or kinks
	Something wrong with the pump.	The water pump may be broken or clogged with a small piece of clothing; call service/maintenance people.
The motor stopped mid-cycle.	The machine's overload protector may have tripped	Take out some of the articles to reduce the load and the protector resets itself. Then restart the machine
Washer does not operate or make any noise when set on any cycle, or it stops when it should continue onto the next cycle	No power	Check power cord, plug. Check for blown fuses or mcb. Do the repair work and restart
	Motor overload or safety shut off	Press reset button on motor or control panel
	Lid switch faulty	Remove detergent build-up in orifice. Make sure switch is making contact, if not- open the lid and check the lid switch and the tab on the lid that it pushes against. Press and release the switch. If it does not click, it is probably broken. Replace or call a service man

(Contd)

Table 16.7 (*Contd*)

Problem	Possible cause	Possible remedies
	Water pump clogged	Call service people
	Motor may be binding	Call service people
	Machine controls defective	Call service people
Washer noisy; it shakes or vibrates too much	Tub unbalance; Out-of-balance wash load	Adjust the load if it has scrunched up on one side of the tub. Follow instruction manual
	The floor surface may not be proper	The floor must be flat, level, and strong enough to support the heavy, water-filled washer without deflecting under the load
	May not be sitting firmly and level on the floor	Check for the leveling of the feet of the washing machine
	Tub bolt loose	Tighten nuts that hold tub
	Motor bearing may be damaged	Check for the condition of motor bearing- replace, if necessary
	Cross bar damaged	Call service people
	Drum brakes may be faulty	Check the drum brakes—replace if necessary
	May be due to over-sudsing	Determine if the machine is over-sudsing. Reduce suds by pouring ½ cup of vinegar with about 1 litre of water into the washer. Then switch to either low-detergent or a low-sudsing option. Reduce the amount of detergent used.
High water level or water overflow	Flow valve washer faulty	Check for any fault—the size may not be also proper. Replace the washer
	Water supply pressure may be too high	Reduce flow by closing water faucets slightly
	Water level control switch may be faulty	Call service people
Washer seems to complete all of its cycles but does not spin	Unbalance tub due to lopsided cloth loading; this activates off-balance shut-off switch	Be sure the load of clothes is not lopsided. Just open the lid and reposition the clothes. This automatically resets shut-off switch. Close the lid and restart.
	A faulty lid switch	Check the lid switch and the tab on the lid that strikes it Press and release the switch: If it doesn't click each time you do this, it's probably broken.
	The drive belt may be loose, worn, or broken)	Replace the belt
	There may be a problem with the controls	Controls such as timer control, centrifugal switch, water level switch, etc. may be defective; call service people
Tub spins during wash cycle	Fault with timer	Check timer; if contacts are bad, replace else call service people
Water fills wash tub too slowly	Water inlet valve may not be fully open	Check the inlet valve and do the needful
	Inlet valve screen may be clogged	Clean the valve screen

(*Contd*)

Table 16.7 (*Contd*)

Problem	Possible cause	Possible remedies
	Water supply pressure at the tap may low	Call plumbing maintenance people
Insufficient water in tub	Water level switch may be defective	Call service people
	Water may drain out through the drain hose	Check that the drain hose is positioned as high as the inner tub
Water fills in wrong cycle	Water inlet control solenoid may be faulty	Clean the valve and gently tap the solenoid for resumption; if there is no result, call service people
	Water level switch faulty- senses no water even when there is water	Call service people
Clothes are too wet after spin cycle	There may be a drain problem	Check that the drain line is clear; check the drain hose for kinks, etc.; inspect water pump for proper functioning.
	There may be a spin problem	Test the motor, coupler, belt (for belt drive units), pulley; test the lid switch; test the timer control; test water level switch.
Clothing comes out damaged	Agitator may be cracked or may have rough spots	Inspect the agitator and do repair work, if found so
	There may be rust inside if clothing have rust stain	Inspect for rust spots
Water leaking—be sure that water appearing to be due to a leak is not drain water from a backed-up standpipe.	Hose connection junctions to the faucets and to the back of the washing machine may be defective	Check the connection fittings—tighten couplings or hose clamps if necessary or replace the hoses, if faulty, along with the fittings altogether
	Could be due to over-sudsing	Check if the machine is oversudsing. Reduce suds by pouring half cup of vinegar with about 1 litre of water into the washer. Then switch to either low detergent or a low-sudsing option. Reduce the amount of detergent used
	Outer tub could be damaged	Inspect outer tub for cracks, holes or corrosion perforations
	Tub lid seal faulty	Replace lid seal
	Pump gasket faulty	Replace gasket
Water does not drain out of the wash tub	May be due to a clog in the drainage system	Check for clogs or kinks
	Pump may be jammed or clogged	Open and clean pump
	Bad drive belt	Tighten belt, if loose. Replace a broken belt.
	Pump impeller loose on shaft	Secure impeller properly onto the shaft with keys etc.
	Timer control may be faulty	Call service people
Water too hot or too cold	Temperature selector switch set wrong	Check in the control panel and set the selection right

(*Contd*)

Table 16.7 (*Contd*)

Problem	Possible cause	Possible remedies
	Supply hoses may be reversed	Check and rectify the error
	Mixing valve faulty	Check for the solenoid valve; replace if faulty
	Hot or cold water valve may be turned off	Check and do the needful
	Temperature selector switch faulty	Call a service people
Washing machine leaves behind a lot of detergent. It is also not taking fabric softener either.	The fillers may be clogged	Fill the detergent and softener cups with warm water let it remain there for 10-15 minutes and run a quick cycle.
Foul odour coming from washed clothing	Drain line may be clogged	Clean the drain line
	Drain hose may be defective	Check for kinks and blockage
	Detergent may build up in the outer tub	Check and remove the detergent

16.7 | CLOTHES DRYER

Clothes dryer or tumble dryer is an appliance that is used to remove moisture from a load of wet clothing and other linens. Generally, a washing machine cleans the textile items and then passes on to the dryer for the purpose of drying. Hospitality industries accumulate a lot of used textiles and linens (such as bed linens, guest garments, furnishing textiles, etc.) and these machines are extensively and regularly used by them.

Clothes dryers are very reliable and yet simple in operation. There is a rotating drum called 'tumbler' that holds the clothes, rotates and tumbles the wet clothing. Hot air is passed through this tumbler. This hot air, in turn, heats up the textile loads and steams out the water in them. An exhaust vent passes out of the dryer and out of the house. The water in the washed clothes goes out through this exhaust vent in the form of steam. The tumbler is rotated by an electric motor usually through a drive belt arrangement and the air is heated by electric/gas heating arrangement.

Care and maintenance

The following procedures may be adopted to keep the dryer in good working conditions. Any major problems must be addressed by competent service personnel.

- Clean the lint trap before each load. This will prevent a fire hazard, save energy, and help dry clothes faster.
- Use the automatic dry cycle to avoid over-drying clothes.
- Once a year or so, use a dry paint brush to clean lint from the corners and racks in the interior of the dryer and around the door.

- If you have a traditional sheet-metal dryer exhaust vent, you should clean it after about every five years.
- Heavier, flexible plastic tubing gets dirty faster and may need to be cleaned every two or three years. Semi-rigid aluminium duct work may have to be cleaned every three or four years.
- Regardless of what kind of exhaust tubing the unit has, clean the outside vent once a year and make sure it opens when the dryer is on and closes properly when the dryer is put off.
- Never exhaust a gas dryer indoors.

Troubleshooting

Drying machine is a sophisticated piece of equipment with modern control system. As such, repairing of it requires service personnel of the manufacturer. However, a few problems can be tried in-house for uninterrupted service. Table 16.8 furnishes a few troubleshooting tips.

Table 16.8 Troubleshooting chart for clothes dryers

Problem	Possible cause	Possible Remedies
Dryer drum does not turn	No power	Power cord, plug base and socket, fuses or tripped miniature circuit breaker (mcb) to be checked
	Drum belt may be loose/torn	Replace drum belt
	Motor may have burnt; no sound of motor running	Check the motor; replace if necessary
Dryer does not heat	No power	Power cord, plug base and socket, fuses or tripped circuit breaker (mcb) to be checked and appropriate measures to be taken
	Motor does not run	Check for motor burn out or circuit breaker problem
	dryer door switch may not be working	Check dryer door switch
	dryer thermostats may not work properly or set wrongly	Check and reset dryer thermostats, if necessary
	Temperature select switch is not set to air fluff	make proper setting
	Dryer element may be broken.	Check dryer element visually to see if it is broken otherwise use multimeter to check continuity- replace, if necessary.
Dryer heats but takes long time to dry	Dryer vent and roof vent may be blocked	Clean dryer vent; also check roof vent. They have screens to keep birds out and often get blocked by debris.
	Dryer thermostats may not work or set wrongly	Check and reset dryer thermostats, if necessary
	Blower may run free on the motor shaft	Make sure blower is not slipping on motor shaft

(Contd)

Table 16.8 (*Contd*)

Problem	Possible cause	Possible Remedies
The dryer door is opened for checking the clothing and then the dryer does not restart	Heat activated thermal fuse are use in modern dryers. Opening the dryer door mid-cycle can trigger a heat spike that blowing the fuse and stopping the motor	Replace the fuse

16.8 | DISHWASHER

A dishwasher is essentially a watertight box that sprays dishes with hot water and soap, drains out the dirty water, and then dries the dishes. For modern dishwashers, rinsing the dishes is unnecessary and a waste of water and time. Most dishes can go from the table to the dishwasher directly without a stop at the sink. Some dishwashers also come equipped with special cleaning cycles that automatically adjust, sensing the soil level of the dishes. This saves water and energy.

Care and maintenance

With proper cleaning and maintenance, a dishwasher will clean dishes to sparkling shine and to the desired sanitation level for many years. Given below are a few routine activities to be done by the operator during the operation of the machines.

- The inside of dishwasher should be cleaned every two weeks.
  - The tub should be cleaned of all debris and then thoroughly wiped.
  - The detergent dispenser should also be cleaned out.
  - The filter screen and basket should be scrubbed.
  - Clear the ports with a pipe cleaner.
  - To remove mineral deposits inside the tub, pour a cup of vinegar into the empty tub and run the machine through a complete wash cycle.
- Cover any nicks on the enamel surface and racks of dishwasher with epoxy coating made especially for such repair. This prevents rust formation which is responsible for creating subsequent leaks.
- If water comes out from under the dishwasher, immediately disconnect the power and call service people.
- Run the machine after it is fully loaded.
- If the dishwasher does not start, first check that the door is properly fastened and then reset the dial or push the buttons. If it does not work, check power connection and circuit breakers to be sure that power supply is on.
- When loading the dishwasher, ensure that no dishes obstruct the rotating spray arms.

- Load more fragile items in the top rack. The highest-pressure jets are directed at the lower rack to help clean pots and utensils.
- If the machine has a garbage disposal, make sure it is run before running the dishwasher.
- If the dishwasher has a strainer or scrap bin in the bottom, clear it before running each cycle of load.
- Use water heating option, if the dishwasher is equipped with it.
- Do not run the dry cycle frequently as it uses a lot of energy. Instead, when the dishwasher stops, open it and pull out both racks; everything will be dry within an hour or so, especially during dry winter season.
- If patches of rust begin to appear on the dishwasher racks, racks can be re-coated with a rubber-like fluid available in hardware stores.
- If the rollers on the racks get damaged, replace them.

Troubleshooting

Dishwashers are now indispensable in modern kitchens. They combine hydraulic, mechanical, and electrical components with electronic control. Maintenance of such equipment is usually carried out by service people. However, like other systems, they are also amenable to in-house troubleshooting for some common problems. Table 16.9 furnishes information about a few common problems, their possible causes and remedies.

Table 16.9   Troubleshooting chart for dishwashers

Problem	Possible cause	Possible remedies
Dishwasher does not work; there is no sound, water, or light	No power	Check the electrical panel for a tripped circuit breaker or blown fuse. Check that the dishwasher is plugged in securely Inspect the wiring connection to the dishwasher for burns or breaks.
	Motor may malfunction; no sound of motor running	Test the motor start relay Test the motor
	If electrical power is available to the dishwasher, motor is fine but the appliance does not run, the problem is likely to be a defective door switch, timer, or selector switch.	Check the door latch Test the door switch Test the timer motor Test the selector switch
Dishwasher does not start but motor hums	Motor-pump may be mechanically jammed	Stop the motor. Remove the driving belt. Try rotating the pump/motor manually or by pipe wrench to make them run free
	Motor operation may be faulty	Test the motor start relay Test the motor for electrical fault- replace if necessary
	Drive belt may have snapped	Check the drive belt

(Contd)

Table 16.9 (*Contd*)

Problem	Possible cause	Possible remedies
Dishwasher does not stop filling	The float switch may be faulty	Check the float switch; call service people, if necessary
	Water inlet valve is stuck open.	Check the inlet water valve and do necessary repair
	The timer may be stuck on 'Fill' status	Check the timer setting;reset or call service people
Dishwasher does not fill with water or it drains while filling	Water supply valve may be closed	Check that the water supply is turned on. The valve may be located under the sink.
	Door latch and door switch interlock relay may not work	Check the door latch Test the door switch
	Float assembly may not function properly	Inspect the float assembly for smooth movement Inspect the fill tube for kinks; replace, if necessary Test the float switch for any electrical fault
	Water inlet line may be faulty	Check the inlet valve filter screens for clogging-clean/replace filter Test the water inlet valve—clean/replace if necessary
	Drain valve operation may be faulty	Test the drain valve lever arm Check the drain valve
Water does not drain from the dishwasher	Excessive water inside means the pump is not pumping water out properly, the drain hose is not carrying it to the drain pipe, or the property drain lines are blocked	Check for proper operation of pump Check sink drain and drain hose for restrictions Check the drain valve
	Drive motor, pump may get jammed, drive belt may snap	Make the motor/pump run free as in some previous step Check the drive belt-replace, if necessary
	Timer motor may not working properly	Test the timer motor—call service people
	If dirty water spews from the air gap, the drain line is kinked or clogged.	Check for blocked or kinked hose—replace, if necessary
	Knock out plug may not have been removed when the connection of a newly installed a garbage disposal was made.	If a garbage disposal is recently installed, be sure the knockout plug for the dishwasher was removed when the connection was made
Dishwasher is leaking water	Water seal provided at various areas may not be working properly	Inspect the seals for the main tub, float, heating element and blower diffuser—replace, if necessary
	The tub may be overfilled	Overfilling is likely a problem with the float or float switch—check these items for proper functioning
	The inlet valve may not be closing properly and leaking internally	Check if valve solenoid is working properly. If water leaks through the valve even when the dishwasher is off, then the valve itself is the problem. Debris may be preventing the valve from closing. The valve needs to be cleaned/ replaced.
Dishwasher is noisy	Spray arm may be damaged and obstructing	Check the spray arm for damage and obstructions
	Motor mountings may be faulty, fan motor blades may touch the casing, creating vibration	Check the motor and mounts Check the fan motor and blade; do the repair work
	Improperly loaded dishes.	Check owner's manual for proper loading methods

(*Contd*)

Table 16.9 (*Contd*)

Problem	Possible cause	Possible remedies
	Drain line may be obstructed	Check the drain for obstructions
	Dishwasher may not be sitting on level ground	Adjust the feet beneath the unit by screwing them up or down. If there are lock nuts on the feet, make sure they are tight.
Door does not close or latch properly	There may be some physical obstruction in the door	Check for obstructions
	Door latch, hinges, gaskets may not work properly	Check the door latch Inspect the door hinges Inspect the door seal gasket
Dishwasher does not complete cycle	Timer motor may not work properly or may have wrong setting	Test the timer motor
	Thermostat may not work properly	Check the thermostat
	improper functioning of heating element	Test the heating element for its electrical health
Cycles times are too long	The setting of timer motor and/or the thermostat may be incorrect	Test the timer motor and thermostat
	improper functioning of heating element	Test the heating element for its electrical conditions
Detergent cup does not open	The system may be gummed up with detergent gum	Clean up the detergent gum by washing the machine
	The bimetal electric switch triggering the open action of the cup may be faulty	Test the bimetal assembly
	The system may not receive the signal to open	This suggests a problem with the timer control or the wiring—check the timer motor
	The actuator arm finally opening the cup may be jammed	Check the actuator arm
Dishwasher does not wash properly	There may not be enough water	Check for adequate water level
	The spray arm may not rotate freely and water supply hose into the spray arm may not be aligned properly	Make sure that the spray arm is free and the inlet pipe is in line with the rack in the spray arm
	During the wash cycle, spray arms may be running poorly	Clean out the spray holes in the spray arm(s)
	The selector switch identifies which cycle is to be used. If a button does not work, that cycle will not function properly.	Test the selector switch and repair/replace, if found defective
	There may be improper loading-dishes block or impede the spray arms or prevent the soap dispenser from opening	Remove blockage and reposition load. Loading should not be haphazard but systematic
	If the dishwasher has chronic problems with washing, the problem may be too low supply water pressure	Check the water supply pressure. In order to fill to appropriate levels, water pressure should be from 20 to 120 pounds per square inch)
	The water hardness and the right amount of detergent combination may not be proper	Water supply authority can supply data for hardness value or in-house hardness testing may be done to arrive at the right amount of detergent for the purpose. In case, water is supplied by in-house treatment plant, better control can be exercised
	Water temperature may be low	Check the water temperature—should be about 60°C

16.9 | PUMPS

Pumps are machines that find one of the widest applications in industrial, sewage disposal, irrigation, municipal water distribution system and residential requirements. Hotels use pumps for various purposes, starting from boiler house, water-treatment plant, drinking and cooking water distribution to gardening, water recreation, to name only a few. This wide use of pumps makes them a vital engineering equipment for which care and proper maintenance assumes a very high degree of priority.

Care and maintenance

The following checklist may aid in keeping pumps in good condition:

1. On a daily basis
   - Inspect the pump for any abnormalities such as excessive heat in the bearing housing or abnormal sound or vibration in the machine
   - Lubricate the pump and motor bearings
   - Check for any leakage through gland packing
2. On a monthly basis
   - Check priming speed
   - Check capacity
   - Check shaft seal leakage of air and water
   - Examine any noise in pump casing
   - Check gaskets and O-rings
   - Examine suction line for leakage, etc.
   - Check suction strainer for obstruction and proper functioning
3. On every 6 month basis
   - Check impeller wear
   - Check clearance between impeller shroud and the volute
   - Check shaft seal wear
   - Check shaft sleeve wear
   - Clean the casing and volute passages

(*Note:* For checking impeller wear and casing-impeller clearances, refer to manufacturer's recommendations.)

Troubleshooting

As pumps form a very common machinery, a few troubleshooting tips for some common problems come in very useful in rectifying them, thus, ensuring a reliable and continuous supply of water. However, most of these problems and their solutions will be undertaken by the maintenance engineering department. Table 16.10 furnishes a troubleshooting chart for pumps.

Table 16.10 Troubleshooting chart for pumps

Problem	Possible cause and remedies
No water delivered	• Priming, casing, and suction pipe not completely filled with liquid • Speed too slow • Discharge head too high—check lift and friction loss • Suction lift too high or suction pipe too small or too long, causing excessive friction loss—check with the reading of suction gauge • Impeller or suction pipe or suction entry completely plugged • Impeller rotating in wrong direction • Air pocket in suction line • Stuffing box packing worn or water seal plugged • Allowing leakage of air into pump casing • Air leak in suction line.
Not enough water delivered	• Priming-casing and suction pipe not completely filled with water • Speed too low—check the voltage and frequency • Discharge head higher than anticipated—check for any excessive number of bends, valves etc. consider increasing the delivery pipe size • Suction lift too high or suction pipe too small or too long, causing excessive friction loss—check with the reading of suction gauge • Impeller or suction pipe or opening partially clogged • Wrong direction of rotation • Air pocket in suction line, perhaps because of sharp vertical bend or concentric reducer in suction line • Stuffing box packing worn or water seal destroyed, allowing leakage of air into pump casing • Air leak in suction line • Intermittent plugging of the suction inlet. Loose rags can do this • A foot valve is stuck • The impeller diameter is too small • The impeller width is too narrow • The impeller is damaged • The impeller has been installed backwards • The wear ring clearance is too large • A wear ring is missing • The suction lift is too high • The fluid is creating vortex at the pump inlet because the sump level is too low • The tank is being heated to de-aerate the fluid, but it is heating the fluid up too much • Two pumps are connected in series. The capacity of first pump is less than the second one and thus not sending enough flow to the second pump • The operating temperature of the pumped fluid has increased

(Contd)

Table 16.10 (*Contd*)

Problem	Possible cause and remedies
	• A bubble is trapped in the eye of the impeller • Foot valve grossly undersized • Foot valve not immersed deep enough • A foreign object is stuck in the piping It was left there when the piping was repaired • A check valve is stuck partially closed in the delivery line
Not enough pressure or head being developed—as shown by the delivery side gauge	• Speed too low—check the voltage of the electric motor. It may be too low • There may be something physically wrong with the motor. Check the bearings etc. • Air entrained in water • Impeller diameter too small • Mechanical defects—wear rings worn, impeller damaged, casing packing defective • Wrong direction of rotation • Pressure measured at incorrect point—measure pressure at top of pump case • The impeller is clogged. This is a major problem with closed impellers while pumping liquid with particulate matter • The double volute casting is clogged with solids • The open impeller to volute clearance is too large • The impeller has become loose on the shaft—key is sheared • The pump is running backwards because the discharge check valve is not holding and system pressure is causing the reverse rotation. This is a common problem with pumps installed in a parallel configuration • The impeller has been installed backwards. This can happen with closed impellers on double suction pumps • Two pumps are in connected in series. The first pump does not have enough capacity for the second pump. They should be running at the same speed with the same width impeller to ensure compatibility • The pump inlet temperature is too high • The tank is being heated to de-aerate the fluid, but it is heating the fluid up too much—a common problem with boiler feed pump applications

Pump consumes too much power	• Speed too high
	• The pump was sized for the maximum operating condition, but does not run anywhere near that point most of the time—head lower than rated, pumps too much water and efficiency becomes low
	• Can be caused by a pump that is too large for the application
	• The closed impeller wear rings are a common source of rubbing
	• The increased amperage can be caused by an increase in bearing loading
	• The increased amperage is caused by two parts rubbing together as a result of shaft displacement
	• Mechanical defects
	   – shaft bent
	   – stuffing boxes too tight
	   – pump and driving unit misaligned
	• The impeller has been installed backwards
	• The shaft is running in the wrong direction

(*Contd*)

Table 16.10 (*Contd*)

Problem	Possible cause and remedies
Pump works for a while then quits pumping	Leakage in suction line Stuffing box packing worn or water seal plugged, allowing leakage of air into pump casing Air pocket in suction line Not enough suction head for hot water or volatile liquids— check carefully as this is a frequent cause of trouble with hot water, etc. Air or gases in liquid Suction lift too high
Pump leaks excessively at the stuffing box	Packing worn or not properly lubricated Packing incorrectly installed or not properly run in Packing type incorrect for liquid handled Shaft scored

Pump makes a lot of noise	• Could be due to cavitation or 'hydraulic noise' —suction lift could be too high, temperature of liquid may have become high, etc. Check with vacuum gauge for very low suction pressure
	• Mechanical defects
	    – bearings worn out
	    – shaft bent
	    – rotating parts are loose or broken
	    – pump and motor shafts misaligned

16.10 | CONCLUSION

It is very clear from the aforesaid discussion regarding care and maintenance of engineering system that a little care and maintenance is expected to be undertaken by the operators for best results from the equipment. It is, therefore, necessary that the operators are given some training related to the daily care and maintenance of the equipment they handle. However, for many of the maintenance work, the specialized technicians of the engineering maintenance department will handle the situation. Care and maintenance of some special equipment are also carried out through maintenance contract with external agencies. Troubleshooting charts furnished are only indicative in nature and may not be exhaustive. Operator usually can do very little if some problem arises and should inform the concerned departmental head who, in turn, will contact the engineering department for troubleshooting.

KEY TERMS

Care and maintenance   It is a routine procedure involving safe and machine-friendly handling and upkeep of equipment.

Defrosting   This is the process of thawing (melting) of ice formed on the outer surface of the evaporator coil in a refrigerator.

Sudsing	The process of forming lather, bubble, or foam during washing is called sudsing.
Troubleshooting	Identifying the cause of problems in a system and suggesting and implementing feasible remedial measures is called troubleshooting.

REVIEW QUESTIONS

16.1 Write about the care and maintenance of refrigerators.

16.2 Discuss the measures taken for the care and maintenance of unit air conditioners.

16.3 List routine care and maintenance steps for cooking ovens.

16.4 Discuss, in brief, the routine care and maintenance of washing machines and dryers.

Index